Lecture Notes in Physics

For information about Vols. 1–90, please contact your bookseller or Springer-Verlag.

Vol. 91: Computing Methods in Applied Sciences and Engineering, 1977, II. Proceedings, 1977. Edited by R. Glowinski and J. L. Lions. VI, 359 pages. 1979.

Vol. 92: Nuclear Interactions. Proceedings, 1978. Edited by B. A. Robson. XXIV, 507 pages. 1979.

Vol. 93: Stochastic Behavior in Classical and Quantum Hamiltonian Systems. Proceedings, 1977. Edited by G. Casati and J. Ford. VI, 375 pages. 1979.

Vol. 94: Group Theoretical Methods in Physics. Proceedings, 1978. Edited by W. Beiglböck, A. Böhm and E. Takasugi. XIII, 540 pages. 1979.

Vol. 95: Quasi One-Dimensional Conductors I. Proceedings, 1978. Edited by S. Barišić, A. Bjeliš, J. R. Cooper and B. Leontič. X, 371 pages. 1979.

Vol. 96: Quasi One-Dimensional Conductors II. Proceedings 1978. Edited by S. Barišić, A. Bjeliš, J. R. Cooper and B. Leontić. XII, 461 pages. 1979.

Vol. 97: Hughston, Twistors and Particles. VIII, 153 pages. 1979.

Vol. 98: Nonlinear Problems in Theoretical Physics. Proceedings, 1978. Edited by A. F. Rañada. X, 216 pages. 1979.

Vol. 99: M. Drieschner, Voraussage – Wahrscheinlichkeit – Objekt. XI, 308 Seiten. 1979.

Vol. 100: Einstein Symposion Berlin. Proceedings 1979. Edited by H. Nelkowski et al. VIII, 550 pages. 1979.

Vol. 101: A. Martin-Löf, Statistical Mechanics and the Foundations of Thermodynamics. V, 120 pages. 1979.

Vol. 102: H. Hora, Nonlinear Plasma Dynamics at Laser Irradiation. VIII, 242 pages. 1979.

Vol. 103: P. A. Martin, Modèles en Mécanique Statistique des Processus Irréversibles. IV, 134 pages. 1979.

Vol. 104: Dynamical Critical Phenomena and Related Topics. Proceedings, 1979. Edited by Ch. P. Enz. XII, 390 pages. 1979.

Vol. 105: Dynamics and Instability of Fluid Interfaces. Proceedings, 1978. Edited by T. S. Sørensen. V, 315 pages. 1979.

Vol. 106: Feynman Path Integrals, Proceedings, 1978. Edited by S. Albeverio et al. XI, 451 pages. 1979.

Vol. 107: J. Kijowski, W. M. Tulczyjew, A Symplectic Framework for Field Theories. IV, 257 pages. 1979.

Vol. 108: Nuclear Physics with Electromagnetic Interactions. Proceedings, 1979. Edited by H. Arenhövel and D. Drechsel. IX, 509 pages. 1979.

Vol. 109: Physics of the Expanding Universe. Proceedings, 1978. Edited by M. Demiański. V, 210 pages. 1979.

Vol. 110: D. A. Park, Classical Dynamics and Its Quantum Analogues. VIII, 339 pages. 1979.

Vol. 111: H.-J. Schmidt, Axiomatic Characterization of Physical Geometry. V, 163 pages. 1979.

Vol. 112: Imaging Processes and Coherence in Physics. Proceedings, 1979. Edited by M. Schlenker et al. XIX, 577 pages. 1980.

Vol. 113: Recent Advances in the Quantum Theory of Polymers. Proceedings 1979. Edited by J.-M. André et al. V, 306 pages. 1980.

Vol. 114: Stellar Turbulence. Proceedings, 1979. Edited by D. F. Gray and J. L. Linsky. IX, 308 pages. 1980.

Vol. 115: Modern Trends in the Theory of Condensed Matter. Proceedings, 1979. Edited by A. Pekalski and J. A. Przystawa. IX, 597 pages. 1980.

Vol. 116: Mathematical Problems in Theoretical Physics. Proceedings, 1979. Edited by K. Osterwalder. VIII, 412 pages. 1980.

Vol. 117: Deep-Inelastic and Fusion Reactions with Heavy Ions. Proceedings, 1979. Edited by W. von Oertzen. XIII, 394 pages. 1980.

Vol. 118: Quantum Chromodynamics. Proceedings, 1979. Edited by J. L. Alonso and R. Tarrach. IX, 424 pages. 1980.

Vol. 119: Nuclear Spectroscopy. Proceedings, 1979. Edited by G. F. Bertsch and D. Kurath. VII, 250 pages. 1980.

Vol. 120: Nonlinear Evolution Equations and Dynamical Systems. Proceedings, 1979. Edited by M. Boiti, F. Pempinelli and G. Soliani. VI, 368 pages. 1980.

Vol. 121: F. W. Wiegel, Fluid Flow Through Porous Macromolecular Systems. V, 102 pages. 1980.

Vol. 122: New Developments in Semiconductor Physics. Proceedings, 1979. Edited by F. Beleznay et al. V, 276 pages. 1980.

Vol. 123: D. H. Mayer, The Ruelle-Araki Transfer Operator in Classical Statistical Mechanics. VIII, 154 pages. 1980.

Vol. 124: Gravitational Radiation, Collapsed Objects and Exact Solutions. Proceedings, 1979. Edited by C. Edwards. VI, 487 pages. 1980.

Vol. 125: Nonradial and Nonlinear Stellar Pulsation. Proceedings, 1980. Edited by H. A. Hill and W. A. Dziembowski. VIII, 497 pages. 1980.

Vol. 126: Complex Analysis, Microlocal Calculus and Relativistic Quantum Theory. Proceedings, 1979. Edited by D. lagolnitzer. VIII, 502 pages. 1980.

Vol. 127: E. Sanchez-Palencia, Non-Homogeneous Media and Vibration Theory. IX, 398 pages. 1980.

Vol. 128: Neutron Spin Echo. Proceedings, 1979. Edited by F. Mezei. VI, 253 pages. 1980.

Vol. 129: Geometrical and Topological Methods in Gauge Theories. Proceedings, 1979. Edited by J. Harnad and S. Shnider. VIII, 155 pages. 1980.

Vol. 130: Mathematical Methods and Applications of Scattering Theory. Proceedings, 1979. Edited by J. A. DeSanto, A. W. Sáenz and W. W. Zachary. XIII, 331 pages. 1980.

Vol. 131: H. C. Fogedby, Theoretical Aspects of Mainly Low Dimensional Magnetic Systems. XI, 163 pages. 1980.

Vol. 132: Systems Far from Equilibrium. Proceedings, 1980. Edited by L. Garrido. XV, 403 pages. 1980.

Vol. 133: Narrow Gap Semiconductors Physics and Applications. Proceedings, 1979. Edited by W. Zawadzki. X, 572 pages. 1980.

Vol. 134: γγ Collisions. Proceedings, 1980. Edited by G. Cochard and P. Kessler. XIII, 400 pages. 1980.

Vol. 135: Group Theoretical Methods in Physics. Proceedings, 1980. Edited by K. B. Wolf. XXVI, 629 pages. 1980.

Lecture Notes in Physics

164

Stability of Thermodynamic Systems

Proceedings of the Meeting
Held at Bellaterra School of Thermodynamics
Autonomous University of Barcelona
Bellaterra (Barcelona) Spain, September 1981

Edited by J. Casas-Vázquez and G. Lebon

Springer-Verlag
Berlin Heidelberg GmbH 1982

Editors
Georgy Lebon
Liège University, Institute of Chemistry
B6, Sart Tilman, B-4000 Liège

José Casas-Vázquez
Universidad Autonoma de Barcelona, Departamento de Termologia
Bellaterra-Barcelona, Spain

ISBN 978-3-540-11581-6 ISBN 978-3-540-39328-3 (eBook)
DOI 10.1007/978-3-540-39328-3

Originally published by Springer-Verlag Berlin Heidelberg New York in 1982

2153/3140-543210

PREFACE

This book contains the manuscripts of the conferences and seminars delivered at the meeting on Stability of Thermodynamic Systems held at the Bellaterra School of Thermodynamics, Autonomous University of Barcelona. The aim of this School is to promote biannual meetings between specialists and scientists wishing to introduce themselves to topics of present interest in thermodynamics.

The purpose of this course is to present not only the general framework but also the more recent progress in the domain of nonequilibrium instabilities.

The general lectures were intended to be a clear, broad and suitable introduction to this fast developing field. A review of the thermodynamic framework, the mathematical methods and the basic phenomenology may be found in the papers of J. Casas-Vázquez, G. Lebon and C. Pérez-García.

The various topics covered by the other lectures deal with instabilities in a wide variety of fields like hydrodynamics (M. Dubois, P. Bergé, M. Zamora, D. Jou, and D. Quemada), electromagnetism (J.M. Rubí), chemistry (M.G. Velarde) and ecology (R. Margalef), as well as some mathematical aspects, such as the development of strange attractors (C. Perelló).

We acknowledge the sponsoring of the Secretaría de Estado de Universidades e Investigación, the Dirección General de Política Científica, the Spanish Ministry of Education, the Dirección General d'Universitats of the Generalitat of Catalonia, the Instituto de Ciencias de la Educación and the Vicerectorado de Extensión Universitaria of the Autonomous University of Barcelona.

Barcelona, April 1982

J. Casas-Vázquez
G. Lebon

CONTENTS

THERMODYNAMIC THEORY OF STABILITY

J. CASAS-VAZQUEZ

Departamento de Termología
Universidad Autónoma de Barcelona
Bellaterra (Barcelona) Spain

1. INTRODUCTION

In these notes we are interested in the study of the stability of thermodynamic systems both in equilibrium and in nonequilibrium states. The first objective of this lecture is to formulate the great lines of the Gibbs theory because it is clearly better adapted to the subject treated here. A second objective is to give a general brief review of the stability problem in different approaches to nonequilibrium thermodynamics, namely TPI or local equilibrium theory, and extended thermodynamics (ET). Stability in rational thermodynamics will be excluded here because it is a subject which has not yet been formulated in a general manner. However, there are several indications pointing to an intimate relationship between second law and stability. Thus in references (2), (5) and (10) we find that the Clausius-Duhem inequality induces stability of equilibrium processes in the sense of Lyapunov for a variety of materials.

2. THE GIBBS THEORY OF STABILITY

2.1 Preliminaries

Two different formulations of equilibrium thermodynamics are usually given in the scientific literature. The two formulations are generally associated with the names of Clausius-Kelvin-Carathéodory (CKC) and Gibbs, this latter appearing in a modern fashion as thermostatics or macroscopic thermodynamics of equilibrium (MTE).

Before entering in its description we now summarize the advantages of the phenomenological Gibbs theory with regard to CKC one. In this theory the thermodynamic system is considered as a "black box", its main achievement being the establishment of the concept of internal energy and entropy from observable quantities. In the Gibbs theory, in contrast, attention is focussed toward the system. The concepts of internal energy and entropy are taken for granted, and are used to provide a more detailed description of the system in equilibrium which includes its chemical and phase structure.

Another distinction can be pointed out turning our attention toward the geometrical methods utilized by each. The application of these methods to thermodynamics is based on a thermodynamic phase space or, in other words, a space spanned by a number of thermodynamic variables. The guiding idea of the CKC theory is that the variables be directly measurable as for example, pressure, volume and the mole numbers of chemical components. Beyond this requirement, the theory is quite insensitive to the specific choice of phase space and, consequently, there is no geometrical significance attached to the distinction between extensive and intensive variables. In the Gibbs theory, on the contrary, a particular phase space spanned by the extensive variables such as internal energy, entropy, volume and mole number plays a privileged role. It is the thermodynamic configuration space or Gibbs space (21)

In Section 2.3 we shall come back to discuss some relevant features of the geometry of the thermodynamic configuration space.

2.2 A brief review of the formal structure

The formal structure of MTE has been well displayed by Callen in his celebrated textbook (1). The basic principles are substituted by the following postulates:

1. There exist particular states (of equilibrium) which are characterized completely by the internal energy u and a set of extensive variables X_1, X_2, ... X_t to be specified later.

2. There exist a function S, called the entropy (alternatively the internal energy U), of the extensive variables, defined for all equilibrium states, and having the following property: the values assumed by the set of extensive variables in the absence of a constraint are those which maximize S (or minimize U) over the manifold of constrained equilibrium states.

In this point it must be underlined that the basic problem of thermodynamics is the determination of the equilibrium state that eventually results after the removal of internal constraints in a closed composite system. Postulate 2 represents an excellent tool for searching the equilibrium state via an extremum principle. The relation that gives the entropy (or the internal energy) as a function of the extensive variables is known as a fundamental equation. It therefore has all conceivable thermodynamic information about the system.

3. The entropy (or the internal energy) of a composite system is additive over the constituent subsystems. The entropy (or the internal energy) is continuous and sufficiently differentiable and is a monotonically increasing function of the energy (the entropy).

The role played by the intensive variables in the formal structure can be revealed from the differential form of the fundamental equation in the energy scheme

$$dU = TdS + \sum_{1}^{t} P_k dX_k = \sum_{0}^{t} P_k dX_k \qquad (2.1)$$

in which

$$P_k = \frac{\partial U}{\partial X_k} \qquad (2.2)$$

Alternatively, in the entropy scheme

$$dS = T^{-1}dU - T^{-1}\sum_{1}^{t} P_k dX_k = \sum_{0}^{t} F_k dX_k \qquad (2.3)$$

where

$$F_k = \frac{\partial S}{\partial X_k} \qquad (2.4)$$

The intensive variables in both schemes are functions of the extensive variables, the functional relations being the equations of state. Furthermore, the condition of equilibrium with respect to a transfer of X_k between two subsystem (the extensive variables obey a conservation law, $\sum_\alpha X_k^{(\alpha)}$ = const) is the equality of the intensive parameters P_k. Note that the definition of an intensive parameter requires that a contact equilibrium be attained.

From the homogeneous first-order property of the fundamental equation follows

$$U = \sum_0^t P_k X_k \tag{2.5}$$

which is known as Euler relation. Combining (2.1) and the differential form of (2.5) one obtains the Gibbs-Duhem relation

$$\sum_0^t X_k dP_k = 0 \tag{2.6}$$

Similar relations can be written in the entropy scheme.

Intensive parameters are usually better measured or controlled than the extensive ones. For this reason, thermodynamics makes widely use of thermodynamic potentials and Massieu-Planck functions obtained from the fundamental equation by means of Legendre transformations. A partial Legendre transformation of $U = U(X_0, X_1 \ldots . X_t)$ can be made by replacing the variables $X_0, X_1, \ldots . X_s$ by $P_0, P_1, \ldots . P_s$, the transformed function being

$$\psi^{(s)} = U - \sum_0^s P_k X_k \tag{2.7}$$

in which

$$\frac{\partial \psi^{(s)}}{\partial P_k} = - X_k, \qquad k = 0, 1, \ldots s \tag{2.8}$$

$$\frac{\partial \psi^{(s)}}{\partial X_k} = P_k, \qquad k = s+1, \ldots t \tag{2.9}$$

since the natural variables of $\psi^{(s)}$ are $P_0, P_1 \ldots . P_s, X_{s+1} \ldots . X_t$. Consequently, the differential form of potential $\psi^{(s)}$ is

$$d\psi^{(s)} = - \sum_0^s X_k dP_k + \sum_{s+1}^t P_k dX_k \tag{2.10}$$

The equilibrium values of any unconstrained extensive variable in a system in contact with reservoirs prescribing constant values of $P_0, P_1 \ldots . P_s$ minimize $\psi^{(s)}$ at constant $P_0, P_1 \ldots . P_s$.

Since (2.10) is an exact differential, its mixed partial derivatives are equal, i.e.

$$\frac{\partial X_j}{\partial P_k} = \frac{\partial X_k}{\partial P_j} \quad , \; j, \; k \leq s$$

$$\frac{\partial X_j}{\partial X_k} = - \frac{\partial P_k}{\partial P_j} \quad , \; j \leq s \quad \text{and} \quad k > s$$

$$\frac{\partial P_j}{\partial X_k} = \frac{\partial P_k}{\partial X_j} \; , \; j, k > s$$

These expressions are the well-known Maxwell relations.

2.3 Geometry of the Gibbs space

The achievements of geometrisation of the Gibbs theory are remarkable when it is realized that the basic geometrical elements of metric and orthogonality are wanting. However, the mathematical foundations were discovered at a later date showing that the geometry of the Gibbs space is an affine differential geometry. Although a metric cannot be defined, it is possible to define a parallel projection which replaces the orthogonal projection in the ordinary Riemann theory of curvature. Furthermore, by representing the entropy (or the energy) as a quadratic form, it is possible to obtain something similar to a metric, i.e. the Gibbs space is one in which "volumes" but not "lengths" are measurable. The volume is represented by the determinant of the matrix associated with the quadratic form. Thus, unlike the Euclidean geometry, parallelism replaces orthogonality, and volumes but not lengths are measurable in the Gibbs space. Also, different spaces can be generated by Legendre transformations, both extensive and intensive variables being spanned.

The fundamental equation introduced by postulate 2 can be represented as a (primitive) surface in the Gibbs space. From an analysis of the curvature of the primitive surface Gibbs obtained his thermodynamic criteria of stability. The alternate use of both entropy and energy representations amounts to a rotation in the Gibbs space. This duality breaks down in non-equilibrium thermodynamics ([13]). Even in equilibrium thermodynamics there are particular situations in which a scheme is preferred to the other. This is the case for the energy in equilibrium stability study.

Stability is related to the positive definite or negative definite forms of the second variations of the internal energy and entropy, respectively. The stability criteria are associated with physically measurable quantities when the quadratic form is reduced to its canonical diagonal form. But, the reduction to diagonal form cannot be performed by the usual eigenvalue methods of metric definable spaces. Since a metric does not exist in Gibbs space, the reduction to diagonal form has to be carried out by the method of "completing square" (21).

2.4 Intrinsic stability of general systems

The problem of stability arises on two distinct levels. There is the problem of mutual stability of two systems (also stability of heterogeneous systems) which concerns the stability of a predicted partition of all extensive parameters between two systems separated by an appropriate wall. But there is also a problem of intrisic stability, which arises even within a single isolated system.

It is evident that an isolated simple system may be subdivided at least mentally in two or more portions transforming thus into a composite system. So a problem of intrinsic stability is reduced to a problem of mutual stability.

We consider a general system with the fundamental equation

$$U^* = U^*(S^*, X_1^*, \ldots X_t^*) \tag{2.11}$$

The X_1^* can be thought of as the volume, the mole numbers and other extensive parameters including the corresponding to gravitational, electric and magnetic fields.

A small subsystem with a constant value of X_t is considered the remainder of the system constituting the complementary subsystem. Since the internal energy is additive

$$U^* = U + U^c \tag{2.12}$$

where U^c is the internal energy of the complementary subsystem. Defining

$$u \equiv U/X_t \ , \quad x_0 = S/X_t \quad \text{and} \quad x_j = X_j/X_t \tag{2.13}$$

$$u^c \equiv U^c/X_t^c \ , \quad x_0^c = S^c/X_t^c \quad \text{and} \quad x_j^c = X_j^c/X_t^c \tag{2.14}$$

the fundamental equation becomes

$$U^* = X_t u(x_0, x_1, \ldots x_{t-1}) + X_t^c u^c(x_0^c, x_1^c, \ldots x_{t-1}^c) \qquad (2.15)$$

We assume that the subsystem is very small with respect to the total system, hence also with respect to the complementary subsystem

$$X_t << X_t^* \text{ or } X_t^c \qquad (2.16)$$

The imaginary wall that separates the subsystem and the complementary subsystem is non-restrictive to all but the X_t extensive parameters. The closure condition on every $X_j^!$ is

$$X_t x_j + X_t^c x_j^c = X_j^* \qquad (2.17)$$

and in virtual changes of these parameters

$$X_t \delta x_j + X_t^c \delta x_j^c = 0 \qquad (2.18)$$

Applying the postulate 2, the total entropy must be held constant

$$X_t x_0 + X_t^c x_0^c = X_0^* \qquad (2.19)$$

and

$$X_t \delta x_0 + X_t^c \delta x_0^c = 0 \qquad (2.20)$$

By virtue of (2.16)

$$|\delta x_j^c| << |\delta x_j| \qquad (2.21)$$

$$|\delta x_0^c| << |\delta x_0| \qquad (2.22)$$

Let us assume a virtual transfer of $X_0, X_1 \ldots X_{t-1}$ across the hypothetical surface. Such a transfer leads to a change in the total energy which can be expressed by a Taylor expansion. According to (2.21) and (2.22), this expansion involves an infinite series of terms relating to the subsystem where all but the first-order terms relating to the complementary subsystem can be neglected. Thus

$$\Delta U^* = X_t\{\delta u + \delta^2 u + \ldots\} + X_t^c \delta u^c \qquad (2.23)$$

in which

$$\delta u = \sum_0^{t-1} (\partial u/\partial x_j)\delta x_j = \sum_0^{t-1} u_j \delta x_j = \sum_0^{t-1} P_j \delta x_j \tag{2.24}$$

$$\delta^2 u = \frac{1}{2}\sum_0^{t-1} (\partial^2 u/\partial x_j \partial x_k)\delta x_j \delta x_k = \frac{1}{2}\sum_0^{t-1} u_{jk}\delta x_j \delta x_k \tag{2.25}$$

$$\delta u^c = \sum_0^{t-1} (\partial u^c/\partial x_j^c)\ \delta x_j^c = \sum_0^{t-1} u_j^c \delta x_j^c = \sum_0^{t-1} P_j^c \delta x_j^c \tag{2.26}$$

At this point, the formalism requires first that the first-order terms vanish and this leads to the equality of each P_j and P_j^c, i.e. the equilibrium conditions. Second, minimum value of the internal energy leads to the second-order terms be positive for any conceivable virtual process. The stability depends upon the form (2.25) being positive definite, $\delta^2 u > 0$.

The stability conditions can be related to physically measurable quantities by putting the quadratic form (2.25) in the canonical diagonal form, provided of course that the detèrminant of the quadratic form is non-singular, i.e.

$$D_{t-1} = |u_{jk}| \neq 0 \tag{2.27}$$

As has already been mentioned, the Gibbs space is one in which volumes and not lengths are measurable, it is required that the linear affine transformation be unimodular (of determinant equal to unity), for only then the volume will be invariant under the transformation [16]. The procedure consists of completing the square and one finally arrives at the canonical form

$$\delta^2 u = \frac{1}{2}\sum_0^{t-1} \lambda_j (\delta y_j)^2 \tag{2.28}$$

where the δy_j are linear combination fo the δx_j. In fact by using the matrix notation (2.25) can be written

$$\delta^2 u = \frac{1}{2}(\delta \underline{x})^T \cdot \underline{\underline{D}} \cdot \delta \underline{x} > 0 \tag{2.29}$$

where $\delta \underline{x}$ is the column vector formed from the components δx_j and $(\delta \underline{x})^T$ is the corresponding transposed vector (row vector). Since the matrix $\underline{\underline{D}}$ is symmetric it can always be changed into a diagonal matrix $\underline{\underline{\Lambda}}$ by a linear affine congruent transformation. We shall denote the element of $\underline{\underline{\Lambda}}$ by λ_j . We have, therefore,

$$\underline{\underline{\Lambda}} = \underline{\underline{Q}}^T \cdot \underline{\underline{D}} \cdot \underline{\underline{Q}} = |\lambda_j \delta_{jk}| \tag{2.30}$$

in which $\underline{\underline{Q}}$ is the transformation matrix. The quantities λ_j above defined are not the eigenvalues of $\underline{\underline{D}}$ since the transformation (2.30) is not orthogonal. Since, on other hand, a linear affine transformation is not unique, the values of the λ_j are also not unique, depending on the initially chosen sequence of variables. However, the Sylvester's law of the inertia of quadratic forms assures that the numbers of positive, negative and vanishing λ_j are not changed by a permutation of the variables. This can be easily verified by means of explicit examples.

The new variables y_j are introduced by the transformation

$$\delta\underline{x} = \underline{\underline{Q}} \cdot \delta\underline{y} \tag{2.31}$$

when substituted in (2.29) and taking into account (2.30), the quadratic form becomes

$$\delta^2 u = \frac{1}{2}(\delta y)^T \cdot \underline{\underline{\Lambda}} \cdot \delta y \tag{2.32}$$

whose explicit form is (2.28). The stability condition can, therefore, be fulfilled only if

$$\lambda_j > 0, \qquad j = 0,1,\ldots t-1 \tag{2.33}$$

It is possible to demonstrate (16) that

$$\lambda_j = D_j / D_{j-1} \tag{2.34}$$

where the D_j are the principal minors of the determinant D and we put $D_{-1} \equiv 1$. From (2.33) and (2.34) we get the equivalent form of the stability conditions

$$D_j > 0, \qquad j = 0,1,\ldots t-1 \tag{2.35}$$

Equation (2.34) gives the important relationship

$$\prod_0^{t-1} \lambda_j = D_{t-1} \equiv D \tag{2.36}$$

Another useful form of the stability conditions can be obtained by starting with Eq. (2.2) and noting that

$$u_{jk} = \partial^2 u/\partial x_j \partial x_k = \partial P_j/\partial x_k = \partial P_k/\partial x_j = u_{kj} \qquad (2.37)$$

The variation fo P_j can be expressed as

$$\delta P_j = \sum_0^{t-1} u_{jk}\delta x_k \qquad (2.38)$$

and in matrix notation

$$\delta \underline{P} = \underline{\underline{D}}\cdot\delta\underline{x} \qquad (2.39)$$

If we now put in this set of inhomogeneous linear equations

$$\begin{aligned} &\delta P_0 = \delta P_1 = \ldots = \delta P_{s-1} = 0 \\ &\delta x_{s+1} = \delta x_{s+2} = \ldots = \delta x_{t-1} = 0 \end{aligned} \qquad (2.40)$$

equation (2.39) becomes

$$\begin{aligned} 0 &= u_{00}\delta x_0 + u_{01}\delta x_1 + \ldots + u_{0s}\delta x_s \\ &\vdots \\ 0 &= u_{s-1,0}\delta x_0 + u_{s-1,1}\delta x + \ldots + u_{s-1,s}\delta x_s \\ \delta P_s &= u_{s0}\delta x_0 + u_{s1}\delta x_1 + \ldots + u_{ss}\delta x_s \\ &\vdots \\ \delta P_{t-1} &= u_{t-1,0}\delta x_0 + u_{t-1,1}\delta x_1 + \ldots + u_{t-1,s}\delta x_s \end{aligned} \qquad (2.41)$$

Confining our attention to the first $s+1$ equations and solving for δx_s we obtain after application of Cramer's rule

$$\delta x_s = (D_{s-1}/D_s)\delta P_s \qquad (2.42)$$

where the D are again the principal minors of the determinant D. Because of (2.40), Eq. (2.42) can be written as

$$(\partial P_s/\partial x_s)_{P_j, x_k} = D_s/D_{s-1}, \quad \begin{aligned} j &= 0,1,\ldots s-1 \\ k &= s+1,\ldots t-1 \end{aligned} \qquad (2.43)$$

Bearing in mind the definitions (2.8)-(2.9) the second derivatives of $\psi^{(s)}$ with respect to extensive variables are

$$\psi^{(s)}_{jk} = \partial^2\psi^{(s)}/\partial x_j \partial x_k = (\partial P_j/\partial x_k)_{P_i, x_{m\neq j}} =$$

$$= (\partial P_k/\partial x_j)_{P_i, x_{m\neq j}} = \psi^{(s)}_{kj} \qquad (2.44)$$

This result together with (2.34) and (2.43) allows us to write

$$\lambda_j = \psi^{(j-1)}_{jj} \qquad (2.45)$$

The stability conditions may thus be expressed in the form

$$\psi^{(j-1)}_{jj} > 0 \,, \quad j = 0,1,\ldots t-1 \quad . \qquad (2.46)$$

Returning now to the expansion (2.28) which can be centred around any point of the primitive surface, we can classify the points of this surface as follows:

i) elliptic points for which all $\lambda_j > 0$
ii) parabolic points where all $\lambda_j > 0$ with at least one $\lambda_j = 0$
iii) hyperbolic points with at least one $\lambda_j < 0$

Elliptic points are stable with respect to small displacements (local stability), but may be either stable or instable with respect to finite processes. Hyperbolic points represent states of essential instability. The case of parabolic points is the most complex and also the most interesting. Whether or not parabolic points are stable depends on the large displacements. While most parabolic points turn out to be instable, there are stable limiting situations. Yet, the stability of parabolic points is of a lower order than the one found in stable elliptic points. The latter is referred as <u>normal</u>, the former as <u>critical equilibrium</u> (21).

For the sake of simplicity, we consider the case of two independent variables. An elliptic point is characterized by the conditions (see 2.45 and 2.46)

$$\lambda_0 = \psi^{(-1)}_{00} = u_{00} = (\partial P_0/\partial x_0)_{x_1} > 0 \qquad (2.47)$$

$$\lambda_1 = \psi^{(0)}_{11} = f_{11} = (\partial P_1/\partial x_1)_{P_0} > 0 \qquad (2.48)$$

According to (2.34)

$$\lambda_1 = D_1/D_0 = (u_{00}u_{11}-u_{01}^2)u_{00}^{-1} = u_{11}-u_{01}^2u_{00}^{-1}$$

$$= (\partial P_1/\partial x_1)_{x_0} - (\partial P_0/\partial x_1)_{x_0}^2/(\partial P_0/\partial x_0)_{x_1} > 0 \qquad (2,49)$$

Hence

$$(\partial P_1/\partial x_1)_{x_0} \geqq (\partial P_1/\partial x_1)_{P_0} > 0 \qquad (2.50)$$

where the equality sign is relevant only if

$$(\partial P_0/\partial x_1)_{x_0} = 0 \qquad (2.51)$$

Relations (2.50) are equivalent to (2.47) and (2.48). They are usually referred as the <u>principle of Le Châtelier</u>.

A system for which we may choose x_0=s and x_1=v relations (2.50) yield

$$-(\partial p/\partial v)_s \geq -(\partial p/\partial v)_T > 0\ ,\ (vk_s)^{-1} \geq (vk_T)^{-1} > 0 \qquad (2.52)$$

where k_s and k_T are the adiabatic and isothermal compressibilities, respectively. If now the choice is x = v and x = s, (2.50) lead to

$$(\partial T/\partial s)_v \geq (\partial T/\partial s)_p > 0\ ,\quad (T/c_v) \geq (T/c_p) > 0 \qquad (2.53)$$

c_v and c_p being the specific heats at constant volume and pressure, respectively. The foregoing results are easily generalized to several variables. In elliptic points we have the generalized Le Châtelier principle

$$(\partial P_k/\partial x_k) \geqq (\partial P_k/\partial x_k)_{P_0} \geq (\partial P_k/\partial x_k)_{P_0,P_1} \geq \ldots \geq (\partial P_k/\partial x_k)_{P_0,P_1..P_{k-1}} > 0 \qquad (2.54)$$

We express this relation in words as follows: a locally stable system is displaced from its equilibrium by the displacement δx_k. The system responds by changing its conjugate intensity by δP_k. This response is the largest if all the other x_i are fixed, and it decreases upon relaxation of each constraint that frees a variable x_i by coupling the system to a reservoir of intensity P_i.

It is well known that if the criteria of stability are not satisfied, a system breaks up into two or more portions called phases (phase transition). Also, the critical phases forming the boundary between full stability and instability are determined by the conditions

$$\psi_{sk}^{(s-1)} = (\partial P_k/\partial x_s)_{P_0,\ldots P_{s-1},\, x_{s+1}\ldots x_{t-1}} = 0 \quad \text{for all } k \geq s \tag{2.55}$$

and

$$\psi_{sss}^{(s-1)} = (\partial^2 P_s/\partial x_s^2)_{P_0\ldots P_{s-1},\, x_{s+1},\ldots x_{t-1}} = 0 \tag{2.56}$$

$$\psi_{ssss}^{(s-\ )} = (\partial^3 P_s/\partial x_s^3)_{P_0\ldots P_{s-1},\, x_{s+1}\ldots x_{t-1}} > 0 \tag{2.57}$$

At such a critical "point" a number of observable parameters become unbounded. In particular

$$(\partial x_k/\partial P_j)_{P_0\ldots P_s,\, x_{s+1}\ldots x_{t-1}} \to \infty \qquad j,k \leq s \tag{2.58}$$

We now give a simple illustrative example: the critical point of vaporization of a one-component simple system. The fundamental equation for molar quantities is

$$u = u(s,v) \tag{2.59}$$

the stiffness matrix

$$\underline{\underline{D}} = \begin{vmatrix} u_{ss} & u_{sv} \\ u_{vs} & u_{vv} \end{vmatrix} = \begin{vmatrix} T/c_v & T\alpha/k_T c_v \\ T\alpha/k_T c_v & c_p/c_v k_T v \end{vmatrix} \tag{2.60}$$

and the compliance matrix

$$\underline{\underline{C}} = \underline{\underline{D}}^{-1} = \begin{vmatrix} u_{vv}/D & -u_{sv}/D \\ -u_{sv}/D & u_{ss}/D \end{vmatrix} = \begin{vmatrix} c_p/T & \alpha \\ \alpha & v_{k_T} \end{vmatrix} \tag{2.61}$$

Transformation into the diagonal form according to (2.45) gives

$$\lambda_0 = \partial^2 u/\partial s^2 = T/c_v \, , \quad \lambda_1 = \partial^2 f/\partial v^2 = -(\partial p/\partial v)_T \tag{2.62}$$

where f is the molar free energy, s and v are the critical parameters. We are thus dealing with the normal case. At the critical point $\lambda_1 = 0$. From (2.55)-(2.56)

$$(\partial p/\partial v)_T = 0, \quad (\partial^2 p/\partial v^2)_T = 0 \tag{2.63}$$

At the same time all the elements of the compliance matrix becomes infinite:

$$c_p = \infty \, , \quad \alpha = \infty \, , \quad k_T = \infty \tag{2.64}$$

2.5 Transformation of stability conditions

We now wish to express the stability conditions with the aid of an arbitrary thermodynamic potential, or a Massieu-Planck function. For the sake of simplicity we confine our calculations to the energy scheme.

Let $\psi^{(k)}$ be the partial Legendre transform of u

$$\psi^{(k)} = u - \sum_0^k P_i x_i = \psi^{(k)} \, (P_0 \ldots P_k, \, x_{k+1} \ldots x_{k-1}) \tag{2.65}$$

and its first order variation

$$\delta\psi^{(k)} = \delta u - \sum_0^k P_i \delta x_i - \sum_0^k x_i \delta P_i = - \sum_0^k x_i \delta P_i + \sum_{k+1}^{t-1} P_i \delta x_i \tag{2.66}$$

in which

$$- x_i = \partial\psi^{(k)}/\partial P_i \qquad (i = 0,1,\ldots k) \tag{2.67}$$

$$P_i = \partial\psi^{(k)}/\partial x_i \qquad (i = k+1,\ldots t-1) \tag{2.68}$$

On the other hand, the second-order variation of u has a definite sign. It is a positive quadratic form and it can be expressed alternatively as

$$\delta^2 u = \sum_{0}^{t-1} \delta P_i \delta x_i = \sum_{0}^{k} \delta P_i \delta x_i + \sum_{k+1}^{t-1} \delta P_j \delta x_j > 0 \tag{2.69}$$

From (2.67) and (2.68) ones sees that $x_i = x_i(P_i, x_j)$ and $P_j = P_j(P_i, x_j)$. Therefore

$$\delta x_i = -\sum_{\ell=0}^{k} \frac{\partial^2 \psi^{(k)}}{\partial P_\ell \partial P_i} \delta P_\ell - \sum_{m=k+1}^{t-1} \frac{\partial^2 \psi^{(k)}}{\partial x_m \partial P_i} \delta x_m \tag{2.70}$$

$$\delta P_j = \sum_{\ell=0}^{k} \frac{\partial^2 \psi^{(k)}}{\partial P_\ell \partial x_j} \delta P_\ell + \sum_{m=k+1}^{t-1} \frac{\partial^2 \psi^{(k)}}{\partial x_m \partial x_j} \delta x_m \tag{2.71}$$

We now substitute in (2.69)

$$- \sum_{i,\ell=0}^{k} (\partial^2 \psi^{(k)} / \partial P_\ell \partial x_j) \delta P_\ell \delta P_i - \sum_{i=0}^{k} \sum_{m=k+1}^{t-1} (\partial^2 \psi^{(k)} / \partial x_m \partial P_\ell) \delta x_m \delta P_i$$

$$+ \sum_{\ell=0}^{k} \sum_{j=k+1}^{t-1} (\partial^2 \psi^{(k)} / \partial P_\ell \partial x_j) \delta x_j \delta P_\ell + \sum_{j,m=k+1}^{t-1} (\partial^2 \psi^{(k)} / \partial x_m \partial x_j) \delta x_j \delta x_m > 0 \tag{2.72}$$

Note that the second and third terms on the l.h.s. cancel out and the remainder can symbolically be written as

$$- |\delta^2 \psi^{(k)}|_{x_j} + |\delta^2 \psi^{(k)}|_{P_i} > 0 \tag{2.73}$$

This inequality can be fulfilled if, and only if, the first quadratic form is negative definite and the second positive definite, i.e.

$$\left|\delta^2 \psi^{(k)}\right|_{x_j} < 0 \tag{2.74}$$

$$\left|\delta^2 \psi^{(k)}\right|_{P_i} > 0 \tag{2.75}$$

An equivalent and more practical formulation is

$$|\partial^2 \psi^{(k)} / \partial P_i \partial P_\ell| \lessgtr 0 \tag{2.76}$$

where the first inequality is to be fulfilled by the principal minors of odd order and the second by the principal minors of even order .

Similarly

$$|\partial^2\psi^{(k)}/\partial x_j \partial x_m| > 0 \quad \text{and all principal minors} > 0 \tag{2.77}$$

Let us now examine a little more closely an important special case of the conditions (2.76): the Gibbs free energy. Putting $\psi^{(k)} \equiv g(T,p)$ we can write (2.76) as

$$\begin{vmatrix} g_{TT} & g_{Tp} \\ g_{pT} & g_{pp} \end{vmatrix} > 0, \quad g_{TT} < 0, \quad g_{pp} < 0 \tag{2.78}$$

It can easily be seen that only two of these conditions are independent e.g. the third inequality follows from the first two. The second inequality gives

$$c_p > 0 \qquad \text{(thermal stability)} \tag{2.79}$$

and the third leads to

$$k_T > 0 \qquad \text{(mechanical stability)} \tag{2.80}$$

Note, however, that for the Helmholtz free energy of a one-component system the conditions (2.76)-(2.77) can only be used in relation to principal minors of odd order, since the even order minor contains mixed terms such as f_{Tv} which have not a definite sign. The stability does not give any information about the mixed terms. In the present case

$$f_{TT} < 0 \quad \text{or} \quad -(\partial s/\partial T)_v < 0, \quad \text{i.e.} \quad c_v > 0 \tag{2.81}$$

$$f_{vv} > 0 \quad \text{or} \quad -(\partial p/\partial v)_T > 0, \quad \text{i.e.} \quad k_T > 0 \tag{2.82}$$

but

$$f_{Tv} = f_{vT} = -(\partial p/\partial T)_v = -\alpha/k_T = ?$$

In the entropy scheme the results are completely analogous

$$|\delta^2\phi^{(k)}|_{x_j} > 0 \tag{2.83}$$

$$|\delta^2\Phi^{(k)}|_{P_i} < 0 \tag{2.84}$$

$\Phi^{(k)}$ being the partial Legendre transform of the entropy s.

2.6 Heterogeneous systems (mutual stability)

The stability conditions for heterogeneous systems can be obtained from the results given in the precedent section. It can be shown that heterogeneous systems are necessarily stable if the stability conditions are fulfilled for each coexistent phase. In the energy representation and following a reasoning similar to given in 2.5 we get for two phases in equilibrium

$$u_{ss}^{(1)} + u_{ss}^{(2)} > 0 \tag{2.79}$$

$$u_{vv}^{(1)} + u_{vv}^{(2)} > 0 \tag{2.80}$$

$$(u_{ss}^{(1)}+u_{ss}^{(2)})(u_{vv}^{(1)}+u_{vv}^{(2)})-(u_{sv}^{(1)}+u_{sv}^{(2)})^2 > 0 \tag{2.81}$$

where $(^1)$ and $(^2)$ denote the phases. Clearly, the stability of the individual phases is a necessary condition for the stability of the entire system. Thus

$$u_{ss}^{(i)} > 0 \tag{2.82}$$

$$u_{vv}^{(i)} > 0 \tag{2.83}$$

$$u_{ss}^{(i)}u_{vv}^{(i)}-(u_{sv}^{(i)})^2 > 0 \tag{2.84}$$

must be true for i=1,2. Obviously, the inequalities (2.79) and (2.80) follow from (2.82) and (2.83). Furthermore, the elementary algebra shows that

$$\begin{aligned}(a_1+a_2)(c_1+c_2)-(b_1+b_2)^2 = \; & \{1+(a_1/a_2)\}(a_2c_2-b_2^2) \\ & +\{1+(a_2/a_1)\}(a_1c_1-b_1^2) \\ & +(a_1b_2-a_2b_1)^2/a_1a_2 \end{aligned} \tag{2.85}$$

and, therefore, that (2.81) follows directly from (2.82)-(2.84). So, it may be concluded that the fulfilment of the stability conditions for each phase separately is the necessary and sufficient condition for the stability of a heterogeneous system or mutual stability.

3. STABILITY IN IRREVERSIBLE THERMODYNAMICS

In this section we shall discuss the stability of nonequilibrium systems. For this goal we are going to study hiefly the linear and non-linear range. We start reviewing the basic equations of continuum mechanics and thermodynamics, remaining in this case in the framework of local equilibrium theory.

For a more detailed account on the foundations, one can be directed to the excellent monographs of De Groot and Mazur [3], Gyarmati [11] and Glansdorff and Prigogine as well to the recent book by Nicolis and Prigogine [8,17] and some interesting reviews [4].

3.1 Basic equations

3.1.1 Balance equations

From continuum mechanics one has

$$\partial\rho_i/\partial t = -\nabla\cdot\rho_i\underline{v}_i+\sigma_{m_i} \qquad (3.1) \quad \text{mass balance}$$
$$d\rho/dt = -\rho\nabla\cdot\underline{v}$$

$$\rho d\underline{v}/dt = -\nabla\cdot\underline{\underline{P}}+\underline{F} \qquad (3.2) \quad \text{momentum balance}$$

$$\rho du/dt = -\nabla\cdot\underline{J}_u+\underline{\underline{P}}:\nabla\underline{v}+\rho r \qquad (3.3) \quad \text{energy balance: local form of first law of thermodynamics}$$

$$\underline{\underline{P}} = \underline{\underline{P}}^T \qquad (3.4) \quad \text{angular momentum balance}$$

where σ_{m_i} is the mass production terms (chemical reactions), $\underline{\underline{P}}$ the pressure tensor, $\underline{J}_u$ the heat flux and r the energy supply density.

From thermodynamics: the irreversibility may be expressed by means of the Clausius inequality [8]

$$dS - d_eS = d_iS \geq 0 \qquad (3.6)$$

and introducing the specific entropy field $s=s(\underline{x},t)$ through

$$S = \int_V \rho s dV \qquad (3.6)$$

the expression (3.5) becomes

$$\rho ds/dt + \nabla \cdot \underline{J}_s - \rho r/T = \sigma \geq 0 \qquad (3.7)$$ local form of second law of thermodynamics

$\underline{J}_s$ and σ being the entropy flux and the entropy production, respectively.

In a given physical problem one can require complete control of $\underline{F}$, r and boundary conditions. The remaining fields ρ, ρ_i, $\underline{v}$, $\underline{v}_i$, $\underline{\underline{P}}$, u, $\underline{J}_u$, T, s, $\underline{J}_s$, σ are to be determined by the theory. Obviously Eqs. (3.1)-(3.3) and (3.7) are not sufficient, and for this reason we must supplement them with information on the nature fo the system of interest and on the type of processes occurring therein. This is usually achieved by the <u>constitutive equations</u>.

3.1.2 Constitutive equations

In the most general case, constitutive equations achieve the closure of the field equations (3.1)-(3.3) and (3.7) by expressing some of the quantities therein as functionals of the others. From a operational point of view one chooses as primitive variables the temperature T and its derivatives, the barycentric velocity $\underline{v}$ and its derivatives, the specific volume $1/\rho$ and the composition variables ρ_i. In the case of anisotropic materials the specific volume is to be replaced by the strain tensor. Now, we can distinguish a group of equations expressing the functional dependence of the fluxes on the primitive variables

$$\underline{v}_i,\ \underline{J}_u,\ \sigma_{m_i} \text{ and } \underline{\underline{P}} \text{ functions of } T,\ \dot{T},\ T,\ \underline{v},\ \underline{v},\ \nabla\underline{v},\ \rho \text{ and } \rho_i \qquad (3.8)$$

from another group that indicates how the state functionals depend on these variables, i.e.

$$u = \mathcal{U}(T,\rho,\{\rho_i\};\ \underline{x},t) \qquad (3.9)$$

$$s = \mathcal{S}(T,\rho,\{\rho_i\};\ \underline{x},t) \qquad (3.10)$$

Once these are specified, the quantities $\underline{J}_s$ and σ in Eq. (3.7) can be determined from the balance equations (3.1)-(3.3). In equilibrium (3.9) and (3.10) become ordinary funtions of the state variables, both in closed and open systems

$$s = s(T,\rho,\{\rho_i\}) \tag{3.11}$$

3.2 Linear irreversible thermodynamics

3.2.1 Gibbs equation and entropy production

Since the fundamental equation $s=s(u,1/\rho,\{\rho_i\})$ contains all possible thermodynamic information about a multicomponent system, its differential form or Gibbs equation

$$T\,ds/dt = du/dt + p\,d\rho^{-1}/dt - \sum_i \mu_i\, d\rho_i/dt \tag{3.12}$$

plays a capital role in the classical description of nonequilibrium processes if the local equilibrium hypothesis is assumed to hold. This equation when combined with balance equations (3.1)-(3.3) writes

$$\rho\, ds/dt = -\nabla\cdot\underline{J}_s + \sigma \tag{3.13}$$

where, in order to be simpler (but no less general), we have neglected the energy supply density r, and the entropy flux and the entropy production can be expressed respectively as

$$\underline{J}_s = \underline{J}_u/T - \sum_i \underline{J}_i(\mu_i/T) \tag{3.14}$$

$$\sigma = \sum_\alpha J_\alpha X_\alpha \geq 0 \tag{3.15}$$

J_i being the diffusion flux of constituent i

$$\underline{J}_i = \rho_i(\underline{v}_i - v) \tag{3.16}$$

J_α and X_α are conjugate variables (generalized fluxes and forces) associated with the various irreversible processes. A suitable choice is the following

Process	Flux, J_α	Generalized force, X_α
Heat conduction	$\underline{J}_u - \sum_i h_i \underline{J}_i$	$\nabla T^{\ 1}$
Diffusion	$\underline{J}$	$-T^{-1}\{(\nabla\mu_i)_T - \underline{F}_i\}$
Viscous flow	$\underline{\underline{P}}$	$-T^{-1}$
Chemical reaction	Reaction rate J_ρ	Affinity $T^{-1}A_\rho$

The affinity A_ρ, which is included in the source term of the balance equation for $\{\rho_i\}$ (3.1), is defined by

$$A_\rho = -\sum_j \mu_j \nu_{j\rho} \tag{3.17}$$

where $\nu_{j\rho}$ is the stoichiometrix coefficient of the constituent j in the chemical reaction ρ. For convection, $\nu_{j\rho} > 0$ if the constituent j is on the r.h.s. of the reaction and $\nu_{j\rho} < 0$ if the constituent is on the l.h.s.

3.2.2 Phenomenological relations

Thermodynamic equilibrium is characterized by

$$X_\alpha^{eq} = 0; \quad J_\alpha^{eq} = 0 \tag{3.18}$$

which suggests the possibility of defining the vicinity of equilibrium by the property that both X_α and J_α remain small and consequently

$$J_\alpha(\{X_\beta\}) = J_\alpha(\{0\}) + \sum_\beta (\partial J_\alpha/\partial X_\beta)_0 X_\beta + \tfrac{1}{2}\sum_{\beta,\gamma}(\partial^2 J_\alpha/\partial X_\beta \partial X_\gamma)_0 X_\beta X_\gamma + \ldots \tag{3.19}$$

the first term of the expansion being $J_\alpha(\{0\}) \equiv J_\alpha^{eq} = 0$

Sufficiently close to equilibrium, the higher-order terms may be neglected so that (3.19) reduces to

$$J_\alpha = \sum_\beta L_{\alpha\beta} X_\beta \tag{3.20}$$

where

$$L_{\alpha\beta} = (\partial J_\alpha/\partial X_\beta)_0 \tag{3.21}$$

are the phenomenological coefficients, which are determined by the internal structure of the system, independently of the applied constraints. They may depend, however on the state variables $T, p, \{\rho_i\}$, etc. : $L_{\alpha\beta} = L_{\alpha\beta}(T, p, \{\rho_i\}, \ldots)$. In virtue of relations (3.20) the balance equations (3.1)-(3.4) become entirely closed, enabling explicit evaluation of the generalized forces or, in an equivalent manner of the state variables. The phenomenological relations (3.20) can be used as a definition of the linear range of irreversible thermodynamics.

We now summarize the restrictions imposed on the matrix of $\{L_{\alpha\beta}\}$. These restrictions proceed from

i. the positive semidefiniteness of the bilinear form

$$\sigma = \sum_{\alpha,\beta} L_{\alpha\beta} X_\alpha X_\beta \geq 0 \tag{3.22}$$

obtained by substitution of (3.20) into (3.16), the inequality remaining valid for all possible values of generalised forces and reducing it to an equality at thermodynamic equilibrium. As we have seen in Section 2.4, the positive definite character of σ involves the non-negativity of the diagonal coefficients, $L_{\alpha\alpha} \geq 0$.

ii. the Curie's principle by means of which one may know the allowed couplings among two irreversible processes. One interpretation of this principle is that, " macroscopic causes cannot have more elements of symmetry than the effects they produce" [22]. Another statement forbids all coupling between quantities of different tensorial character as long as the medium is isotropic at equilibrium [19]. This is no longer true when second-order terms are taken into account in (3.19). It follows that relation (3.15) can be split in

$$\sigma = \sigma_{ch}^{(s)} + \sigma_{v}^{(t)} + \sigma_{th}^{(v)} + \sigma_{d}^{(v)} \tag{3.23}$$

where the entropy production for viscous phenomena $\sigma_v^{(t)}$ can be also decomposed in the following way

$$\sigma_v^{(t)} = \sigma_v^{(s)} + \sigma_v^{(a)} + \sigma_v^{(0)} \tag{3.24}$$

i.e. in its scalar, axial vector and traceless 2nd. order tensor parts.

iii. the time-reversal invariance which implies that the matrix

$\{L_{\alpha\beta}\}$ is symmetrical (3)

$$L_{\alpha\beta} = L_{\beta\alpha} \tag{3.25}$$

These equations, which play a similar role in irreversible thermodynamics as the Maxwell relations in thermostatics, are the celebrated Onsager's reciprocal relations.

3.2.3 A variational criterion: minimum entropy production

It is well-known from classical thermodynamics the important role played by the thermodynamic potentials in the description of equilibrium and stability conditions of systems in contact with reservoirs prescribing one or more intensive parameters $P_0^{(r)}$, $P_1^{(r)}$... For example, a system mantained at a constant temperature T through contact with a heat reservoir can evolve to a state of minimum free energy F if phase transitions are excluded. On the contrary, if all contact is absent (isolated system) $d_eS = 0$ and (3.5) implies that $dS \geq 0$, and if the system exists in a single stable phase, it follows that evolution will lead to a unique equilibrium state for which entropy S is maximum.

At first glance, the extension of these variational properties away from equilibrium may appear difficult. An open system subject to time-independent constraints has the possibility of evolving to a regime known as a (non-equilibrium) steady state for which all state variables are time-independent. From (3.5)

$$- d_eS = d_iS > 0$$

or (3.26)

$$\nabla \cdot J_s = \sigma > 0$$

Because of this, the entropy variation dS (and similarly dF) need not have a definite sign. Hence S or F cannot serve as thermodynamic potentials.

However, away from equilibrium a new thermodynamic potential emerges in the linear range of dissipative phenomena, namely the entropy production. Consider

$$P = \int_V \sigma \, dV \geq 0 \qquad (3.27)$$

One can evaluate the time derivative dP/dt using the explicit form of $\sigma|Eq.(15)|$, the definition of J_α and X_α as given in §3.2.1 and the balance equations for u, ρ, $\underline{v}$, $\{\rho_i\}$. In doing this one applies the reciprocity relations (3.25) and besides one assumes that $L_{\alpha\beta}$ can be treated as constants (strict linearity condition). One is led in this way to an expression of dP/dt which is quadratic in the derivatives of the state variables. For instance, the part of the expression coming from heat conduction

$$\frac{dP}{dt} = -2\int_V \rho \, c_v T^{-2} (\partial T/\partial t)^2 \, dV \qquad (3.28)$$

It can be show, that if equilibrium is stable against heat conduction, i.e. if the system at equilibrium is capable of damping any thermal disturbance that may arise, then $c_v > 0$. For diffusion and chemical reactions

$$\frac{dP}{dt} = -\frac{2}{T}\int_V \sum_{ij} (\partial\mu_i/\partial\rho_j) \partial\rho_i/\partial t \; \partial\rho_j/\partial t \; dV \qquad (3.29)$$

where the coefficient matrix $(\partial\mu_i/\partial\rho_j)$ of this quadratic form is known from the analysis of equilibrium state stability. In this case, if the system at equilibrium is able of absorbing any material inhomogeneity that may emerge $\partial\mu_i/\partial\rho_j > 0$.

We must bear in mind that we are restricted to a local theory where the various functions are given in terms of the state parameters by the same formal expressions as in equilibriu. This means that u or $\partial u/\partial T$ is the same function of T is the same function of as in equilibrium. Similarly, this means also that μ_i or $\partial\mu_i/\partial\rho_j$ is the same function of $\{\rho_i\}$ as in equilibrium. atlthough its numerical value for a non-equilibrium distribution of $\{\rho_i\}$ will differ slightly from its value at the corresponding equilibrium state. Since positive definiteness is an intrinsic property that remains valid for all possible stable equilibria, it follows that $\{\partial\mu_i/\partial\rho_j\}$ will keep this property away from equilibrium as well. This fact is a consequence of the following: the values of state variables for all stable equilibria define a domain in state space which is larger than or equal to the domain of the values of the variables at the nonequilibrium steady state. From (3.28) we arrive to

$$\frac{dP}{dt} \leq 0 \qquad (3.30)$$

in which inequality is verified by states away from the steady state while the time derivative of P becomes zero at the steady state. These two relations (3.27) and (3.30) imply that at nonequilibrium steady state entropy production becomes a minimum, compatible with the constraints applied on the system, provided equilibrium itself is a stable state. This situation is shown in Fig. 1 where P_{in} and P_{st}

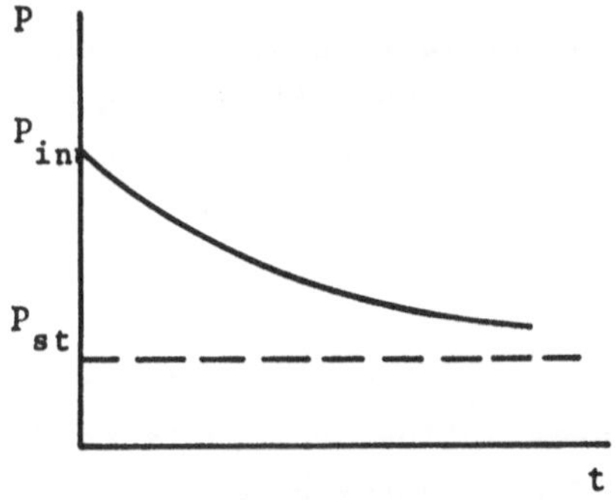

Fig. 1

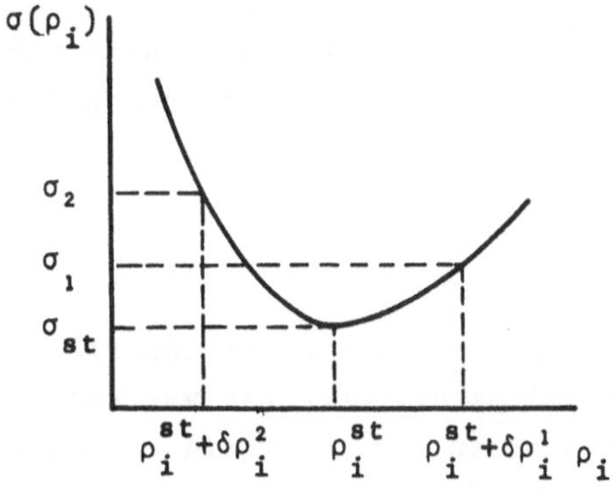

Fig. 2

stand for the initial value of the entropy production and the corresponding value at the steady state, respectively. Observe that if acting externally on the system, this is removed from the steady state, it will be directed again toward this state according to the minimum entropy production criterion. This behaviour is sketched in Fig.2, where $\delta\rho_i$ represent arbitrary deviations in composition variables.

As we see the stability of equilibrium ensures the stability of stationary states near equilibrium and therefore any qualitatively new patterns of behaviour emerging spontaneously, such as temporal oscillations, are excluded in this range.

3.3 Non-linear thermodynamics

The equations of evolution of a system undergoing dissipative processes in the linear range are themselves linear in the state variables. In fact, if one substitutes the phenomenological equations (3.20) into the balance equations (3.1)-(3.3) and expresses the thermodynamic forces X in terms of the state variables x_i , one then finds a linear set of first-order equations

$$\partial x_i/\partial t = - \sum b_{ij} x_j \tag{3.31}$$

where $\{b_{ij}\}$ is in general a matrix which phrases the effects of the diverse dissipative phenomena occurring in the system, for instance, heat conduction, chemical reactions and so on. On the other hand, by virtue of the criterion of minimum entropy production there exists a suitable choice of variables for which these equations take a variational form, i.e. they derive from a kinetic potential related to the entropy production.

A variety of reasons advises the extension of irreversible thermodynamics away the linear range. In particular, chemical reactions are very poorly described by linear phenomenological laws. Leaving aside the details of the new phenomenological relations, the balance equations (3.1)-(3.3) will now take the form of a set of non-linear equations

$$\partial x_i/\partial t = f_i(\{x_j\};\lambda) \tag{3.32}$$

in which x_i represent the state variables, f_i the rates and λ a set of parameters that may enter in the description. For instance, in the absence of convective motion the mass and energy balance equations would give the following structure for f_i

$$\partial \rho_i/\partial t = -\nabla\cdot\underline{J}_i + \sum_\rho \nu_{i\rho} J_\rho \tag{3.33}$$

$$\partial(\rho_u)/\partial t = -\nabla\cdot\underline{J}_u + \rho r \tag{3.34}$$

The non-linear phenomenological relations imply that $\underline{J}_i$, J_ρ, $\underline{J}_u$ are given in terms of the state variables and their space derivatives. Hence the balance equations become closed.

3.3.1 A general criterion of evolution

Starting from the well-known expression of total entropy production of the system

$$P = \int_V \sum_\alpha J_\alpha X_\alpha(\{x_j\})\, dV \tag{3.35}$$

where x_j obey the evolution equations (3.32), we evaluate the quantity

$$\frac{d_X P}{dt} \equiv \int_V \sum_\alpha J_\alpha \partial X_\alpha / \partial t \; dV = \int_V \sum_\alpha J_\alpha \frac{\partial X_\alpha}{\partial X_i} \frac{\partial X_i}{\partial t} \; dV \qquad (3.36)$$

subject to time-independent boundary conditions. Using now the explicit forms of J_α, X_α and f_i (see Eq. (3.32)-(3.34)) and Table given in §3.2.1 one obtains a quadratic form in the time derivatives of x_i of the same kind as in Eq. (3.28). Assuming that equilibrium is stable it can be demonstrated that this quadratic form is negative definite and then

$$\frac{d_X P}{dt} \leq 0 \qquad (3.37)$$

equality being valid at the stationary state. It may be pointed out that this inequality gives no information on the sign of dP/dt itself or on the sign of ([3])

$$\frac{d_J P}{dt} \equiv \int_V \sum_\alpha \frac{\partial J_\alpha}{dt} X_\alpha dV \qquad (3.38)$$

From this result, we may now begin to understand why in general the non-linear rate laws (3.32) cannot derive from a potential.

3.2 Thermodynamic stability criteria

Let us first write the results on minimum entropy production in somewhat different terms. Introducing the function $\Delta P = P - P_{st}$ we see (Fig.1) that

$$\Delta P = P - P_{st} \geq 0 \qquad (3.39)$$

with equality at the stationary state. For given values x_{st}, ΔP is a function of the displacements $\delta x = x - x_{st}$, or more precisely a positive definite functional because it is positive if $\delta x \neq 0$ and vanishes only at $\delta x = 0$. On the other hand, from (3.30) we arrive at

$$\partial \Delta P / \partial t \leq 0 \qquad (3.40)$$

where the time derivative of ΔP vanishes at the stationary state.

Such a functional is called a Lyapunov functional. By virtue of a theorem of analysis fue to Lyapunov properties (3.39) and (3.40) guarantee that the system will evolve back to the reference state x_{st} if perturbed initially from this state (asymptotic stability).

We see that the criteria of minimum entropy production is essentially a Lyapunov stability theorem. Now, the properties of the differential form $d_X P$ suggest that in the non-linear range of irreversible phenomena Lyapunov stability cannot be expressed in terms of entropy production.

The search of a new Lyapunov functional giving information on stability in this range was carried out by Glansdorff and Prigogine(8). As a first step, these authors introduced the excess entropy around the reference stationary state

$$\Delta S = S(\{x_i\}) - S(\{x_i\}_{st}) \tag{3.41}$$

expanded $S(\{x_i\})$ around $S(\{x_i\}_{st})$ and obtained

$$\Delta S = (\delta S)_{st} + \tfrac{1}{2}(\delta^2 S)_{st} + \dots \tag{3.42}$$

in which

$$(\delta^2 S)_{st} = \int_V \sum_{ij} \left(\frac{\partial^2(\rho s)}{\partial x_i \partial x_j}\right)_{st} \delta x_i \delta x_j \; dV \tag{3.43}$$

derivatives veing evaluated at the stationary state. If this were true equilibrium state, the stability against phase changes would imply

$$(\delta^2 S)_{st} \equiv (\delta^2 S)_{eq} \leq 0$$

Later, they extended these results away from equilibrium

$$(\delta^2 S)_{st} \leq 0 \tag{3.44}$$

equality remaining at the reference state. The next step consists of evaluating $d(\delta^2 S)_{st}/dt$ from (3.43) together with the evolution equations (3.32). The obtained result is

$$\frac{1}{2}\frac{d}{dt}(\delta^2 S)_{st} = \int_V \sum_\alpha \delta J_\alpha \delta X_\alpha \; dV \tag{3.45}$$

called the excess entropy production, It can be shown that this quantity is identical to $\delta_X P$, defined in a way similar to (3.36)

$$\delta_X P \equiv \int_V \sum_\alpha J_\alpha \delta X_\alpha dV = \int_V \sum_\alpha J_{\alpha,st} \delta X_\alpha dV + \int_V \sum_\alpha \delta J_\alpha \delta X_\alpha dV \qquad (3.46)$$

because the first term on r.h.s vanishes identically (17). The Lyapunov theorem ensures then the stability of the reference state,

$$\delta_X P \geq 0 \qquad (3.47)$$

along the solutions of the dyanmical equations of the thermodynamic system. At the reference state the excess entropy production becomes zero. Relations (3.45) and (3.47) constitute a thermodynamic stability condition for nonequilibrium states. It must be emphasized that, in contrast to (3.45), inequality (3.47) is not satisfied identically unless the system operates in the linear range, constituting rather a sufficient condition for stability. When this condition is violated away from equilibrium the possibility to transitions driving the system toward new and qualitatively different regimes is created. In this case the transition threshold could be expressed by

$$\delta_X P = \delta_X P(\{x_j\},\lambda_c) = 0 \qquad (3.48)$$

This relation determines the threshold values λ_c of λ driving the system away from the reference state. Among these parameters the strength of generalized thermodynamic forces measuring the distance from equilibrium plays an important role.

3.3 Nonequilibrium phase transitions. Dissipative structures

In the preceding section we saw that non-equilibrium phenomena in the non-linear range can lead to new regimes differing qualitatively from the reference stationary state.

Imagine now a process causing a systematic displacement from equilibrium, for instance, the increase of a state variable or of a parameter λ. The composition variables are subject to the changes described qualitatively in Fig. 3. By virtue of the criterion of minimum entropy production, the steady states close to equilibrium remain asymptotically stable. These states are depicted on the branch (a) in Fig. 3. By reasons of continuity this branch of states (the

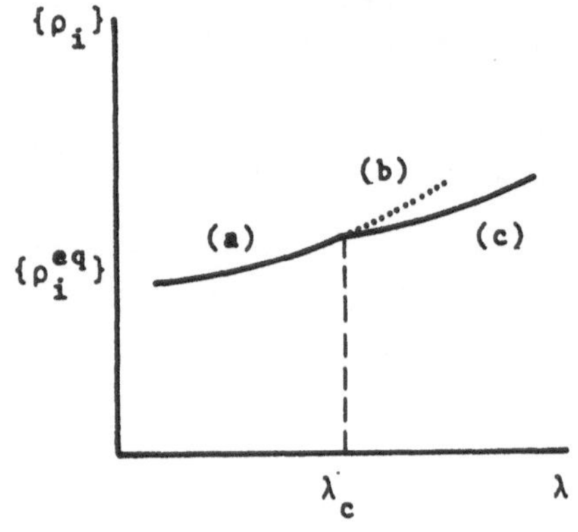

Fig. 3

thermodynamic branch) keeps up its stability in a finite neighborhood of the equilibrium state. However is a c critical value λ_c is reached we could not exclude the possibility that the states on the thermodynamic branch become unstable, situation sketched as branch (b) on Fig. 3. When this point is reached, the least disturbance on the system forces it to evolve away f from this branch. An ordered configuration can characterize the new regime represented by branch (c). At $\lambda=\lambda_c$ it occurs a bifurcation to a new branch of solutions in which the system is stable.

Now we summarize the different situations that can emerge in a given system. Near equilibrium the excess entropy production $\delta_X P$ is positive definite as an immediate consequence of the second law, but far from equilibrium this property need not be positive definite. The possibilities arising in this more general case are depicted in Fig. 4. Bearing in mind that the time derivative of $\frac{1}{2}(\delta^2 S)$ is the excess entropy production, the nonequilibrium steady state becomes unstable as soon as $\delta_X P$ becomes negative for $t \geq t_0$. In consequence we say that one has an unstable reference state if $\delta_X P<0$ for $t \geq t_0$ where as if $\delta_X P>0$ for $t \geq t_0$ we refer to this reference state as asymptotically stable, The sign of this last inequality may be inverted for a critical value of λ and the reference state looses its stability. This situation is known as one of marginal stability, for which $\delta_X P(\lambda_c)=0$.

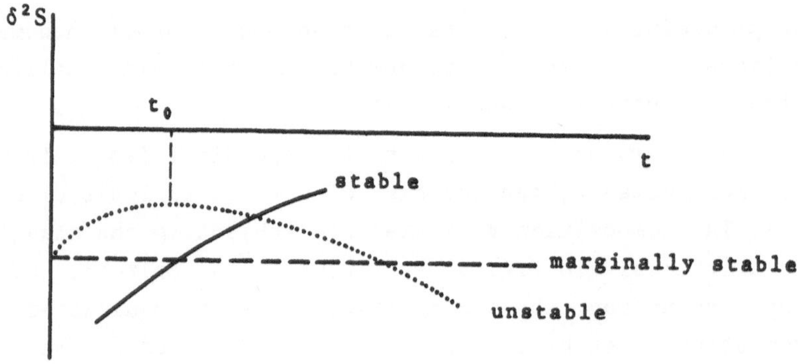

Fig. 4

As an example we relate the above stability conditions to the kinetics of chemical reactions. Let us consider first an unimolecular reaction

$$Y \underset{k_2}{\overset{k_1}{\rightleftarrows}} C \qquad (3.49)$$

where k_1 , k_2 are the kinetic constants of the reaction. The rate of this reaction and its affinity can be written

$$J = k_1\rho_y - k_2\rho_c \qquad (3.50)$$

$$A = k_BT \ln(k_1\rho_y/k_2\rho_c) \qquad (3.51)$$

k_B being the Boltzmann's constant. Assuming ρ_c and T constant, the variations around the steady state are

$$\delta J = k_1\delta\rho_y \qquad (3.52)$$

$$\delta A = k_BT(\delta\rho_y)/\rho_y^o \qquad (3.53)$$

and consequently the excess entropy production is

$$\delta_X\sigma = \delta J\delta(A/T) = k_Bk_1(\delta\rho_y^o)^2/\rho_y > 0 \qquad (3.54)$$

with ρ_y^o the value of the state variable at the steady state. According to this result we can conclude that reaction (3.49) tends to stabilize the system. By the contrary in the autocatalytic reaction

$$C + Y \underset{k_2}{\overset{k_1}{\rightleftarrows}} 2Y \qquad (3.55)$$

the expressions corresponding to (3.50)-(3.54) are respectively

$$J = k_1\rho_c\rho_y - k_2\rho_y^2 \qquad (3.56)$$

$$A = k_BT \ln(k_1\rho_c\rho_y/k_2\rho_y^2) \qquad (3.57)$$

$$\delta J = (k_1\rho_c - 2k_2\rho_y^o)\delta\rho_y \qquad (3.58)$$

$$\delta A = -k_BT(\delta\rho_y)/\rho_y^o \qquad (3.59)$$

and

$$\delta_X \sigma = - \frac{k_B}{\rho_y^\circ} (k_1 \rho_c - 2k \rho_y^\circ)(\delta \rho_y^\circ) \tag{3.60}$$

where as before T and ρ_c have been mantained constant. This expression has not a definite sign and under certain conditions, it can become negative. Therefore, we arrive to the following conclusion: autocatalytic reactions or, in general, reactions involving nonlinear steps tend to destabilize the system. Naturally, a single reaction such as (3.55) cannot produce an instability, as it will always evolve to equilibrium. On the other hand, such a process can be a part of an open system undergoing a whole sequence of reactions (9).

We have seen that the distance from equilibrium and the nonlinearity may both be sources of order capable of driving the system to an ordered configuration. It may be said that order, stability and to indicate clearly this convection the ordered configurations that arise beyond instability of the thermodynamic branch are called dissipative structures.

4. STABILITY IN GENERALIZED (EXTENDED) THERMODYNAMICS

It is ell known that the classical theory of irreversible thermodynamics (3) (8) rests on the local equilibrium hypothesis, which states that locally the Gibbs equation remains valid. In the last decade, some authors (6) (12) (15) have tried to build a theory of irreversible thermodynamics which is not based on that hypothesis. Indeed, in the usual development of irreversible thermodynamics one is led to the customary Fourier's law for heat conduction in solids, which gives a parabolic equation for the evolution of the temperature and therefore predicts the propagation of the thermal distrubances at infinite velocity, and which is not able to describe, for instance, the phenomenon of second sound observed in some solids at low temperatures.

In our approach to nonequilibrium thermodynamics, the existence of a nonequilibrium entropy is assumed whose dependence on the dissipative fluxes as well as on the classical thermostatic variables can be expressed through a generalized Gibbs equations. In the next subsections we present some examples of the stability problem in the framework of extended thermodynamics.

4.1 Heat conduction ([14])

The starting point is a generalized Fourier law in the form

$$q_i + \tau \overset{\circ}{q}_i = LT^{-1}_{,i} \qquad \text{(isotropic systems)} \tag{4.1}$$

T is the temperature, q_i the ith Cartesian coordinate of the heat flux vector, τ the relaxation time and L the phenomenological coefficient related to the thermal conductivity λ by $L=\lambda T^2$. The subscript, i denotes partial derivation with respect to the coordinate x_i. The phenomenological law (4.1) emerges from the generalized Gibbs equation

$$ds_v = T^{-1}du_v - \alpha_i T^{-1} dq_i \tag{4.2}$$

in which $\alpha_i = -T\tau T^{-1}_{,i}$. This Gibbs equarion is the differential form of an extended fundamental equation

$$s_v = s_v(u_v, q_i) \tag{4.3}$$

For fixed temperature, or heat flux, at the boundary of the body, Glansdorff and Prigogine ([8]) have shown that heat conduction is always stable with respect to infinitesimally small temperature disturbances. Based on the Lyapunov analysis, the Lyapunov functional is

$$\delta^2 S = \int_V \delta^2 s_v dV = \int_V \delta u_v \delta T^{-1} dV = -\int_V \frac{c_v}{T}(\delta T)^2 dV \tag{4.4}$$

In the framework of the local equilibrium hyphothesis, $c_v>0$ from which follows that $\delta^2 S<0$, i.e. a negative definite quantity. Moreover, these authors show that

$$\frac{d}{dt}\,\delta^2 S \geq 0 \tag{4.5}$$

for any dependence of L on the temperature. Consequently stability is concluded.

From (4.2) one can obtain the second variation of s_v

$$\delta^2 s_v = \delta T^{-1} \delta u_v - \delta(T^{-1}\alpha_i)\delta q_i \tag{4.6}$$

and after some lengthy but elementary calculations, one gets

$$\delta^2 S = \int_V \delta^2 s_v dV = - \int_V \{\frac{c_v}{T^2}(\delta T) + \frac{1}{T\chi}\delta q_i \delta q_i\} \, dV \tag{4.7}$$

where

$$c_q = T(\partial s_v/\partial T)_q \quad \text{and} \quad \chi = (\partial q_i/\partial \alpha_i)_T \tag{4.8}$$

Since the local equilibrium hypothesis cannot be involved at this point, no conclusion can be drawn about the signs of c_q and χ.

Assuming that the coefficients c_q/T^2 and $1/\chi T$ in (4.7) are frozen at a given arbitrary time t_0, one defines another functional which is no longer $\delta^2 S$. For constant values of these coefficients the time derivative of (4.7) is given by

$$\frac{1}{2}\frac{d}{dt}\int_V \delta^2 s_v dV = \int_V \{ \delta T^{-1}\delta q_{i,i} + \delta(\tau T^{-1}_{,i})\delta \mathring{q}_i\} dV \tag{4.9}$$

if the energy conservation law is used in the form

$$\delta \mathring{u}_v = -\delta q_{i,i} \tag{4.10}$$

Keeping T fixed at the boundary, interchanging the symbols δ and $,i$ and integrating by parts the first term under the integral results in

$$\frac{1}{2}\frac{d}{dt}\delta^2 S = \int_V \{\delta q_i(\delta T^{-1})_{,i} + \tau\delta T^{-1}_{,i}\delta\mathring{q}_i + T^{-1}_{,i}\delta\tau\delta q_i\} \, dV \tag{4.11}$$

If one assumes that τ is constant, (4.11) reduces to

$$\frac{1}{2}\frac{d}{dt}\delta^2 S = \int_V \{(\delta q_i + \tau\delta\mathring{q}_i)\delta T^{-1}_{,i}\} \, dV \tag{4.12}$$

and by virtue of (4.1)

$$\frac{1}{2}\frac{d}{dt}\delta^2 S = \int_V \delta(LT^{-1}_{,i})\delta T^{-1}_{,i} \, dV \tag{4.13}$$

For constant and positive L, the r.h.s. of (4.13) is positive definite and the system is stable if $S\ 0$, which implies that the sufficient stability conditions are

$$c_q > 0 \quad \text{and} \quad \chi > 0 \tag{4.14}$$

If L depends on T, (4.13) becomes after integration by parts

$$\frac{1}{2}\frac{d}{dt}\delta^2 S = \int_V \{L(\delta T^{-1}_{,i}) - \frac{1}{2}L_{,ii}(\delta T^{-1})^2\}\, dV \tag{4.15}$$

Clearly, if $L_{,ii}<0$, the stability conditions are still given by (4.14) However, for $L_{,ii}>0$ and for the more general case that T is not constant, no general information can be derived. The stability conditions depend then on the particular characteristics of the material and the process involved.

Therefore and contrary to what happens in the classical situation ([8]), it cannot be concluded that heat conduction is always stable.

4.2 Coupled heat and mass transfer

In order to generalize the results given in 4.1, Sieniutycz ([29]) has used an alternative method, complementary in realtion to that described above. The author starts from the matrix phenomenological equation

$$\underline{J} + \underline{\underline{T}}\cdot\dot{\underline{J}} = \underline{\underline{L}}\cdot\text{grad}\ \underline{u} \tag{4,16}$$

in which

$$\underline{\underline{T}} = -\underline{\underline{L}}\cdot\underline{\underline{C}}^{-1}(\rho c_0^2)^{-1} \tag{4.17}$$

$$\underline{J} = \text{col}(\underline{J}_1, \underline{J}_2, \ldots \underline{J}_n = \underline{J}_q)$$

$$\text{grad}\ \underline{u} = \text{col}(\text{grad}\ \frac{\mu_n - \mu_1}{T}, \text{grad}\ \frac{\mu_n - \mu_2}{T}, \ldots, \text{grad}\ \frac{1}{T})$$

If y_i and h denote the independent mass fractions and the specific enthalpy

$$\underline{z} = \text{col}(y_1, y_2 \ldots y_{k-1}, h)$$

$$\underline{u} = \text{col}\left(\frac{\mu_n - \mu_1}{T}\ , \ \ldots \frac{1}{T}\right)$$

then

$$\underline{\underline{C}} = \partial \underline{z}/\partial \underline{u} \ = (\partial^2 s/\partial \underline{z}\partial \underline{z})^{-1} \tag{4.18}$$

In the linear theory $\underline{\underline{L}}$, $\underline{\underline{C}}$ and $\underline{\underline{T}}$ are usually constant. Taking the divergence operator for (4.16) and using (4.17) and (4.18) yields the wave matrix equation of change

$$\rho \underline{\underline{C}} \cdot \frac{\partial u}{\partial t} + \underline{\underline{L}} \cdot \left(\nabla^2 \underline{u} - \frac{\partial^2 \underline{u}}{c_0^2 \partial t^2}\right) = 0 \tag{4.19}$$

which generalizes the well-known parabolic equation

$$\rho \underline{\underline{C}} \cdot \frac{\partial u}{\partial t} + \underline{\underline{L}} \cdot \nabla^2 \underline{u} = 0 \tag{4.20}$$

for the case of the finite propagation speed of disturbances.

Consider now the small transfer potential vector perturbation $\delta \underline{u}$ with respect to the definite special solution, $\underline{u}^\circ(\underline{x},t)$ of (4.19), i.e.

$$\delta \underline{u}(\underline{x},t) = \underline{u}(\underline{x},t) - \underline{u}^\circ(\underline{x},t) \tag{4.21}$$

The perturbed equation is

$$\rho \underline{\underline{C}} \cdot \frac{\partial \delta u}{\partial t} + \underline{\underline{L}} \cdot \left(\nabla^2 \delta \underline{u} - \frac{\partial^2 \delta \underline{u}}{c_0^2 \partial t^2}\right) \tag{4.22}$$

and applying the second method of Lyapunov it is shown that a functional V exists such that

$$V(\delta \underline{u}, \text{grad } \delta \underline{u}, \frac{\partial \delta u}{\partial t}, \underline{x}, t) \geq 0 \tag{4.23}$$

$$\partial V/\partial t \leq 0$$

being

$$V = \frac{1}{2} \int \{\underline{\underline{L}} : \left|(\text{grad } \delta \underline{u})(\text{grad } \delta \underline{u}) + \frac{\partial \delta \underline{u}}{c_0^2 \partial t} \frac{\partial \delta \underline{u}}{\partial t}\right|\} \, dV$$

and

$$\partial V/\partial t = \int_V \{\rho \underline{\underline{C}}: \frac{\partial \delta \underline{u}}{\partial t} \frac{\partial \delta \underline{u}}{\partial t}\} dV \leq 0 \quad \text{if} \quad \underline{\underline{C}} \leq 0$$

As stated by Glansdorff and Prigogine ([8]) the inequality C <0 is true even for steady states far from equilibrium. No conclusion can be , however, made here for any nonlinear models generalizing (4.19).

4.3. Local stability

Stability conditions can be stated from a generalized Gibbs equation. Following one of the diverse approaches to the extension of irreversible thermodynamics, it has been shown ([7]) that the extended Gibbs relation

$$Tds = du + pdv - \Sigma_i \hat{\mu}_i dc_i + \Sigma_{i\alpha} X_i^\alpha d\Phi_i^\alpha \tag{4.25}$$

can be derived from the Boltzmann equation for a mixture of dilute (ideal) gases. The X_i^α denote the generalized thermodynamic force conjugate to the irreversible flux Φ_i^α which may be a stress tensor, a heat flux or the material flux of i. The extended Gibbs relation implies that the entropy density may be looked upon as a hypersurface in a Euclidean space spanned by the variables u, v, c_i and Φ_i^α (i=1,2,... r; = 1,2,...) whose temporal and spatial evolutions are determined by the evolution equations: ordinary balance equations (3.1)-(3.3) and

$$d\Phi_i^\alpha/dt = Z_i + \Sigma_j(\overrightarrow{\Lambda}_{ij}^\alpha - \overleftarrow{\Lambda}_{ij}^\alpha) \tag{4.26}$$

the last term being the collisional contribution to the change in Φ_i^α and they are intimately related to the local entropy production due to dissipative irreversible processes Φ_i^α. Therefore, they cannot be arbitrary, if the evolution of irreversible processes is to be consistent with the second law of thermodynamics.

Let us note that for an infinitesimal variation we may write the extended Gibbs relation as ([7])

$$T\delta s = \delta u + p\delta v - \Sigma_i \hat{\mu}_i \delta c_i + \Sigma_{i\alpha} X_i^\alpha \delta \Phi_i^\alpha \tag{4.27}$$

and the second order variation in s

$$T\delta^2 s = -\delta T\delta s + \delta p\delta v - \Sigma_i \delta\hat{\mu}_i \delta c_i + \Sigma_{i\alpha}\delta X_i^\alpha \delta\Phi_i^\alpha \quad (4.28)$$

After lengthy calculations, it becomes

$$T\delta^2 s = -(c_v T^{-1})(\delta T)^2 - (v\kappa)^{-1}(\delta v)^2_\Phi - \Sigma_{ij}(\partial\hat{\mu}_i/\partial c_j)_{v,c'}(\delta c_i)(\delta c_j)$$

$$+ \Sigma_{ij}\Sigma_{\alpha\beta} S_{ij}^{\alpha\beta}\delta\Phi_i^\alpha \delta\Phi_j^\beta \quad (4.29)$$

in which

$$S_{ij}^{\alpha\beta} = (\partial X_j^\beta/\partial\Phi_i^\alpha)_{t,p,\mu_i,\ \Phi_{j\neq i}^\beta\ ,\ \beta\neq\alpha} \quad (4.30)$$

Since $\delta^2 s$ must be negative for thermodynamic equilibrium to be locally stable, the necessary and sufficient conditions for stability are

$$c_v \geq 0 \quad (4.31)$$

$$\kappa \geq 0 \quad (4.32)$$

$$\Sigma_{ij}\left(\frac{\partial\hat{\mu}}{\partial c_j}i\right)(\delta c_i)_\Phi(\delta c_j)_\Phi \geq 0 \quad (4.33)$$

and

$$\Sigma_{ij}\Sigma_{\alpha\beta} S_{ij}^{\alpha\beta}\delta\Phi_i^\alpha\ \delta\Phi_j^\beta \leq 0 \quad (4.34)$$

The first three inequalities (4.31)-(4.33) are the usual inequalities in the Gibbs-Duhem stability theory except for the additional constraint Φ_i^α which must be kept fixed, but (4.34) is an additional condition that must be satisfied by the nonequilibrium local entropy around the thermodynamic equilibrium.

Defining the function

$$\Xi = Ts - pv - u + \Sigma_i \hat{\mu}_i c_i = \Sigma_{i\alpha} X_i^\alpha \Phi_i^\alpha \quad (4.35)$$

the second order variation in Ξ is easiliy seen to be

$$\delta^2\Xi = \delta T\delta s - \delta v\delta p + \Sigma_i \delta\hat{\mu}_i \delta c_i + \Sigma_{i\alpha}\delta X_i^\alpha \delta\Phi_i^\alpha \quad (4.36)$$

and, in particular, for T, p and $\hat{\mu}_i$ constant

$$(\delta^2\Xi)_{T,p,\hat{\mu}_i} = \Sigma_{i\alpha}\delta X_i^\alpha \delta\Phi_i^\alpha = \Sigma_{ij}\Sigma_{\alpha\beta}\ S_{ij}^{\alpha\beta}\delta\Phi_i^\alpha\delta\Phi_j^\beta \leq 0 \qquad (4.37)$$

Therefore, by the stability criterion (4.34) we see

$$(\delta^2\Xi)_{T,p,\hat{\mu}_i} \leq 0 \qquad (4.38)$$

This means that Ξ is a quantity which increases for all spontaneous processes and is equal to zero at equilibrium. Ξ is the nonequilibrium part of the entropy and its change is related to the local uncompensated heat which must be positive for any irreversible process.

REFERENCES

(1) Callen, H.B. Thermodynamics , Wiley, New York, 1960

(2) Dafermos, C.M., Arch.Rat.Mech.Anal.70, 167 (1979)

(3) De Groot S.R. and Mazur, P. Non-Equilibrium Thermodynamics, North-Holland, Amsterdam, 1962

(4) Domingos, J., Nina R. and Whitelaw J. (eds) Foundations of Continuum Thermodynamics, MacMillan, London 1974 ; Hutter K., Acta Mech. 27,1 (1977); Nicolis G., Rep.Progr.Phys.42,225 (1979).

(5) Eriksen, J.L, Int. J. Solids Structures 2, 573 (1966).

(6) Eu B.C., J. Chem. Phys. 73, 2958 (1980); Bampi F. and Morro A., J. Math. Phys. 21, 1201 (1980); Nonnenmacher T., J. Non-Equilib., Thermodyn. 5, 361 (1980).

(7) Eu B.C., J. Chem. Phys. 74, 2998 (1981).

(8) Glansdorff P. and Prigogine I., Thermodynamic Theory of Structure, Stability and Fluctuations, Wiley-Interscience, London, 1971.

(9) Glansdorff P. Nicolis G. and Prigogine I. Proc. Nat. Acad. Sci. (U.S.A.) 71, 197 (1974).

(10) Gurtin M.E., Arch. Rational Mech. Anal. 59 63 (1975).

(11) Gyarmati I., Non-Equilibrium Thermodynamics, Springer, Berlin, 1970.

(12) Jou D., Casas-Vázquez J. and Lebon G., J. Non-Equilib. Thermodyn. 4 349 (1974); Lebon G., Jou D. and Casas-Vázquez J., J. Phys A 13, 275 (1980); Casas-Vázquez J., Jou D. and Rubí J.M. in Systems Far From Equilibrium, L. Garrido, ed., Springer, Berlin, 1980

(13) Lavenda B.H., Thermodynamics of Irreversible Processes, MacMillan, London, 1978.

(14) Lebon G. and Casas-Vázquez J., Phys. Lett.A 55, 393 (1976).

(15) Müller I., Z. Phys. 198, 329 (1967); Israel W., Ann. Phys. N.Y. 100, 310 (1976); Kranys M., J. Phys.A 10, 689 (1977); Kluitenberg G.A., Physica A 93, 273 (1978); Lebon G., Bull Acad. Roy Belg. Classe Sci. 64, 456 (1978); Robles-Dominguez J.A., Silva B. and García-Colín L.S., Physica A 106, 539 (1981).

(16) Münster A., Classical Thermodynamics, Wiley-Interscience, London, 1970.

(17) Nicolis G. and Prigogine I., Self-Organization in Nonequilibrium Systems, Wiley, New York, 1977.

(18) Onsager L., Phys. Rev. 37, 405 and 38, 2265 (1931)

(19) Prigogine I and Mazur P., Physica 17, 661 (1951).

(20) Sieniutycz S., Phys. Lett.A 78, 433 (1980) and J. Non-Equilib. Thermodyn. 6, 79 (1981).

(21) Tisza L., Generalized Thermodynamics, M.I.T. Press,Cambridge, Mass, 1966.

(22) Yourgrau W., van der Merwe A. and Raw G.,Treatise on Irreversible and Statistical Physics, MacMillan, New York, 1966.

MATHEMATICAL METHODS IN STABILITY THEORY

G. LEBON
Department of Mechanics
Liège University
B6 Sart Tilman
B-4000 Liège (Belgium)

1. INTRODUCTION

The general problem of stability may be stated as follows. Consider a material system in a given reference state : the latter is either an equilibrium state, defined by space and time independent variables, a steady state, characterized by time independent variables, or even more generally, a non-steady state. Assume that under the action of a disturbance, the system leaves its reference state. These disturbances may be caused deliberately or induced by irregularities and imperfections of the apparatus; in absence of external agents, they may be produced by internal fluctuations. We call perturbation of a given physical parameter, the deviation with respect to its value in the basic unperturbed state. In any stability problem, the initial value of the perturbation is considered to be given.

The fundamental problem is then whether, starting from its initial value, the perturbation will decay, or, on the contrary, grow in the course of time. In the first case, the reference state is said stable, otherwise it is unstable. In the latter eventuality, the mathematical solution cannot survive in the physical world and the behaviour of the physical system will be governed by another mathematical solution. In the modern langage, it is said that the solution bifurcates from the basic one. The limiting case between stability and unstability is called neutral or marginal stability, it occurs when the perturbation remains constant in time.

The purpose of these lectures is to provide a rather large survey of the most significant theories of stability. As a preliminary, it is necessary to state precisely the notion of stability; this is done in chapter two wherein Lyapounov's concept of stability is recalled. Since the amplitude of the perturbations may be fixed at will, one must distinguish between perturbations of small and large amplitudes.

The study of small perturbations forms the subject of the classical linearized theory. This theory is developed in chapter three, where as illustrative exemple, the Rayleigh-Bénard problem is treated. As soon as the disturbances are important in size, the linearized theory reveals unadequate and must be replaced by non-linear approaches. Among them, one may distinguish the iterative and the global methods. In the latter, the details of the motion and the flow geometry are omitted; instead, attention is focused on the behaviour of global quantities, generally chosen as some positive definite functionals of the disturbance. A typical global method is Reynolds-Orr's theory [1,2] wherein the functional is the total kinetic energy of the perturbation. The Reynolds-Orr technique has been improved by Serrin [3] and more recently by Joseph [4] . These formalisms are examined in chapter four. They may be viewed as particular examples of Lyapounov's general theory, presented in chapter five. The next chapter concerns iterative approaches to the non-linear stability problem. These techniques were initiated by Lindsted [5] , Poincaré [6] and in a different form by Landau [7] , who was interested by the transition to turbulence (see also Hopf [8]). This process was considered as a succession of stability losses of flows of more and more complicated structure. In chapter six, two perturbative schemes are discussed, namely Gorkov-Malkus-Veronis' power series development [9,10] and Stuart's amplitude method [11]. A related and powerful technique used for studying non-linear stability is the theory of bifurcation [12-14]. This approach is briefly discussed in chapter seven, wherein the basic features are presented and discussed.
A final chapter is devoted to variational formulations of the stability problems. The principles are essentially aimed at providing the neutral stability solution : both perturbations of small and large sizes are investigated. The proposed criteria are rather natural extensions of Chandrasekhar's principle [15].

All the methods discussed in this work are systematically illustrated by the Rayleigh-Bénard convective instability problem.

It must be realized that no attempt has been made to give a complete review of the subject. Moreover, the references listed at the end of these notes must be considered as an arbitrary and small fraction of the numerous papers published in the field.

2. DEFINITION OF STABILITY IN THE SENSE OF LYAPOUNOV

The intuitive idea that a system is stable if, in the motion following a sufficiently small, but otherwise arbitrary perturbation, the resulting disturbances remain arbitrary small, is relatively vague. In order to make the above definition more precise, one has to define the distance to the reference configuration as well as the measure for the magnitude of the initial disturbance. These requirements have been met by Lyapounov's theory [16] , at least for discrete systems. Lyapounov's formalism has been generalized in a form particularly well suited for continuous systems by Movchan [17], Zubov [18] and Koiter [19].

2.1. Discrete systems

Assume that the motion of a discrete system is defined by the set of variables $\underset{\sim}{q}(t)$ $(q_1(t), q_2(t) \dots q_n(t))$. In the space E_q^n, the state of the system at time t is represented by the vector $\underset{\sim}{q}(t)$. During the evolution, the end of the vector $\underset{\sim}{q}(t)$ describes a trajectory in the space E_q^n (see figure 2.1).

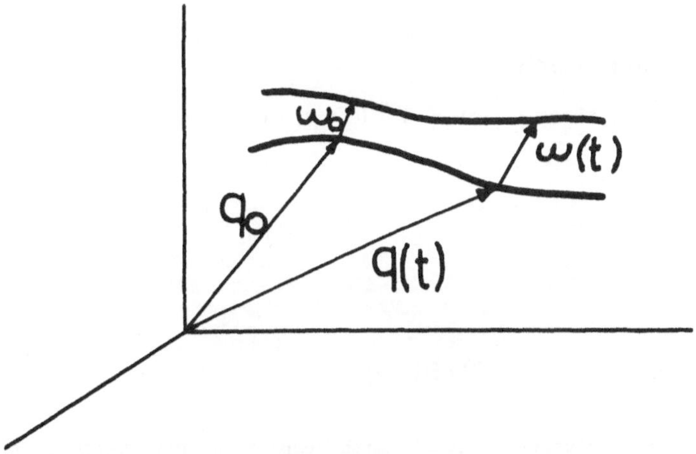

Figure 2.1. Basic and perturbed trajectories.

Let $\underset{\sim}{q}(t)$ be a solution of the differential system

$$\dot{\underset{\sim}{q}} = \underset{\sim}{X}(\underset{\sim}{q}, t) \tag{2.1}$$

Here and later on, an upper dot denotes derivation with respect to the time. The solution $\underset{\sim}{q}(t)$ taking the value $\underset{\sim}{q}_o$ at $t = t_o$ may be written as

$$\underset{\sim}{q} = \underset{\sim}{f}(\underset{\sim}{q}_o, t) \tag{2.2}$$

If t does not explicitely appear, the system (2.1) is called autonomous.

After perturbation of the initial value $\underset{\sim}{q}_o$ by a quantity $\underset{\sim}{\omega}_o$, the evolution of the system is governed by

$$\underset{\sim}{q} = \underset{\sim}{f}(\underset{\sim}{q}_o + \underset{\sim}{\omega}_o, t)$$

and one observes a perturbed trajectory, different from the basic (unperturbed) solution. Denote by $\underset{\sim}{\omega}(t)$, the distance between the perturbed and basic trajectories :

$$\underset{\sim}{\omega}(t) = \underset{\sim}{f}(\underset{\sim}{q}_o + \underset{\sim}{\omega}_o, t) - \underset{\sim}{f}(\underset{\sim}{q}_o, t) \qquad (2.3)$$

Clearly, $\underset{\sim}{\omega}_o$ and $\underset{\sim}{\omega}$ are measures of the disturbance at time t_o and time t respectively. The norm $|\underset{\sim}{\omega}|$ is defined by

$$|\underset{\sim}{\omega}| = \left(\sum_{i=1}^{n} \omega_i^2\right)^{1/2} \qquad (2.4)$$

The basic trajectory is <u>stable in Lyapounov's sense</u>, if it fulfills the following conditions :

1) $\underset{\sim}{\omega}$ is a continuous function of t.

2) Given $\varepsilon > 0$ and $t_o > 0$, there exists a $k^*(\varepsilon, t_o) > 0$ such that any solution $\underset{\sim}{\omega}(t)$, for which

$$|\underset{\sim}{\omega}_o| < k^*(\varepsilon, t_o) \qquad (2.5)$$

satisfies also

$$|\underset{\sim}{\omega}| < \varepsilon \qquad \text{for} \qquad \forall\, t \in (t_o, \infty) \qquad (2.6)$$

Otherwise stated, the basic solution is stable with respect to perturbations of size less than k^* , if given an area of radius ε , there corresponds an area of radius k^*, such that any trajectory issuded from k^* at time t_o crosses the area ε at time t . If no such k^* exists, the solution is unstable.

The solution is said <u>unconditionally</u> or globally stable if $k^* \to \infty$; for finite values of k^* , it is said <u>conditionally stable</u>.

If k^* is independent of t_o , the reference trajectory is called <u>uniformly stable</u>.

The basic solution is said <u>asymptotically stable</u> if

$$\lim_{t \to \infty} |\underset{\sim}{\omega}(t)| = 0 \qquad (2.7)$$

In other terms, the solution is asymptotically stable if all perturbed solutions approach it asymptotically.

2.2. Continuous systems

The variables are now depending on the space variables $\underset{\sim}{x}$ in addition to t. Let $\underset{\sim}{q}(\underset{\sim}{x},t)$ $\left[q_1(\underset{\sim}{x},t) \ldots q_n(\underset{\sim}{x},t)\right]$ be the values taken by the set of variables in a reference configuration. If the latter is an equilibrium or a steady state, $\underset{\sim}{q}$ is time-independent. Introduce a perturbation $\underset{\sim}{\omega}(\underset{\sim}{x},t)$ whose initial value is $\underset{\sim}{\omega}(\underset{\sim}{x},o)$ $= \underset{\sim}{\omega}_o(\underset{\sim}{x})$. This perturbation obeys non-linear partial differential equations as well as initial and boundary conditions.

In Hilbert's functional space L^n, the norm of $\underset{\sim}{\omega}$ is defined by

$$||\underset{\sim}{\omega}|| = \left[\int \sum_{i=1}^{n} \left(\omega_i(\underset{\sim}{x},t)\right)^2 d\underset{\sim}{x}\right]^{1/2}$$

By definition, the basic solution is called <u>stable in the sense of Lyapounov</u> if and only if it is possible to find for every given positive number ε, a second positive number $k^{*}(\varepsilon)$ such that inequality

$$||\underset{\sim}{\omega}_o(\underset{\sim}{x})|| < k^{*}$$

implies

$$||\underset{\sim}{\omega}(\underset{\sim}{x},t)|| < \varepsilon \qquad \forall t \in [t_o,\infty)$$

The system is <u>asymptotically stable</u> if

$$\lim_{t \to \infty} ||\underset{\sim}{\omega}(\underset{\sim}{x},t)|| = 0$$

The above definition of stability employs the L^n norms as a measure for the distance from the basic solution. An extension of Lyapounov's definition, applicable to continuous systems in a more general context have been proposed by Movchan [17] who uses two different metrics. However, as far as we are aware, applications of this general type of theory in fluid mechanics have been restricted to small amplitude perturbations [20].

3. THE LINEARIZED THEORY

3.1. Normal modes superposition

It is assumed that the field variables undergo perturbations of infinitesimally small size and that the equations governing the disturbances are linearized; all the non-linear terms are neglected.

Let $q(\underset{\sim}{x},t)$ represent a typical parameter, like the temperature or the velocity field, and $\omega(\underset{\sim}{x},t)$ the corresponding disturbance :

$$\omega(\underset{\sim}{x},t) = q(\underset{\sim}{x},t)_{\text{perturbed}} - q(\underset{\sim}{x},t)_{\text{basic}} \tag{3.1}$$

Stability means stability with respect to all arbitrary infinitesimal disturbances. As a consequence, one must examine the reaction of the system to all possible perturbations. This is accomplished by expressing $\omega(\underset{\sim}{x},t)$ as a superposition of basic modes and examining the stability of the system with respect to all these modes. If the system possesses invariance properties in the plane x,y, it is convenient to express the perturbation as

$$\omega(x,y,z,t) = \sum_k \Omega_k(z,t)\phi_k(x,y) \tag{3.2}$$

k is the horizontal wave number, Ω_k the amplitude of the disturbance and ϕ_k a function of x and y reflecting the invariance properties.

Since the perturbations are arbitrary, it is essential that the ϕ_k's form a complete system. This is realized by selecting for instance ϕ_k, as solution of the eigenvalue problem

$$\left(\frac{\partial^2}{\partial x^2} + \frac{\partial^2}{\partial y^2}\right)\phi_k(x,y) = -k^2\phi_k(x,y) \tag{3.3}$$

from which,

$$\phi_k(x,y) = \exp\left(i(k_x x + k_y y)\right) \quad , \quad (k_x^2 + k_y^2 = k^2) \tag{3.4}$$

With solutions of the form (3.4), one evades the difficulty of writing boundary conditions at infinity. Here, this problem is circumvented by taking perturbations which are periodic in the directions where the system extends to infinity. Clearly, (3.4) is bounded at $x = y = \infty$. Substitution of (3.4) in (3.2) yields

$$\omega(x,y,z,t) = \sum_k \Omega_k(z,t) \exp\left(i(k_x x + k_y y)\right) \tag{3.5}$$

If one single mode Ω_k increases with time, the system is unstable. On the contrary even if all the modes decrease with respect to the time, one cannot conclude in favour of stability. Indeed, it cannot be excluded that the system becomes unstable with respect to disturbances of finite size. It is therefore worth to point out that a linearized theory can only provide <u>sufficient conditions of instability</u>.

The dependence on time in (3.5) can be eliminated by seeking solutions of the form

$$\Omega_k(z,t) = W_k(z)\exp(\sigma_k t) \tag{3.6}$$

where σ_k is a complex quantity to be determined. Expression (3.6) is appropriate for basic steady solutions. If the basic flow is time-periodic, (3.6) is replaced by

$$\Omega_k(z,t) = W_k(z,t)\exp(\gamma_k t) = W_k(z,t + T)\exp \gamma_k t$$

where γ_k are the Floquet exponents and T the period. With (3.6), the relation (3.5) reads as

$$\omega(x,y,z,t) = \sum_k W_k(z)\exp\left(i(k_x x + k_y y) + \sigma_k t\right) \tag{3.7}$$

Despite its arbitrariness, expression (3.7) is very useful for handling a lot of practical problems. However, it must be kept in mind that the above analysis is confined to the study of stability with respect to the class of disturbances represented by (3.7).

The stability is determined by the sign of the real part of σ_k .

If all the $\underline{\mathrm{Re}\sigma_k < 0}$, the system is <u>stable</u>.

If one single $\underline{\mathrm{Re}\sigma_k > 0}$, the system is <u>unstable</u> (a sufficient albeit not a necessary condition).

The limiting case corresponding to $\underline{\mathrm{Re}\sigma_k = 0}$ is called <u>marginal or neutral stability</u>.

In most problems, it is postulated that besides $\mathrm{Re}\sigma_k = 0$, one has $\mathrm{Im}\sigma_k = 0$. This conjecture is called the <u>principle of exchange of stability</u>. The term was introduced by Jeffreys [21] but was certainly not the most suitable. Although the basic solution becomes unstable when $\sigma_k = 0$, it cannot be excluded that beyond $\sigma_k = 0$, the physical system works according to another mathematical solution which may itself be stable. In some circumstances, the principle of exchange of stability receives a demonstration. This is the case when the set of equations governing the disturbances is self-adjoint.

If $\mathrm{Im}\sigma_k \neq 0$, the onset of instability is initiated by oscillatory perturbations; it is said that one has <u>overstability</u>.

By setting $\mathrm{Re}\sigma_k = \mathrm{Im}\sigma_k = 0$ in the perturbed equations of motion, one is faced with an eigenvalue problem for some relevant dimensionless parameters, like the Rayleigh, the Reynolds or the Marangoni numbers.

3.2. The Bénard problem

The problem of Bénard will serve as illustration of the linear stability theory. The study of Bénard's thermal convection has been a subject of increasing interest during the last two decades. This was motivated because thermal convection occurs in a great variety of natural phenomena and industrial applications. A detailed description can be found in such excellent review papers as those of Segel [22], Finlayson [23], Palm [24], Normand et al [25] and Busse [26].

Consider a horizontal incompressible fluid layer of thickness d, extending to infinity and initially at rest. When the temperature difference between the lower and upper surfaces reaches a certain value, a spontaneous motion of the fluid is observed. These cells take the form of rolls or hexagonal polygons. In liquids, wherein viscosity decreases with temperature, the fluid flows upward at the center of the cell, spreads across the surface, cools until it reaches the edges of the cell where it descends towards the bottom where it is warmed. In contrast, in gases, for which the viscosity increases with the temperature, one observes convection cells with a downward motion in the centre.

According to Rayleigh, who was the first to produce a physical interpretation, only buoyancy effects are responsible for the onset of instability. The fluid adjacent to the lower heated plate becomes warmer and consequently, less dense. Therefore, it starts rising, but this flow is inhibited by viscous damping and thermal conduction. The latter effects resist to buoyancy, but at the critical point, the energy liberated by the uprising of the less dense fluid overcomes the rate at which energy is dissipated, and convection starts.

However, this interpretation is certainly not complete. Indeed, one has observed Bénard's cells in fluids heated from above or cooled from below. It is now established that buoyancy is not the sole agency responsible for the instability but that convection is also induced by temperature variations of the surface tension [27]. Generally, the surface tension in a liquid decreases when the temperature increases. As a consequence the temperature drop between the center and the wall of the cells generates an increase of surface tension, and hence promotes circulation. Since both buoyancy and surface effects are cooperative, it is natural to examine how the one influences the other. This is accomplished within the foregoing analysis.

Assume that the temperature drop between the lower and upper faces of the layer is ΔT. Cartesian axes are selected with their origin at the lower face and the vertical axes $\underset{\sim}{e}_z$ pointing upwards (see figure 3.1.).

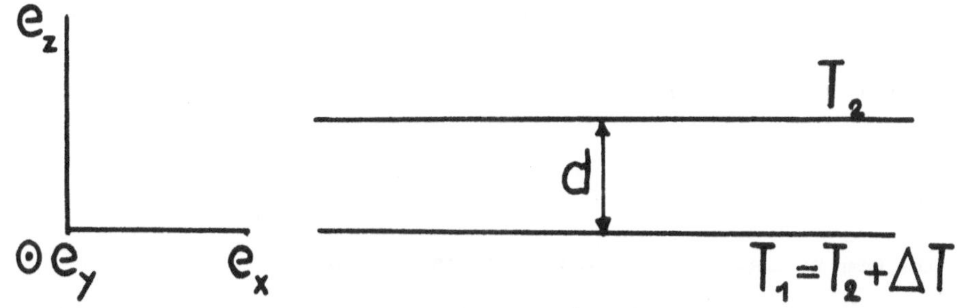

Figure 3.1. Fluid layer heated from below.

In the Oberbeck-Boussinesq approximation, the motion is governed by the following equations :

$$\nabla.\underset{\sim}{v} = 0 \qquad \text{(balance of mass)} \tag{3.8}$$

$$\frac{\partial \underset{\sim}{v}}{\partial t} + \underset{\sim}{v}.\nabla\underset{\sim}{v} = -\frac{1}{\rho_o}\nabla p + \frac{\rho}{\rho_o}\,\underset{\sim}{g} + \nu\nabla^2\underset{\sim}{v} \tag{3.9}$$

(balance of momentum)

$$\frac{\partial T}{\partial t} + \underset{\sim}{v}.\nabla T = \kappa\nabla^2 T \tag{3.10}$$

(balance of energy)

The notation is classical, $\underset{\sim}{v}$ is the velocity field, T the temperature field, p the pressure, ν the kinematic viscosity, κ the thermal diffusivity and ρ the density, supposed to te linear with respect to the temperature

$$\rho = \rho_o\left(1 - \alpha(T - T_o)\right) \tag{3.11}$$

ρ_o and T_o are reference quantities, α is the coefficient of thermal expansion ($\alpha \approx 10^{-3}$ for ordinary fluids).

In the basic unperturbed state, the solutions of the balance equations are

$$\underset{\sim}{v}_b = 0 \tag{3.12}$$

$$T_b = -\beta z + T_1 \qquad (\beta = \frac{\Delta T}{d} > 0) \tag{3.13}$$

$$\frac{dp_b}{dz} = -\rho_o g(1 + \alpha\beta z) \tag{3.14}$$

T_1 is the temperature, assumed uniform, of the lower face.

The disturbances are designated by

$$\underset{\sim}{u} = \underset{\sim}{v} - \underset{\sim}{v}_b \quad , \quad \theta = T - T_b \quad , \quad \pi = p - p_b \tag{3.15}$$

By using $d, d^2/\nu, \frac{\nu}{\kappa\beta d}$ and $\rho \frac{v^2}{d^2}$ as scales for length, time, temperature and pressure, the balance equations for the disturbances can be written in non-dimensional form as

$$\nabla.\underset{\sim}{u} = 0 \tag{3.16}$$

$$\frac{\partial}{\partial\tau}\underset{\sim}{u} + \underset{\sim}{u}.\nabla\underset{\sim}{u} = -\nabla\pi + Ra\theta\, \underset{\sim}{e}_z + \nabla^2\underset{\sim}{u} (\tau = t\nu/d^2) \tag{3.17}$$

$$Pr(\frac{\partial}{\partial\tau} + \underset{\sim}{u}.\nabla)\theta = w + \nabla^2\theta \tag{3.18}$$

$\underset{\sim}{u}(u,v,w)$, π and θ are now the dimensionless velocity, pressure and temperature disturbances, Pr and Ra are the Prandtl and the Rayleigh numbers :

$$Pr = \frac{\nu}{\kappa} \qquad\qquad Ra = \frac{g\alpha d^4}{\kappa\, \nu}\beta \tag{3.19}$$

The Rayleigh number is a measure of the relative importance of the buoyancy and the dissipative effects.

After elimination of π in (3.17) by application of the rot operator, one is left with the two following linearized equations

$$\frac{\partial}{\partial\tau}\nabla^2 w = Ra\Delta_1\theta + \nabla^4 w (\Delta_1 = \frac{\partial^2}{\partial x^2} + \frac{\partial^2}{\partial y^2}) \tag{3.20}$$

$$Pr\frac{\partial}{\partial\tau}\theta = w + \nabla^2\theta \tag{3.21}$$

Their solutions are of the form

$$w = W(z)\, \exp\left(i(k_x x + k_y y) + \sigma t\right) \tag{3.22}$$

$$\theta = \Theta(z)\, \exp\left(i(k_x x + k_y y) + \sigma t\right) \tag{3.23}$$

Substitution in (3.20) and (3.21) yields

$$(D^2 - k^2)\ (D^2 - k^2 - \sigma)W = \mathrm{Rak}^2\ \phi \qquad (3.24)$$

$$(D^2 - k^2 - \mathrm{Pr}\sigma)\phi = -\ W \qquad (3.25)$$

where D stands for d/dz.

Solution to these equations must be found subject to appropriate boundary conditions.

Regardless the nature of the bounding surfaces, one must require

$$\text{at}\quad z = 0 \quad\text{and}\quad z = 1\ :\ W = 0 \qquad (3.26)$$

If moreover,

i) the surface is rigid : $DW = 0$ (3.27)

ii) the surface is stress free : $D^2W = 0$ (3.28)

iii) the surface tension ξ is temperature dependent : $D^2W = -\ k^2\mathrm{Ma}\ \phi$ (3.29)

where Ma is the Marangoni number

$$\mathrm{Ma} = -\ \frac{(\partial\xi/\partial T)\beta d^2}{\rho_0 \nu \kappa}$$

Ma measures the importance of the surface effects generated by temperature inhomogeneities with respect to the dissipated energy.

A supplementary boundary condition on the temperature is needed, namely

$$D\theta = -\ h\theta \qquad (3.31)$$

(3.31) is frequently called the radiation boundary condition, it expresses that the heat flux at the boundary is equal to the rate of heat loss from the surface, h is Biot's heat transfer coefficient :

$h = 0$ corresponds to an insulated surface $(D\theta = 0)$;

$h = \infty$ describes a perfectly heat conducting surface $(\theta = 0)$.

By assuming exchange of stability ($\mathrm{Im}\sigma = 0$), one obtains the solution corresponding to neutral stability by setting $\sigma = 0$ in (3.24) and (3.25). After elimination of ϕ between these two equations, one has

$$(D^2 - k^2)^3\ W = -\ k^2\mathrm{Ra}W \qquad (3.32)$$

We are now faced with a characteristic eigenvalue problem defined by the equation (3.24) and (3.25) (or 3.32) and the adequate boundary conditions (3.26) to (3.31). It is only for some particular values of Ra (and Ma) that the system will admit non trivial solutions, for a given k .

Let us consider as particular example, the case of two stress-free surfaces, perfectly conducting $(Ma = 0 \ , \ h = \infty)$.

The exact solution of (3.32) is

$$W = A \sin(n \pi z) \qquad n = 0,1,2, \ldots \qquad A = \text{constant} \tag{3.33}$$

The marginal stability curve (Ra versus k) is obtained after substitution of (3.33) in (3.32) and is given by (see figure 3.2).

$$Ra = \frac{(n^2\pi^2 + k^2)^3}{k^2}$$

The lowest point on the curve $Ra(k)$ is obtained from

$$\partial Ra/\partial k = 0$$

which yields

$$R_c = \frac{27\pi^4}{4} = 657.5 \qquad k_c = \frac{\pi}{\sqrt{2}} = 2.2215$$

For the more realistic case of two rigid perfectly conducting boundaries, numerical calculations have given

$$R_c = 1707.7 \qquad k_c = 3.117$$

The increase of R_c compared with the previous result indicates that rigid boundaries have stabilizing influence on the motion. At the bifurcation point $Ra = R_c$, the state of conduction is marginally unstable and a convective motion with wave number k_c is induced. For $R > R_c$, a continuous spectrum of modes becomes unstable and one should expect a very complicated structure. But practically, one observes rather simple cellular forms like rolls, or hexagonal cells. This ordering originates clearly in the non-linear terms.

The advantages of the normal mode expansion are that it determines the critical values of the parameters for small size disturbances and provides preliminary informations in view of a non-linear approach. Moreover, for most situations met in engineering, a linear analysis may suffice.

Among the deficiencies, we notice :

1. That the method does not predict what happens beyond the bifurcation point $R > R_c$. In particular, it does not allow to compute the amplitude of the disturbance or to determine whether all, or only certain modes remain unstable beyond the critical point.

2. That some flows which are enacted stable by the linear theory may actually be unstable with respect to perturbations of finite size. As clearly pointed out, the linearized theory leads only to sufficient conditions of instability.

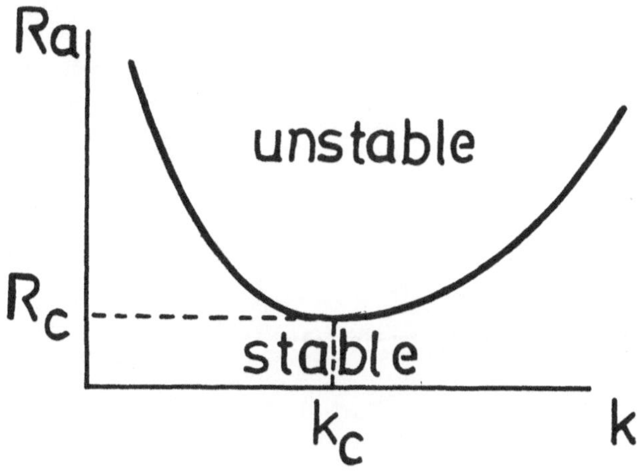

Figure 3.2. Linear theory : Rayleigh number versus the horizontal wave number.

4. THE ENERGY METHODS

These methods provide one way to take account of the non-linear effects. They lead to global statements about stability taking the form of sufficient conditions for stability. The essence of the method consists of finding positive definite functionals of the disturbances, which decrease in the course of time.

The simplest energy method is incontestably Reynolds-Orr's formalism (2) , but it yields critical values that are too conservative. A modern extension of this method was proposed by Serrin (3) , but it gives still unsatisfactory results. Serrin's technique has been recently improved by Joseph (4,28) who introduced a supplementary parameter which is extremalized to obtain more realistic critical values. In this chapter, we analyse successively the theories of Reynolds-Orr, Serrin and Joseph, and apply them to the Bénard problem.

4.1. The Reynolds-Orr method

We are no longer interested in the detailed local evolution of the disturbances but instead, we examine how some global properties, closely related to the disturbances, behave. In absence of temperature effects, it is usual to take as typical global quantity, the kinetic energy K of the disturbance, integrated on the whole system. If dK/dt decreases, the basic state is called stable in the mean.

In Bénard's problem, where thermal effects play a leading part, it is essential to introduce a more general functional, like

$$E = K + Pr\,\Theta \qquad (> 0) \tag{4.1}$$

where

$$K = \frac{1}{2} \langle \underset{\sim}{u} \cdot \underset{\sim}{u} \rangle \tag{4.2}$$

$$\Theta = \frac{1}{2} \langle \theta^2 \rangle \tag{4.3}$$

brackets denote integration on a cell, explicitely

$$\langle \underset{\sim}{u} \cdot \underset{\sim}{u} \rangle = \int \underset{\sim}{u} \cdot \underset{\sim}{u} \; dx\, dy\, dz$$

E is defined as the total energy of the disturbance. The basic convection-free state is stable in the mean, if

$$dE/dt < 0 \tag{4.4}$$

Let us now determine the expression of dE/dt .

Assume that the lower plate is rigid and perfectly heat conducting, and that the upper surface is free (with a temperature dependent surface tension), but exchanges heat according to Biot's law (3.31). Take the dot product of (3.17) by $\underset{\sim}{u}$, integrate over the volume of a cell and make use of (3.16) and the boundary conditions $\underset{\sim}{u} = 0$ at $z = 0$ and $z = 1$. This yields

$$\frac{dK}{dt} = - \langle \nabla \underset{\sim}{u} : \nabla \underset{\sim}{u} \rangle + Ra \langle \theta w \rangle - Ma \langle \theta\, Dw \rangle_1 \tag{4.5}$$

A double dot stands for the double scalar product while index 1 affecting the brackets $\langle\ \rangle$ means that the corresponding quantity must be evaluated at the upper surface, $z = 1$.

By multiplying (3.18) by θ , one obtains after integration on the volume

$$Pr \frac{d\theta}{dt} = - <\nabla\theta.\nabla\theta> + <\theta w> - <h\theta^2>_1 \qquad (4.6)$$

From (4.1)-(4.5) and (4.6), dE/dt may be written as

$$\frac{dE}{dt} = - \left(<\nabla \underset{\sim}{u} : \nabla \underset{\sim}{u}> + <\nabla\theta.\nabla\theta> + <h\theta^2>_1\right)$$
$$+ \left((Ra + 1) <\theta w> - Ma<\theta Dw>_1\right) \qquad (4.7)$$

The first three terms in the r.h.s. represent the dissipated energy by viscosity and heat conduction; the two remaining terms are related to the buoyancy and the surface tension effects and contribute to the positive production.

The idea is now to replace the r.h.s. eq. (4.7) by upper bounds in order to secure that $dE/dt < 0$.

Assume that the following inequalities are verified :

$$<\nabla \underset{\sim}{u} : \nabla \underset{\sim}{u}> \geq a^2 <\underset{\sim}{u}.\underset{\sim}{u}> \geq a^2 <w^2> \qquad (a > 0) \qquad (4.8)$$

$$<\nabla\theta.\nabla\theta> \geq b^2<\theta^2> \qquad (b > 0) \qquad (4.9)$$

The problem of finding the most appropriate values of a and b is a very subtle one and has been solved for some particular situations [28].

According to Schwarz's inequality,

$$(Ra + 1) <\theta w> \leq \frac{1}{4b^2} (Ra + 1)^2 <w^2> + b^2<\theta^2>$$

and

$$Ma<\theta Dw>_1 \geq - \frac{Ma^2}{4h} < (Dw)^2>_1 - <h\theta^2>_1$$

Collecting all these results in (4.7), one obtains after some reductions

$$\frac{dE}{dt} \leq - \frac{1}{b^2} <w^2> \left(a^2b^2 - (\frac{Ra + 1}{2})^2\right) + \frac{1}{4h} Ma^2 < (Dw)^2 >_1$$

This result indicates that whatever Ra , small values of h and large values of Ma promote instability. In absence of surface tension effects $(Ma = 0)$, or for a perfectly conducting boundary $(h = \infty)$, stability is assured as long as

$$Ra < 2ab - 1 \qquad (4.10)$$

For the particular case of two rigid and perfectly conducting plates, the values of a and b have been determined by Joseph (28) :

$$a = 1.93\,\pi \qquad b = \pi \tag{4.11}$$

By replacing in (4.10), one finds

$$Ra < 37$$

to be compared with the value 1707 of the linear theory.

Since the disturbances are not imposed to be solution of the equations of motion, it is not surprising to obtain critical values that are too conservative. Indeed, the method guarantees stability with respect to a wider class of disturbances than those allowed by the governing equations.

The results are not more exciting for other problems. For Poiseuille flow, Serrin (3) has found a critical Reynolds number equal to 5.71 while the linearized theory predicts 5780. Clearly, the method produces values which are too low, even for the most simple flows. This is the reason why the method has been ignored for several years. It is only recently that it knew a revival, thanks to the works of Serrin and Joseph.

4.2. Serrin-Joseph theory

In order to derive sharper estimates of the stability limit, Serrin (3) proposed to maximalize the r.h.s. of (4.7) by a variational principle. Serrin's theory has been recently improved by Joseph (4). We shall directly discuss the latter approach, since it contains Serrin's formalism as a particular case. As before, the Rayleigh-Bénard problem serves as a support.

A sufficient condition of stability

One introduces a total disturbance energy E_λ by

$$E_\lambda = \frac{1}{2} \langle \underset{\sim}{u} . \underset{\sim}{u} \rangle + \frac{1}{2} \lambda \, Pr \, \langle \theta^2 \rangle \tag{4.12}$$

wherein λ is a positive parameter. Expression (4.12) is similar to (4.1) by carrying out the change of variable

$$\theta' = \lambda^{1/2} \theta \tag{4.13}$$

In terms of $\underset{\sim}{u}$ and θ', it is directly checked that the rate of change of energy may be written as

$$\frac{dE_\lambda}{dt} = -D + RP \tag{4.14}$$

R is a new stability parameter

$$R = \sqrt{Ra^2 + Ma^2} \tag{4.15}$$

D is the positive dissipated energy

$$D = \langle \nabla \underset{\sim}{u} : \nabla \underset{\sim}{u} \rangle + \langle \nabla\theta' . \nabla\theta' \rangle + h \langle \theta'^2 \rangle_1 \tag{4.16}$$

and P is the positive production term

$$P = \frac{1 + \lambda/Ra}{\lambda^{1/2}} \frac{Ra}{R} \langle \theta' w \rangle - \frac{1}{\lambda^{1/2}} \frac{Ma}{R} \langle \theta' Dw \rangle_1 \tag{4.17}$$

Denoting by $R_{\lambda m}^{-1}$ the maximum value of the ration P/D

$$R_{\lambda m}^{-1} = \max \left(\frac{P}{D}\right) \tag{4.18}$$

it follows from (4.14) that

$$\frac{dE_\lambda}{dt} \leq - D\left(1 - \frac{R}{R_{\lambda m}}\right) \tag{4.19}$$

The problem is to find the greatest Rayleigh and Marangoni numbers and the best value of λ for which the inequality

$$\frac{d}{dt} E_\lambda \leq 0 \tag{4.20}$$

holds.

Clearly, the system is stable if $R < R_{\lambda m}$, and marginally stable if $R = R_{\lambda m}$. We are now faced with the determination of $R_{\lambda m}$.

A variational criterion

The problem amounts of finding the (most unfavourable) disturbance which guarantees stability. Or stated in other terms, seek for the disturbance $(\underset{\sim}{u}, \theta')$ which makes P maximum when the following constraints are satisfied :

$$D = 1 \qquad (4.21)$$

$$\nabla . \underset{\sim}{u} = 0 \qquad (4.22)$$

$$\underset{\sim}{u} = \theta' = 0 \quad \text{on} \quad z = 0 \qquad (4.23\ a)$$

$$w = 0 \quad \text{on} \quad z = 1 \qquad (4.23\ b)$$

Such a formulation is typically of variational nature and can be expressed by the variational equation

$$\delta\left(P - \Lambda_\lambda D + 2\langle\mu\nabla . u\rangle\right) = 0 \qquad (4.24)$$

δ is the variation symbol, Λ_λ and $\mu(x,y,z)$ are Lagrange multipliers. The solutions $(\underset{\sim}{u},\theta')$ which render P extremum are solutions of the linearized stationary Euler-Lagrange equations :

$$\nabla^2 \underset{\sim}{u} + \frac{(1 + \lambda/Ra)}{2\lambda^{1/2}\Lambda_\lambda}\frac{Ra}{R}\,\theta' \underset{-}{e}_z - \frac{1}{\Lambda_\lambda}\nabla\mu = 0 \qquad (4.25)$$

$$\nabla^2\theta' + \frac{(1 + \lambda/Ra)}{2\lambda^{1/2}\Lambda_\lambda}\frac{Ra}{R}\,w = 0 \qquad (4.26)$$

with at $z = 0$: $\underset{\sim}{u} = \theta' = 0$ (4.27)

at $z = 1$: $w = 0$ (4.28)

$$Du + \frac{1}{2\lambda^{1/2}\Lambda_\lambda}\frac{Ma}{R}\frac{\partial\theta'}{\partial x} = 0 \qquad (4.29)$$

$$Dv + \frac{1}{2\lambda^{1/2}\Lambda_\lambda}\frac{Ma}{R}\frac{\partial\theta'}{\partial y} = 0 \qquad (4.30)$$

$$D\theta' + h\theta' + \frac{1}{2\lambda^{1/2}\Lambda_\lambda}\frac{Ma}{R}\,Dw = 0 \qquad (4.31)$$

If $\frac{\mu}{\Lambda_\lambda}$ is identified with the pressure disturbance π and Λ_λ with R^{-1} , it is seen that (4.25) and (4.26) bear strong resemblances with the stationary linearized balance equations (3.17) and (3.18). Both sets are strictly identical if in addition, λ is equal to the Rayleigh number Ra . In contrast, (4.29)-(4.30) are not the correct boundary conditions, even for $\lambda = Ra$. Serrin was the first to propose a variational principle of the form (4.24); however, in Serrin's approach, λ is fixed from the start and put equal to Ra .

The problem set up by equations (4.25) to (4.31) is an eigenvalue problem in Λ_λ. It is easily verified that the solution of the above equations satisfy

$$\langle \nabla u : \nabla u \rangle = \frac{1 + \lambda/Ra}{2\lambda^{1/2}\Lambda_\lambda} \frac{Ra}{R} \langle \theta' w \rangle - \frac{1}{2\lambda^{1/2}\Lambda_\lambda} \frac{Ma}{R} \langle \theta' Dw \rangle_1 \tag{4.32}$$

and

$$\langle \nabla\theta' . \nabla\theta' \rangle + h \langle \theta'^2 \rangle_1 = \frac{1 + \lambda/Ra}{2\lambda^{1/2}} \frac{Ra}{R} \langle \theta' w \rangle - \frac{1}{2\lambda^{1/2}\Lambda_\lambda} \frac{Ma}{R} \langle \theta' Dw \rangle_1 \tag{4.33}$$

After addition of (4.32) and (4.33) and use of the condition $D = 1$, one finds that the eigenvalues are given by

$$\Lambda_\lambda = \frac{(1 + \lambda/Ra)}{\lambda^{1/2}} \frac{Ra}{R} \langle \theta' w \rangle - \frac{Ma}{R\lambda^{1/2}} \langle \theta' Dw \rangle_1 \tag{4.34}$$

This expression is nothing but the production term P. As a consequence, the fields (θ', w) which maximize P are the eigenvectors corresponding to the greatest eigenvalue Λ_λ^{max}. Recalling that $R_{\lambda m}^{-1}$ has been defined as the maximum of P/D and using the condition that $D = 1$, it follows that

$$R_{\lambda m}^{-1} = \Lambda_\lambda^{max} \tag{4.35}$$

The determination of $R_{\lambda m}^{-1}$ is thus equivalent to the search of the greatest eigenvalue Λ_λ^{max} in the eigenvalue problem (4.25)-(4.31).

Of course, to each given value of λ, there corresponds a maximum value Λ_λ^{max}. The parameter λ must be selected in order to give the best possible limits for stability. Since the stability condition $R < R_{\lambda m}$ is a sufficient condition, the method will provide a value for $R_{\lambda m}$, smaller than the actual one. Thus, the best limit of stability is clearly defined by the largest value of $R_{\lambda m}$: otherwise stated, we shall seek that $\tilde{\lambda}$ for which

$$\tilde{R}_{\lambda m} = \max_{\lambda > 0} R_{\lambda m} \tag{4.36}$$

This value is obtained from

$$0 = \frac{\partial R_{\lambda m}^{-1}}{\partial \lambda} = \frac{1}{2\lambda^{3/2}} \left(\frac{Ma}{R} \langle \theta' Dw \rangle_1 - (1 - \frac{\lambda}{Ra}) \frac{Ra}{R} \langle \theta' w \rangle \right) \tag{4.37}$$

$$\frac{\partial^2 R_{\lambda m}^{-1}}{\partial \lambda^2} > 0 \tag{4.38}$$

It is readily checked that

$$\tilde{\lambda} = Ra - Ma \frac{\langle\theta' Dw\rangle_1}{\langle\theta' w\rangle} \tag{4.39}$$

and that

$$\left.\frac{\partial^2 R_{\lambda m}^{-1}}{\partial\lambda^2}\right|_{\lambda = \tilde{\lambda}} = \frac{\langle\nabla\theta'\rangle^2 + h \langle\theta^2\rangle}{2\lambda^{3/2} R} > 0$$

ensuring that $\tilde{\lambda}$ corresponds to a minimum of $R_{\lambda m}^{-1}$. It is interesting to observe that one recovers Serrin's result ($\lambda = Ra$) when the tension effects are omitted ($Ma = 0$).

While the essential merit of the energy method is to avoid the details of the phenomena, it demands further remarks.

1. The method gives stability conditions with respect to a wider class of disturbances than the linear theory, which excludes non-linear transport terms and supposes that everywhere, the growth rate is the same. This is the reason why the energy method gives critical values which are never larger than the critical values of the linear theory (4) :

$$R \text{ (energy)} \leq R \text{ (linear)} \tag{4.41}$$

2. The method does not tell anything about instability. The instability is absolutely not guaranteed beyond the critical treshold. When greater than critical, the energy of the disturbance will certainly increase for a time, but it is not excluded that ultimately, the energy decays.

3. The determination of the critical value of the characteristic parameter implies the solution of an eigenvalue problem like (4.25)-(4.31). In certain circumstances, this problem may reveal as intricate as the original equations.

5. LYAPOUNOV'S THEORY

5.1. Lyapounov's theorems

The energy methods, discussed in the previous chapter may be considered as particular applications of Lyapounov's theory (16). The latter has known an increasing interest in the last years, specially in the field of non-equilibrium thermodynamics (29).

As in chapter two, consider the set of non-linear differential equations

$$\dot{\underset{\sim}{q}} = \underset{\sim}{X}(\underset{\sim}{q},t)$$

We introduce the following definition

Positive (negative) definite functions

Let $\phi(q_1,q_2, \ldots q_n)$ be a function of the n variables $q_1(t) \ldots q_n(t)$. The function $\phi(\underset{\sim}{q})$ is said positive definite if :

1. ϕ is a continuous function of t , as well as its first derivative in an open domain Ω around the origin of the space E^n_q ;

2. ϕ vanishes at the origin : $\phi(0) = 0$;

3. outside the origin, ϕ is overall positive (negative) : $\phi(\underset{\sim}{q}) > 0$.

ϕ is positive semi-definite if it vanishes at other points than the origin : $\phi(\underset{\sim}{q}) \geq 0$.

When only two variables q_1 and q_2 are involved, one can represent the positive definite function $\phi(q_1,q_2)$ by a reversed bowl (see figure 5.1.).

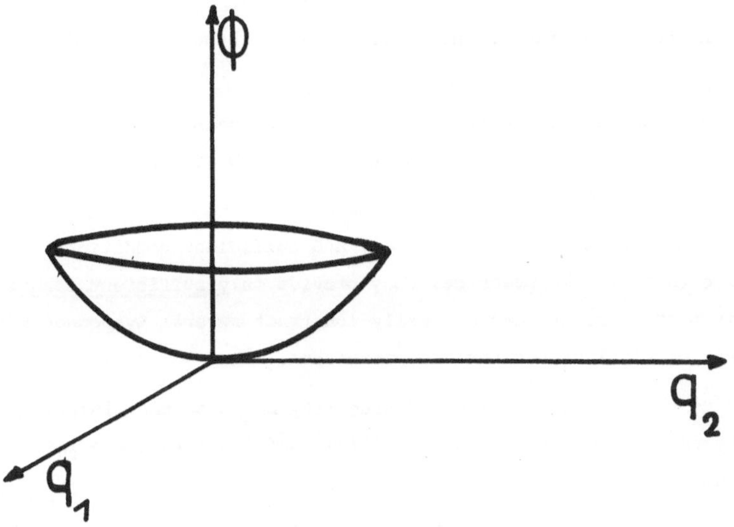

Figure 5.1. Example of Lyapounov's function.

If, in addition, ϕ depends explicitely on the time, $\phi(\underset{\sim}{q},t)$ is said positive definite if :

1. $\phi(\underset{\sim}{q},0)$ is positive definite;

2. there exists a positive definite function $\psi(q)$ such that

$$\phi(\underset{\sim}{q},t) \geq \psi(\underset{\sim}{q}) \qquad \forall t \geq 0$$

We shall state without proof the *fundamental theorems of Lyapounov.*

<u>Theorem 1</u> : The reference state ($\underset{\sim}{q} = 0$) is stable in a domain around the origin if there exists a positive definite function ϕ whose time derivative(*) is negative semi-definite or equal to zero :

$$\frac{d\phi}{dt} = \underset{\sim}{X}.\nabla\,\phi \leq 0 \qquad \left(\nabla\,\phi \equiv (\frac{\partial\phi}{\partial q_i})\right) \tag{5.2}$$

<u>Theorem 2</u> : The state ($\underset{\sim}{q} = 0$) is asymptotically stable if $d\phi/dt$ is negative definite :

$$\phi > 0 \qquad \frac{d\phi}{dt} < 0 \tag{5.3}$$

<u>Theorem 3</u> : The state ($\underset{\sim}{q} = 0$) is unstable if there exists a positive definite function whose time derivative is positive definite :

$$\phi > 0 \qquad \frac{d\phi}{dt} > 0 \tag{5.4}$$

Of course, the above theorems remain valid by interchanging the words positive and negative.

The extension to partial differential equations has been developed by Zubov [18]. In that case, one deals with Lyapounov's functionals which are integrals over the space coordinates, instead of functions.

Although the above theorems give necessary and sufficient conditions, it is instructive to notice that, in the practice, they provide only <u>sufficient conditions</u>. Indeed, for a given problem, one can generally construct several Lyapounov's functions yielding different critical values for the parameters.

In contrast to the classical theory of stability based on the linearization of the equations, Lyapounov's method yields stability conditions that are independent of the integration of the balance equations.

(*) If the system is non-autonomous, the time derivative is given by :

$$\frac{d\phi}{dt} = \frac{\partial\phi}{\partial t} + \underset{\sim}{X}.\nabla\,\phi$$

Another advantage is that it is applicable to disturbances of large amplitude.

The main difficulty is the construction of the most appropriate Lyapounov function. If the choice is not fortunate, one obtains conditions that are too conservative. Moreover, the method is disarmed to describe the behaviour of the system beyond the critical point.

5.2. Examples of Lyapounov's functions

i) in classical mechanics

Consider a conservative system like a frictionless material point attached to a spring. The motion can be described by a generalized coordinate q and a generalized momentum p. Let K designate the kinetic energy and $V = \frac{1}{2} q^2$ the potential energy, which is minimum at $q = 0$. Let us show that the hamiltonian

$$H(q,p) = K(p) + V(q) \tag{5.5}$$

is an appropriate Lyapounov function in a domain around the origin $p = q = 0$.

There is no lost of generality by taking

$$H(0,0) = 0$$

Since $V(q)$ is minimum at $q = 0$, one has in the vicinity of origin

$$H(p,q) > 0 \tag{5.6}$$

Classical mechanics tells us that for a conservative system

$$\frac{dH}{dt} = 0 \tag{5.7}$$

As a consequence of theorem 1, the origin $q = p = 0$ is stable.

ii) in continuum mechanics

Consider a diffusion process described by

$$\frac{\partial q}{\partial t} = \frac{\partial^2 q}{\partial x^2} \qquad 0 \le x \le 1 \tag{5.8}$$

$$\frac{\partial q}{\partial x} = 0 \qquad \text{at} \quad x = 0 \quad \text{and} \quad x = 1 \tag{5.9}$$

q represents for instance a dimensionless temperature or concentration disturbance. The question is asked whether the solution $q = 0$ is stable. The answer is affirmative as seen by constructing the positive definite functional

$$\phi = \frac{1}{2} \int_0^1 (\frac{\partial q}{\partial x})^2 \, dx > 0 \tag{5.10}$$

whose time derivative is

$$\frac{d\phi}{dt} = \int_0^1 \frac{\partial q}{\partial x} \frac{\partial}{\partial t} \frac{\partial q}{\partial x} \, dx$$

After integration by parts and use of the boundary conditions (5.9), one has

$$\frac{d\phi}{dt} = - \int_0^1 (\frac{\partial^2 q}{\partial x^2}) \, dx < 0 \qquad \text{Q.E.D.} \tag{5.11}$$

Another example of Lyapounov function associated to (5.8) and (5.9) is

$$\phi = \frac{1}{2} \int q^2 dx > 0$$

for which it is proved that

$$\frac{\partial \phi}{\partial t} = - \int (\frac{\partial q}{\partial x})^2 \, dx < 0$$

iii) in non-equilibrium thermodynamics

Glansdorff and Prigogine (29) have established a Lyapounov function describing the stability of thermo-mechanical systems characterized by non-linear constitutive equations.

According to the classical theory of irreversible processes, it is admitted that locally, the entropy depends on the same set of variables than in equilibrium (local equilibrium hypothesis). For a multi-component mixture, these variables are the concentrations $C_i (i = 1,2, \dots r)$, the temperature T and the density ρ , so that

$$s = s(T, \rho, C_1, C_2, \dots C_r) \tag{5.12}$$

The global entropy S is linked to the specific entropy s by

$$S = \int \rho \, s \, d\Omega \tag{5.13}$$

Assume that after application of a disturbance of infinitesimal size, the entropy deviates from its reference value by a quantity ΔS. Expanding ΔS in the form (*)

$$\Delta S = \delta S + \frac{1}{2}\delta^2 S + \dots \tag{5.14}$$

it has been calculated [28] that

$$\delta^2 S = - \int \Big(\frac{1}{\rho T} C_v (\delta T)^2 + \frac{\rho}{T\chi_T} (\delta\rho^{-1})^2 + \sum_{i,j} \frac{\partial\mu_i}{\partial C_j} \delta C_i \delta C_j\Big) d\,\Omega < 0 \tag{5.15}$$

δT, $\delta\rho$, δC_i measure deviations with respect to the reference state, C_v is the specific heat, χ_T is the isothermal compressibility, μ_i is the chemical potential. From the local equilibrium hypothesis, it is inferred that C_v, χ_T and the last term in (5.15) are positive definite. As a consequence, the quadratic form $\delta^2 S$ is negative definite and from Lyapounov's theorems, stability of the reference state is ensured at the condition that

$$\frac{d}{dt} \delta^2 S \geq 0 \tag{5.16}$$

For situations close to equilibrium, that is for linear constitutive equations, the quantity (5.16) is the total entropy production [29] which is positive according to the second law of thermodynamics. It follows that such reference states are automatically stable. It must be pointed out that this is no longer true for states far from equilibrium. The criterion (5.16) has been largely used in the literature, specially in chemistry.

Observe that the above Lyapounov function (5.15) is not suitable for fluid flows. Indeed, (5.15) may vanish for non-zero values of the velocity perturbations $\delta \underset{\sim}{v}$, and is therefore not negative definite. Glansdorff and Prigogine have suggested to use as correct Lyapounov's function, the quantity

$$\delta^2 Z = \delta^2 S - \frac{1}{2T_o} \int \delta \underset{\sim}{v} \cdot \delta \underset{\sim}{v} \; d\,\Omega \tag{5.17}$$

which is undoubtedtly negative definite, T_o is a positive reference temperature.

(*) Observe that here δ is not the variation symbol. The notation $\delta^2 S$ is largely used in Prigogine and collaborators works and therefore, has been maintained in this section.

6. ITERATIVE METHODS IN NON-LINEAR HYDRODYNAMICS

The linear theory states that when the dimensionless Rayleigh (or Marangoni) number is greater than its critical value, the amplitude of the disturbance will increase exponentially with the time, untill the non-linear terms become important. When the latter are introduced, the behaviour of the disturbance may change drastically; in most cases, the exponential growth is modified in favour of a steady final amplitude situation. There is another reason for taking non-linear terms into account. The linear theory predicts that a whole spectrum of horizontal wave numbers becomes unstable. But observation shows a tendency towards simple cellular patterns, indicating that only a simple wave-number (or perhaps a small band of them) is selected.

Moreover, we have seen that the energy methods do not bring more light in explaining the mechanisms occuring beyond the linear instability point.

The problem that is set up is a non-linear eigenvalue problem. Unfortunately, no method for solving non-linear differential equations in closed form have been presented. This has motivated the development of perturbation techniques.

A widely used approach is the power series method. This technique has started with the works of Lindsted [5] , Poincaré [6] and Hopf [8] and has been extended to fluid mechanics by Gorkov [9] and independently by Malkus and Veronis [10] . It consists essentially of expanding the steady convective state in terms of a small parameter. An extension has been proposed recently by Schlüter, Lortz and Busse [30] and is analysed in the next section. We next discuss another approach due to Stuart [11]. Stuart assumes that the non-linear disturbance has the same form as the marginal solution of the linear theory with an unknown time-dependent amplitude. Stuart's method leads to an amplitude equation of the type conjectured by Landau [31] in his theory on the transition to turbulence.

6.1. The Gorkov-Malkus-Veronis power series method

To fix the ideas, we shall examine how the formalism works on the Bénard problem with two stress free boundaries. The Gorkov-Malkus-Veronis method proceeds in two steps. Firstly, one seeks the steady solutions governing the problem near the onset of convection. Secondly, the preferred form of convection is selected by examining the stability of the solutions with respect to disturbances of infinitesimal size.

The method is based on an expansion in powers of a small parameter ε , which is a measure of the amplitude of the convection. The steady non-linear equations are given by

$$\nabla . \underset{\sim}{u} = 0 \tag{6.1}$$

$$\underset{\sim}{u} . \nabla \underset{\sim}{u} = - \nabla \pi + Ra\ \theta\ \underset{\sim}{e}_z + \nabla^2 \underset{\sim}{u} \tag{6.2}$$

$$Pr\ \underset{\sim}{u} . \nabla \theta = w + \nabla^2 \theta \tag{6.3}$$

with

$$w = D^2 w = 0 \tag{6.4}$$

at the boundaries.

The solution of these equations are assumed to be of the form

$$w = w^{(o)} + \varepsilon\ w^{(1)} + \varepsilon\ w^{(2)} + \ldots \tag{6.5}$$

$$Ra = R_c + \varepsilon R^{(1)} + \varepsilon R^{(2)} + \ldots \tag{6.6}$$

and expansions of the same type for the other dependent variables; $w^{(o)}$ is the unperturbed vertical component of velocity (here $w^{(o)} = 0$) and R_c the minimum value of the marginal curve $Ra(k)$:

$$R_c = 27\ \pi^4/4$$

Substitution of the series (6.5) and (6.6) in the non-linear set (6.1)-(6.4) yields a sequence of non-homogeneous partial differential equations :

$$L\left(w^{(1)}\right) = 0 \tag{6.7}$$

$$L\left(w^{(2)}\right) = N_1\left(w^{(1)}, R^{(1)}\right) \tag{6.8}$$

$$L\left(w^{(3)}\right) = N_2\left(w^{(1)}, w^{(2)}, R^{(1)}, R^{(2)}\right) \tag{6.9}$$

$$L = (\nabla.\nabla)^3 - R_c\nabla_1 \tag{6.10}$$

L, N_1, N_2 are differential operators but in contrast to N_1 and N_2, L is linear and self-adjoint (22) :

$$\int w^{(1)} L w^{(n)} d\,\Omega = \int w^{(n)} L w^{(1)} d\,\Omega$$

Let us in particular examine the behaviour of the finite amplitude rolls in the direction y. The linearized solution of (6.7) is given by

$$w^{(1)} = \cos\,(kx)\,\sin(\pi z) \tag{6.11}$$

It is well known that the inhomogeneous equation (6.8) has a solution only if its r.h.s. is orthogonal to the solution $\overset{*}{w}{}^{(1)}$ of the adjoint linear problem (Fredholm's theorem). This is expressed by

$$\int \overset{*}{w}{}^{(1)} N_1(w^{(1)}, R^{(1)})\,dxdz = 0 \tag{6.12}$$

Here $\overset{*}{w}{}^{(1)} = w^{(1)}$ due to the self-adjointness property of the operator. The existence requirement (6.12) fixes the value of $R^{(1)}$ which is found to be zero. More generally $R^{(n)} = 0$ for odd integers when the boundary conditions are symmetric. With that result in mind, (6.9) can be solved in the form

$$w^{(2)} = \left(d_1 + d_2 \sin(2\pi z)\right)\cos(2kx) \tag{6.13}$$

wherein the constants d_1 and d_2 are given by

$$d_1^{-1} = 64\pi^3\ , \quad d_2^{-1} = 60\pi^3\left(\frac{k^2}{\pi 2} + 1\right)^3$$

The condition of solvability applied to (6.9) gives

$$\int w^{(1)} N_2(w^{(1)}\ ,\ w^{(2)}\ ,\ 0\ ,\ R^{(2)})\ dxdz = 0 \tag{6.14}$$

hence,

$$R^{(2)} = \gamma/k^2 \tag{6.15}$$

with

$$\gamma = \frac{1}{4}\pi^3(2d_1 + d_2)$$

It follows from (6.6) that

$$Ra = R_c + \varepsilon^2 R^{(2)} \tag{6.16}$$

Since Ra is an externally given parameter, (6.16) defines ε , namely

$$\varepsilon = \frac{k}{\gamma 1/2}\ (Ra - R_c)^{1/2} \tag{6.17}$$

This indicates that the steady solution is proportional to $(Ra - R_c)^{1/2}$.

Instead of treating separately first approximations $w^{(1)}$ appropriate to roll, hexagons, etc, Schlüter et al [30] and Busse [26] considered a whole manifold of steady solutions at once. More specifically, they took for $w^{(1)}$, the general form

$$w^{(1)} = \sum_{n=-N}^{+N} c_n \exp i(\underset{\sim}{k}_n\cdot\underset{\sim}{r})f(z) \tag{6.18}$$

where c_n are arbitrary complex numbers and $\underset{\sim}{r} = (x,y)$. The $\underset{\sim}{k}_n$'s are arbitrary except that they have the same absolute value

$$(k_x^2)_n + (k_y^2)_n = (k_x^2)_m + (k_y^2)_n = \dots = k^2$$

$N = 1$ corresponds to two dimensional rolls and $N = 3$ to hexagonal solutions.

Although the number of solutions (6.18) is considerably restricted by the solvability conditions (6.12), (6.14), ..., there remains still an infinite number of solutions. In order to find which of them are physically expected, one must examine their stability, even though such a criterion does not guarantee an unique solution. Since the disturbances $\tilde{w}, \tilde{\theta}$ of the steady-state solutions w, θ are assumed of infinitesimal amplitude, they satisfy linear homogeneous differential equations. According to the results of chapter III, their time-dependence can be written in the form exp (σt) . Moreover, the expansions (6.5) and (6.6) of the steady solution suggest to develop $\tilde{w}, \tilde{\theta}$ and σ in power series of ε , namely

$$\tilde{w} = \tilde{w}^{(o)} + \varepsilon\tilde{w}^{(1)} + \varepsilon^2\tilde{w}^{(2)} + \dots\ ,\quad \tilde{\theta} = \tilde{\theta}^{(o)} + \varepsilon\theta^{(1)} + \varepsilon^2\theta^{(2)} + \dots \tag{6.19}$$

$$\sigma = \sigma^{(o)} + \varepsilon\sigma^{(1)} + \varepsilon^2\sigma^{(2)} + \ldots \qquad (6.20)$$

For the sake of simplicity, we examine the stability of the solution (6.11) rather than (6.18). After linearization of the equations of motion, one obtains at the various order of ε ,

$$-\sigma^{(o)}\tilde{w}^{(o)} + L\left(\tilde{w}^{(o)}\right) = 0 \qquad (6.21)$$

$$-\sigma^{(o)}\tilde{w}^{(1)} + L\left(\tilde{w}^{(1)}\right) = L_1\left(\sigma^{(1)}, \tilde{w}^{(o)}, w^{(1)}\right) \qquad (6.22)$$

$$-\sigma^{(o)}\tilde{w}^{(2)} + L\left(\tilde{w}^{(2)}\right) = L_2\left(\sigma^{(2)}, \tilde{w}^{(o)}, \tilde{w}^{(1)}, w^{(1)}, w^{(2)}\right) \qquad (6.23)$$

At the zeroth-order approximation, a particular solution of (6.21) is

$$\tilde{w}_o = \cos(kx)\sin(\pi z) \qquad (\text{with } \sigma^{(o)} = 0) \qquad (6.24)$$

Clearly (6.24) has the same form as the steady solution whose stability is investigated. Moreover, by taking for k the same wave number as that of the steady solution, the disturbance with $\sigma^{(o)} = 0$ is the most critical.

At the first order, it is found that $\sigma^{(1)}$ vanishes just as $R^{(1)}$ did. That means that at the first order, no steady solution is preferred. At the next order, the existence requirement yields

$$\sigma^{(2)} = -2\gamma \qquad (6.25)$$

Hence

$$\sigma = \varepsilon^2\sigma^{(2)} = -2k^2(Ra - R_c) < 0 \qquad (6.26)$$

where (6.17) has been used.

We conclude that when non-trivial solutions exist, they are locally stable. When a more complicated steady solution like (6.18) is selected, the analysis is more involved. As shown by Schlüter et al [30] , all the solutions with $N > 1$ are unstable, while for $N = 1$, all the $\sigma^{(2)}$ values are negative, which is a confirmation of (6.26). Further investigations have brought out the following results.

1. For a fluid of constant viscosity, confined between rigid or free boundaries, the rolls are stable while hexagons are always unstable.

2. For a fluid of temperature-dependent viscosity, confined between rigid or free boundaries, hexagons are stable.

3. For a fluid of constant viscosity, with a temperature dependent surface tension,

hexagons are stable.

6.2. The Stuart method

A different, but closely related procedure is the Stuart scheme. The solutions of Bénard's non-linear problem are expressed as

$$\theta = \theta^{(o)} + \theta^{(1)} + \theta^{(2)} + \dots \tag{6.27}$$

$$w = w^{(o)} + w^{(1)} + w^{(2)} + \dots \tag{6.28}$$

θ^{o}, w^{o} is the reference solution ($w^{o} = 0$ in Bénard's problem) and upper indices correspond to higher order of approximations.

The various approximations obey equations of the form

$$L\left(w^{(1)}\right) = 0 \tag{6.29}$$

$$L\left(w^{(2)}\right) = M_1\left(w^{(1)}\right) \tag{6.30}$$

$$L\left(w^{(3)}\right) = M_2\left(w^{(2)}\right) \tag{6.31}$$

L is the linear time-independent operator (6.10), M contains not only non-linear terms, but also <u>time derivatives</u> and some linear terms of the same order of magnitude.

The basic idea is to assume that the velocity and temperature fields keep the same form as in the linear analysis, except that the time dependence is given by an unknown function $A(t)$. The latter grows exponentially when non-linear terms are dropped but remains bounded in presence of non-linearities.

Equation (6.29) and the boundary conditions (6.4) are satisfied if

$$w^{(1)} = A(t)\cos(kx)\sin(\pi z) \quad \text{and} \quad R_c = \frac{27}{4}\pi^4 \tag{6.32}$$

The spatial dependence of $w^{(1)}$ is clearly inspired by the corresponding solution of the linear theory while $A(t)$ is the unknown time-dependent amplitude.

At the next-order of approximation, one has

$$\begin{aligned} L\left(w^{(2)}\right) &= M\left(w^{(1)}\right) \\ &= (\dot{A} - \sigma A)\cos(kx)\sin(\pi z) \\ &\quad + \pi^3 A^2(1 + \cos 2kx)\sin(2\pi z) \end{aligned} \tag{6.33}$$

where

$$\sigma = k^2 (Ra - R_c) \tag{6.34}$$

A particular solution

$$w^{(2)} = A^2 (d_1 + d_2 \cos 2kx) \sin(2\pi z) \tag{6.35}$$

is obtained by removing in (6.33), the term proportional to the solution of the homogeneous problem, i.e. by setting

$$\dot{A} - \sigma A = 0 \tag{6.36}$$

Like in the linear theory, the perturbation grows exponentially :

$$A \sim \exp(\sigma t) \tag{6.37}$$

Due to non-linear self-interactions of the first order modes in cos(kx), it is seen in (6.35) that second order modes proportional to A^2 are produced.

At the next order of approximation, the first and second order modes interact to generate third order modes proportional to A^3. At this order of approximation, A satisfies a Landau-like equation (31)

$$\dot{A} - \sigma A + \gamma A^3 = 0 \qquad \left(\gamma = \frac{1}{4} \pi^3 (2d_1 + d_2)\right) \tag{6.38}$$

Observe that only odd power terms in A occur. This follows from the property that an equivalent pattern is obtained by reversing the time, i.e. by transforming A into - A. In contrast to (6.36), (6.38) possesses a steady solution given by equation

$$A_s = \left(\frac{\sigma}{\gamma}\right)^{1/2} = \frac{k}{\gamma 1/2} (Ra - R_c)^{1/2} \tag{6.39}$$

in complete agreement with (6.17).

The stability of (6.39) is derived by following the classical procedure : set

$$A = A_s + \tilde{A}$$

where $\tilde{A}$ is the disturbance. By substituting in (6.38) and omitting non-linear terms, one obtains

$$\dot{\tilde{A}} = (\sigma - 3\gamma A_s^2)\,\tilde{A} = -\,2k^2 (Ra - R_c)\tilde{A} \tag{6.40}$$

Since $Ra > R_c$ under supercritical conditions, it is clear that the local steady solution (6.39) is stable, in compliance with Gorkov-Malkus-Veronis theory.

The above analysis has been extended to the case that the first order approximation $w^{(1)}$ contains various modes. By analogy with (6.18), Palm (24) considers solutions like

$$w^{(1)} = \sum_n A_n(t)\, e^{i\underset{\sim}{k}_n \cdot \underset{\sim}{r}}\, f(z) \qquad (6.41)$$

where it is assumed that all the modes have the same wave-number k . Instead of (6.36) and (6.38), one is then led to a set of coupled non-linear equations

$$\dot{A}_n = \sigma A_n - A_n \sum_m b_{mn} A_m^2 - b\, A_n^3 \qquad (6.42)$$

It is important to recall that the perturbation methods outlined in this chapter are restricted to Ra values near R_c ; this restriction is however relaxed by using the more general development (6.41) or other methods requiring extensive numerical computations (26).

7. BIFURCATION THEORY

7.1. Introduction

Unlike non-linear theories, bifurcation theory is not concerned with explicit calculation of the details of the flow. Instead, bifurcation theory describes the stability problem from a spectral analysis of the linear theory. More precisely, the theory provides a catalogue of the various solutions which branch from a basic solution when the latter gives up its stability to infinitesimally small disturbances.

The first study of bifurcation was done by Euler who studied the buckling of a column subjected to axial compression. Recent and extensive bibliography can be found in the books of Joseph (4) , Sattinger (13), Iooss and Joseph (14), Stakgold (32) or in the monographs edited by Keller and Antman (33) , Haken (34) , Swinney and Gollub(35), Bardos and Bessis (36).

Our purpose is to give only a brief and simplified sketch of the bifurcation theory. Readers interested in more detailed analyses are recommanded to consult the above mentioned references.

7.2. Regular and singular points

Consider a disturbance u of a basic equilibrium or steady solution u_b obeying the evolution equation

$$\frac{du}{dt} = F(R,u) \tag{7.1}$$

For the sake of simplicity, u is supposed to be a scalar, R is a real scalar parameter, for instance Rayleigh's or Reynold's number, F is a non-linear function of u and R, whose first and second derivatives are continuous. The null solution $u = 0$ satisfies (7.1) for all R, i.e

$$F(R,0) = 0 \tag{7.2}$$

A solution is said to bifurcate from $u = 0$ at the value R_c of the parameter if there exist two or more solutions which tend to $u = 0$ as R approaches R_c.

The main questions raised in the theory of bifurcation are the following :

1. What are the solutions u and parameter values R_c at which bifurcations occur?
2. How many solutions emanate from the bifurcation point?
3. What can be said about their stability?

To start the analysis, let us consider the linear eigenvalue problem.

$$Lu = Ru \tag{7.3}$$

where L is a linear operator acting in some normed linear space. For every value of R, $u = 0$ is solution of (7.3). Assume that there exist a set of eigenvalue $R_1, R_2, R_3, \ldots$ with $u_1, u_2, u_3, \ldots$ as normalized eigenfunctions

$$Lu_j = R_j u_j \qquad ||u_j|| = 1$$

If A is a real value, other solutions of (7.3) are

$$u = Au_j$$

with

$$||u|| = A$$

These solutions are represented on figure 7.1. ; it is seen that the solution $u = 0$

bifurcates into two branches at each eigenvalue R_i : the points $(R_i,0)$ are the bifurcation points of the problem.

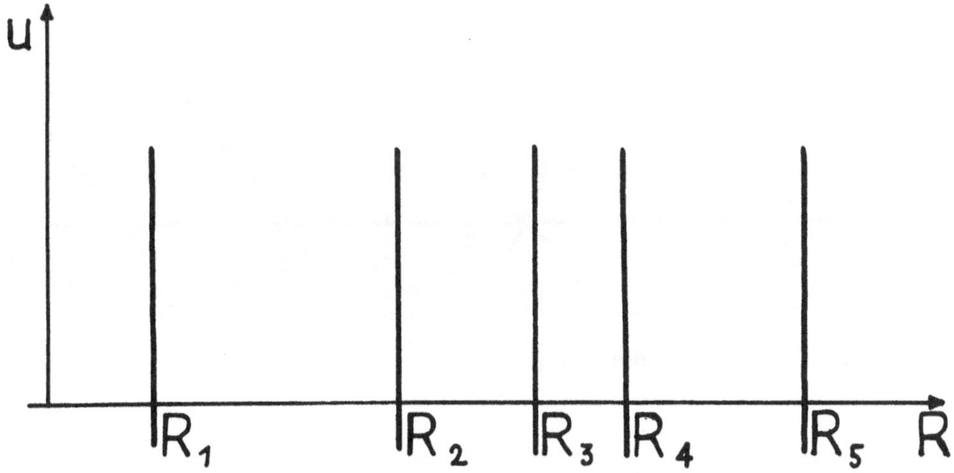

Figure 7.1. Bifurcated solutions for the linear problem.

Let us now examine the non-linear eigenvalue problem

$$F(R,u) = 0 \tag{7.4}$$

which has (7.3) as its linearization. A graph of the possible solutions u is shown on figure 7.2. The following comments can be afforded :

1. The branches issued from the eigenvalues of the linear problem are curved.

2. There may be one or several branches emanating from an eigenvalue (see R_1 and R_2 respectively).

3. It may happen that there is no branching from an eigenvalue of the linear problem (see R_3).

4. There may exist secondary bifurcations (see R_4).

5. Some branches may not emanate from the eignevalues of the linear problem (see R_5). They are called isolated solutions.

7.3. Stability of solutions

Since $u = 0$ represents the basic solution, the stability of the latter amounts

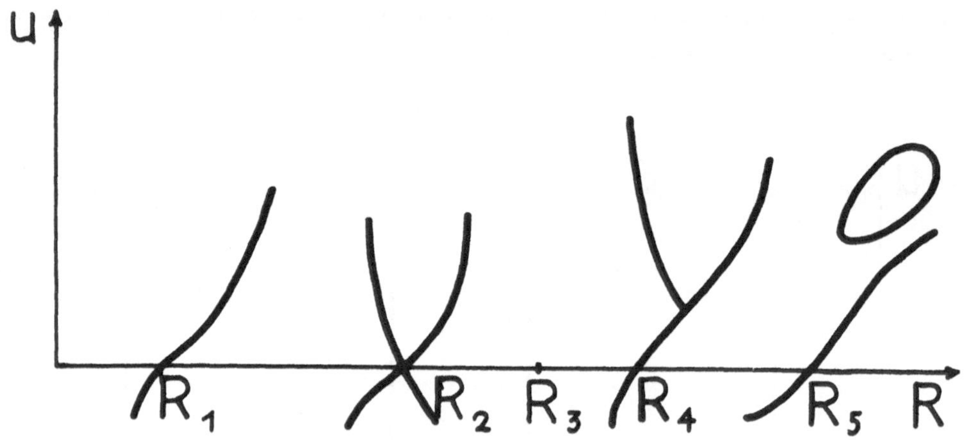

Figure 7.2. Bifurcating solutions.

to the discussion about the stability of the null solution. This is done by setting

$$u = \varepsilon \nu \tag{7.5}$$

where ε is a small parameter, and by substituting in (7.1). Dropping all the second order terms in ε^2, one obtains an equation of the form

$$\frac{\partial \nu}{\partial t} = F_L(R|\nu) \tag{7.6}$$

where $F_L(R|\nu)$ is a linear operator, identified as the derivative of $F(R,u)$ with respect to u, calculated at $u = 0$[4] :

$$F_L(R|\nu) = F_u(R,u) \tag{7.7}$$

Solutions of the form

$$\nu = e^{\sigma t} V(x)$$

satisfy the differential equation (7.6) at the condition that σ and V solve the spectral problem

$$\left(\sigma - F_u(R|\ )\right) V = 0 \tag{7.8}$$

σ is a complex quantity,

$$\sigma = \text{Re}\,\sigma + i\,\text{Im}\,\sigma \tag{7.9}$$

The basic solution is stable with respect to infinitesimally small disturbances if $\text{Re}\,\sigma < 0$ and unstable if $\text{Re}\,\sigma > 0$. The critical value R_c is the value of R for which $\text{Re}\,\sigma$ changes its sign as R is varied across R_c .

For values of R smaller than R_c , the representative eigenvalues lie in the l.h.s. of the complex σ-plane : the basic solution u_b is then unique.

When R is increased $(R > R_c)$, some eigenvalues cross the y-axis and are found in the r.h.s. of the σ-plane . There exists than new solutions $u_b + u$ which bifurcate from the basic flow u_b .

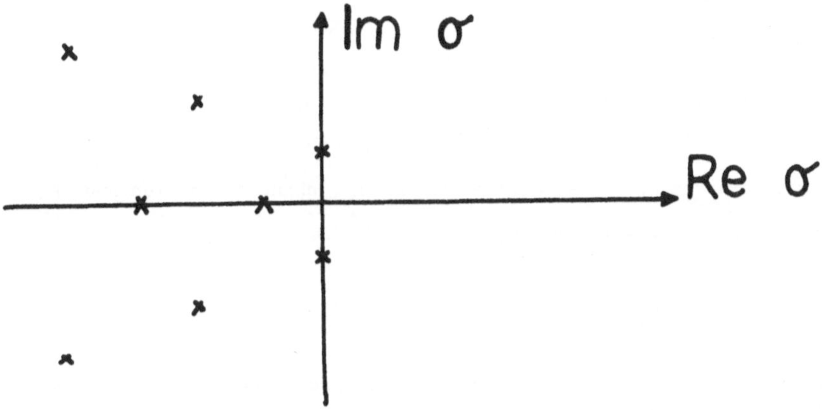

Figure 7.3. Complex σ-plane .

At the critical $R = R_c$, one has

$$\text{Re}\,\sigma = 0 \qquad\qquad \text{Im}\,\sigma = i\,\omega_o \tag{7.10}$$

We distinguish two cases :

1. $\omega_o = 0$: then, a single eigenvalue crosses to the r.h.s. at the origin. One observes symmetry breaking and the solution which bifurcates is a steady solution.

2. $\omega_o \neq 0$: a complex pair $\pm\, i\omega_o$ crosses to the r.h.s. as R is increase. The solution which bifurcates is a time-periodic solution.

We need now to discuss the stability of the new solutions observed as R is increased beyond R_c. To simplify the discussion, we assume that the solutions branching from the bifurcating point are steady. Introducing the notation $u = w(\frac{dw}{dt} = 0)$ for steady solutions, one has from (7.1)

$$F(w,R) = 0 \tag{7.11}$$

For later purpose, it is convenient to introduce the following classification of points of a plane curve.

a. Regular point of $F(w,R) = 0$.

It is a point for which

$$F_R \neq 0 \qquad\qquad F_w \neq 0 \tag{7.12}$$

Subscripts like R and w mean derivation with respect to R and w respectively. When (7.12) is satisfied, there exists a unique curve $w(R)$ or $R(w)$ passing through the point. The curve $w(R)$ is called a bifurcation branch.

b. Regular turning point : a point at which R_w changes its sign and $F_R \neq 0$.

c. Singular point : a point at which

$$F_R = 0 \quad , \qquad F_w = 0 \tag{7.13}$$

d. (Singular) double point : a singular point through which pass two branches of $F(R,w) = 0$, with two distinct tangents.

e. Singular turning (double) point : a double point where R_w changes its sign.

The next task is to relate bifurcation and stability. This is achieved by the following two theorems.

Theorem 1 (Factorization theorem) : For every steady solution $F(w,R) = 0$ located on the bifurcation branch, one has

$$\sigma(w) = - R_w(w)F_R\big(w,R(w)\big) \tag{7.14}$$

Demonstration. It follows from (7.8) that

$$\sigma(w) = F_w\big(w,R(w)\big) \tag{7.15}$$

Moreover, one has the identity

$$\frac{d}{dw} F\left(R(w),w\right) = \frac{\partial F}{\partial w} \left(R(w),w\right) + \frac{\partial F}{\partial R} R_w = 0 \qquad (7.16)$$

Combining (7.15) and (7.16) yields the required results(7.14).

The theorem implies that σ (w) changes its sign as w is varied across a regular turning point, for which R_w changes its sign. This mean that the solution is stable on one side of a regular turning point and unstable on the other side. (See figure 7.4., where a dotted line represents an unstable solution while a solid line represents a stable solution).

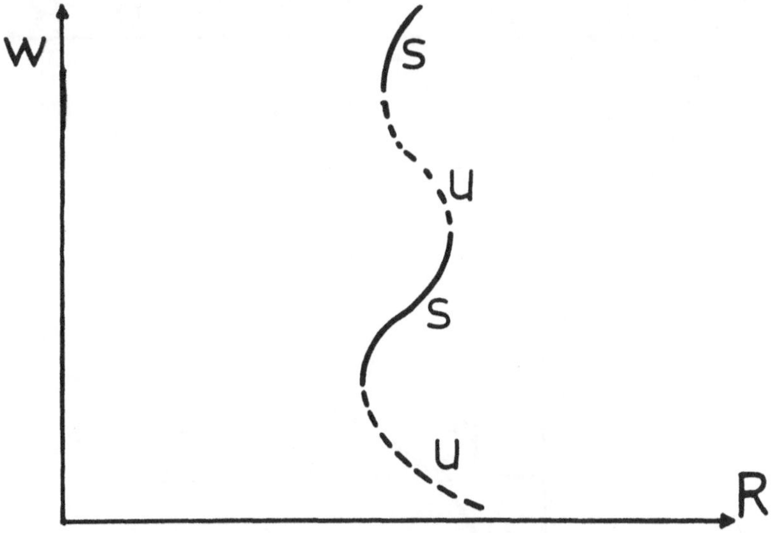

Figure 7.4. Change of stability at turning points.

We now formulate a theorem for double-point bifurcations

Theorem 2 : If one assumes that all singular solutions of $F(R,w) = 0$ are double points, the stability of such solutions changes at each regular turning point and at each singular point (which is not a turning point) and only at such points.

Illustrations of this theorem are found on figure 7.5.

Bifurcations solutions which exist for $R > R_c$, and are consequently unstable by the criteria of the linear theory, are called supercritical (figure 7.5 a). Bifurcation solutions which exist for $R < R_c$ are called subcritical (figure 7.5 b). When the amplitude of the disturbance is small, it has been shown (4) that solutions which bifurcate supercritically are stable, those which bifurcate subcritically are unstable. An intermediate situation is represented by the transcritical bifurcation (figure 7.5. c) : the bifurcation is supercritical when $R > R_c$ and subcritical when $R < R_c$.

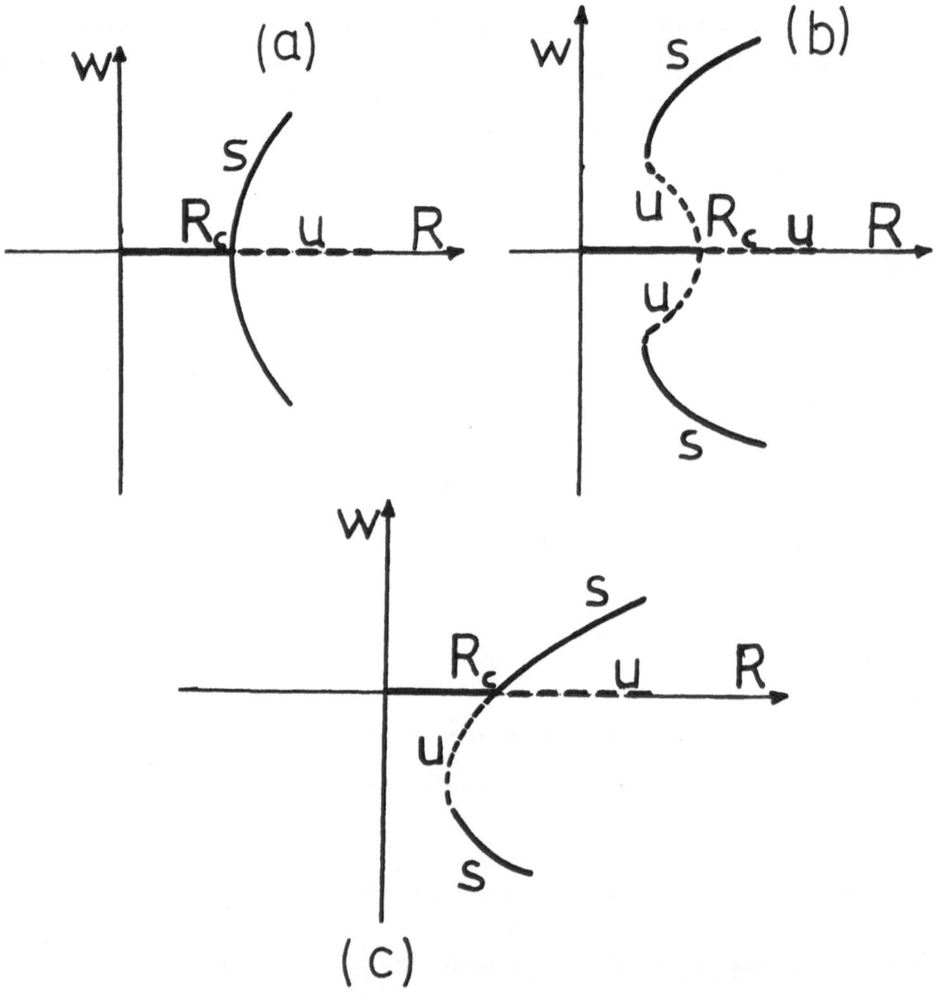

Figure 7.5. Typical bifurcations of steady solutions.
a) supercritical , b) subcritical , c) transcritical.

When the bifurcation is supercritical, the transition between the successive solutions is continuous and there is no abrupt change when the bifurcation point is crossed. In contrast, subcritical bifurcation is characterized by a discontinuous transition : the basic solution "snaps" through the bifurcation to some flow with a larger amplitude.

The above bifurcation diagrams (figure 7.5.) are frequently met in hydrodynamics. In particular (7.5.a) is typical of problems with high degree of symmetry like the Bénard layer between two rigid (or free) surfaces. Diagram (7.5.c) is characteristic of Bénard-Marangoni problem where a region of subcriticality is observed.

7.4. An illustration

An illustration of the results embodied in theorem 2 is provided by Benjamin's experiment [37], sketched on figure 7.6. and reported in Joseph's paper [35]. Benjamin's apparatus is a board with two holes through which passes a viscoelastic wire like a bicycle brake cable. The wire forms an arch above the board of length l and undergoes buckling under the action of gravity. Denote by θ the angle between the vertical plane and the plane of the wire arch : $\theta = 0$ corresponds to the wire arch in the vertical upright position. The equation of motion of the wire is assumed to be given by

$$\frac{d\theta}{dt} = F(l,\theta)$$

with steady solutions of the form

$$F\big(l(\theta),\theta\big) = 0$$

There is a one-to-one correspondance between the behaviours of the wire and the bifurcation diagram represented on figure 7.7.

For small $l (< l_o)$, the only stable solution is the upright one corresponding to $\theta = 0$. This position becomes unstable when $l > l_c$ and the arch falls to the right or the left as seen in figure (7.6.b). The point $(l_c, 0)$ is a singular turning point and according to theorem 2, the stability changes while crossing this point (see figure 7.7). A new solution $(l_c, \pm\theta_c)$ appears that corresponds to the bent arch. For $l < l_c$ the bent solution of the arch may also be stable : in fact, for $l_o < l < l_c$, there exist three stable steady solutions : $\theta = 0$ (the upright one) and $\theta = \pm\,\theta_o$ (the symmetric left and right bent ones). The points $(l_o, \pm\,\theta_o)$ are regular turning points at which the stability changes. In the region $l_o < l < l_c$, the system is said to exhibit hysterisis : if starting from (l_c, θ_c), the length of the wire is decreased, the bent configuration is observed to remain stable until $l = l_o$: this point is a regular turning point at which stability changes and the bent position becomes unstable.

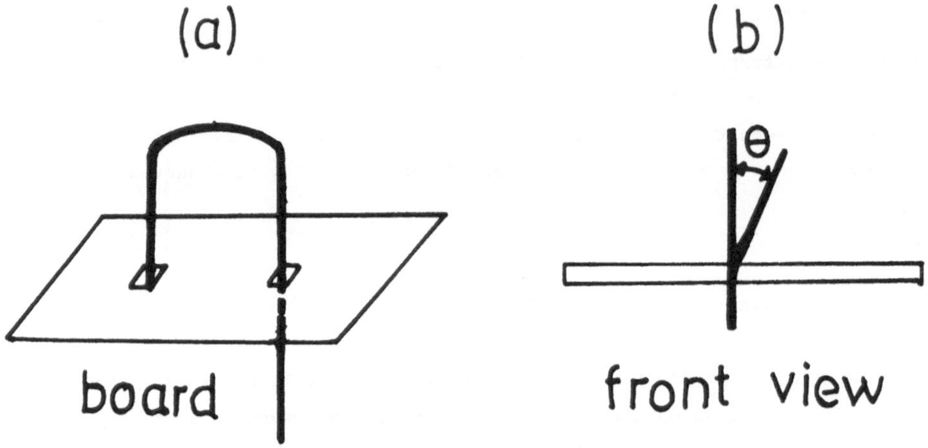

Figure 7.6. Buckling of a viscoelastic wire under action of gravity.

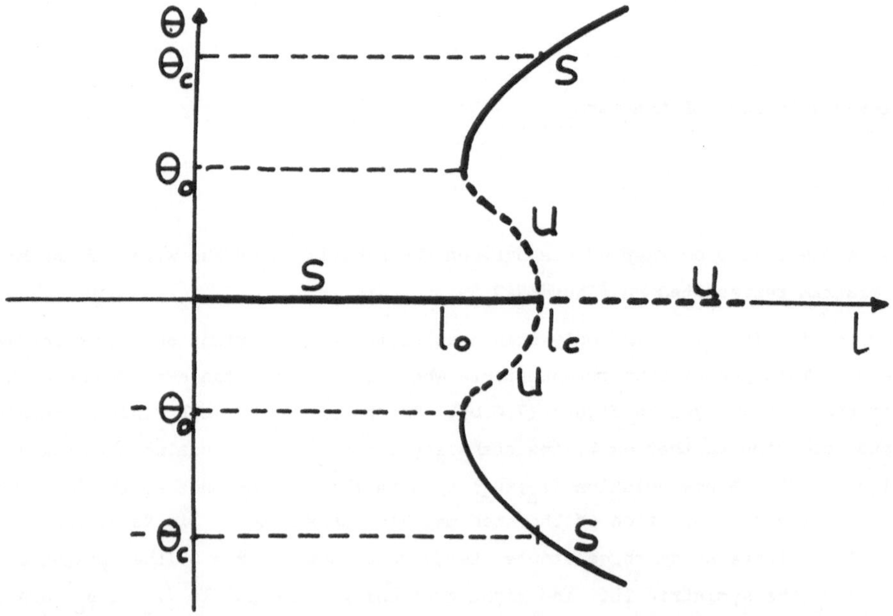

Figure 7.7. Bifurcation diagram for the buckling of the viscoelastic wire.

In short, by increasing l , the arch remains stable in the vertical position up to l_c . When $l > l_c$, the upright solution is unstable and the arch occupies a right or left stable position. Now decreasing l from l_c to l_o , the bent positions remain stable until $l = l_o$: at this point, the arch snaps through the vertical position since only the solution $\theta = 0$ is stable.

The extension of the above simplified discussion to hydrodynamic problems, described for instance by Navier-Stokes equations, is not a trivial matter. It is based on the method of projections whose a detailed analysis may be found in Joseph's paper [35].

8. VARIATIONAL METHODS

8.1. General condition of existence of a variational principle

Variational methods provide specific and powerful techniques for solving stability problems. The general problem is the following : given a differential equation, (or a set of differential equations), find the variational principle whose Euler-Lagrange equations are precisely the original set of differential equations. The solution of this problem is important in that it furnishes an opportunity to use the variational techniques, like the Rayleigh-Ritz or the Kantorovitch partial integration methods, for obtaining approximate solutions.

It must be realized that generally it is not possible to construct a variational principle corresponding to a given set of equations.

A general condition of existence of a variational principle has been established by Vainberg [38]. Vainberg's important result can be briefly stated as follows.

Let

$$N(u) = 0 \tag{8.1}$$

be a non-linear differential equation and

$$B(u) = 0 \tag{8.2}$$

the corresponding boundary condition. The Fréchet derivative $N'(u)$ in the direction η is defined by

$$N'(u) = \lim_{\varepsilon \to 0} \frac{N(u + \varepsilon\eta) - N(u)}{\varepsilon} \tag{8.3}$$

The Fréchet derivative is said to be symmetric in the directions η and ϕ if

$$\int\phi\, N'(u)\eta\, d\,\Omega = \int\eta\, N'(u)\phi\, d\,\Omega \tag{8.4}$$

where $d\Omega$ is the elementary volume of integration.

Theorem : The necessary and sufficient conditions that there exists a variational principle corresponding to $N(u) = 0$ is that the Fréchet derivative $N'(u)$ be symmetric. The variational principle writes explicitely as

$$\delta I(u) = \delta\int_\Omega\int_0^1 uN(su)ds\, d\,\Omega = 0 \tag{8.5}$$

where s is a real parameter and δ the variation symbol.

The existence of a variational principle is thus subordonated to the symmetry property of the Fréchet derivative. No variational principle can be formulated if the Fréchet derivative is not symmetric.

It is instructive to examine the particular case of a linear operator L. By definition,

$$L(\alpha u + \beta v) = \alpha\, L(u) + \beta\, L(v) \tag{8.6}$$

where α and β are arbitrary scalar quantities. Its Fréchet derivative is

$$L'(u) = \lim_{\varepsilon\to 0} \frac{L(u+\varepsilon\eta) - L(u)}{\varepsilon} = \lim_{\varepsilon\to 0} \frac{L(\varepsilon\eta)}{\varepsilon} = L(\eta) \tag{8.7}$$

The symmetry condition (8.4) reduces now to

$$\int\phi L(\eta) d\,\Omega = \int\eta L(\phi) d\,\Omega \tag{8.8}$$

expressing that L must be a self-adjoint operator. The variational principle giving

$$L(u) = 0 \tag{8.9}$$

as Euler-Lagrange equation is by (8.5),

$$\delta\int_\Omega uL(u)d\Omega = 0 \tag{8.10}$$

In short, there exists always a variational principle associated to a linear operator, at the condition that the latter is self-adjoint. If the operator is non-linear, the conditions of existence of a variational criterion is the symmetry of its Fréchet derivative.

The equations governing Bénard's problem are generally not linear and do not possess symmetric Fréchet derivatives. Even the linearized problem is not self-adjoint, except if $Ma = 0$. It follows that solely the linearized Bénard problem with rigid or stress-free surfaces can receive a variational formulation.

In most problems of fluid mechanics, it is not possible to associate a variational principle to a differential equation. The term variational principle is here understood in its classical sense. This means that all the quantities appearing in the integrant of the functional are submitted to variation and that the functional is made stationary.

To circumvent this difficulty, some authors have proposed to relax somewhat this definition by allowing some terms in the integrant to be kept constant during the variational procedure. Such criteria are called quasi-variational or restricted principles (23). An example is given in the next section.

8.2. The variational principle for the Bénard linearized problem

Since we are essentially interested by the marginal stability solution, let us recall the stationary linearized balance equations; in non-dimensional form,

$$\nabla . \underset{\sim}{u} = 0 \tag{8.11}$$

$$\nabla^2 \underset{\sim}{u} + Ra\, \theta\, \underset{\sim}{e}_z - \nabla \pi = 0 \tag{8.12}$$

$$\nabla^2 \theta + w = 0 \tag{8.13}$$

In terms of the amplitudes of the normal modes, one has

$$(D^2 - k^2)^2 W = Rak^2 \phi \qquad (8.14$$

$$(D^2 - k^2)\phi = - W \tag{8.15}$$

As boundary conditions, one takes

at $z = 0$ $\quad W = \phi = 0 \qquad (8.16)$

(rigid surface) $\quad DW = 0 \qquad (8.17)$

at $z = 1$ $\quad W = 0 \qquad (8.18)$

(stress-free surface)

$$D\phi = - h\phi \tag{8.19}$$

$$D^2 W = 0 \tag{8.20}$$

The conditions (8.16) to (8.18) are called essential boundary conditions while (8.19) and (8.20) are natural boundary conditionss. Only the former must be satisfied by the trial functions, because the natural ones are automatically met by the variational principle.

Our problem is to formulate a variational equation of the form

$$\delta I(W,\phi) = 0 \tag{8.21}$$

producing (8.14) and (8.15) as Euler-Lagrange equations and the natural boundary conditions (8.19) and (8.20).

The construction of the functional I is rather simple (39). Multiply equation (8.14) by W and equation (8.15) by ϕ , substract and integrate over z ; these operations yield

$$I = \int_0^1 \Big(\frac{1}{Rak^2} (k^4W^2 + WD^4W - 2k^2WD^2W) - \phi(D^2\phi) + k^2\phi^2 - 2W\phi\Big) dz \tag{8.22}$$

After integration by parts of the terms involving WD^4W and $\phi D^2\phi$, one finds after use of the boundary conditions (8.16) to (8.20),

$$I = \int_0^1 \Big(\frac{1}{Rak^2} (D^2W - k^2W)^2 + (D\phi)^2 + k^2\phi^2 - 2W\phi\Big) dz + \Big(h\phi^2\Big)_{z=1} \tag{8.23}$$

It is easily checked that the Euler-Lagrange equations corresponding to $\delta I = 0$ are the required equations (8.14) and (8.15), and that equations (8.19) and (8.20) are recovered as natural boundary conditions. The above expression of the principle remains also valid for a rigid upper surface. A principle analogous to (8.23) was also proposed by Chandrasekhar (15), who assumed that $h = 0$.

If the surface tension is temperature dependent, the boundary conditions (8.20) is replaced by

$$D^2W = - k^2Ma\ \phi \tag{8.24}$$

This condition renders the problem non self-adjoint and excludes the possibility of producing a classical variational principle.

The quasi-variational principle equivalent to the set of equations (8.14)-(8.19) and (8.24) can be written as

$$\delta\left[\int_0^1\left(\frac{1}{\mathrm{Rak}^2}(D^2W-k^2W)^2+(D\phi)^2+k^2\phi^2-2W\phi\right)dz + \left(h\phi^2\right)_{z=1}\right] + 2\,\frac{\mathrm{Ma}}{\mathrm{Ra}}\left(\phi\delta(DW)\right)_{z=1} = 0 \qquad (8.25)$$

It differs from the exact criterion $\delta I = 0$ only by the last term, wherein it is worth noting that ϕ is not submitted to variation. Clearly (8.25) must be classified as a quasi-variational principle. Many principles, like the Hamilton principle in analytical mechanics, the Glansdorff-Prigogine local potential (28) , the Biot (40) and Lebon-Lambermont variational criteria (41,42) in continuum physics pertain to this category. Despite their quasi-variational character, such formulations revealed very successful for handling problems of heat transfer and fluid mechanics.

The variational equation corresponding to the steady equations (8.11)-(8.13), is expressed by

$$\delta\left(2\,\langle\omega\theta\rangle - \frac{1}{\mathrm{Ra}}\langle\nabla\underset{\sim}{u} : \nabla\underset{\sim}{u}\rangle - \langle\nabla\theta . \nabla\theta\rangle - \langle h\theta^2\rangle_1 + \langle\mu\nabla . \underset{\sim}{u}\rangle\right) + 2\,\frac{\mathrm{Ma}}{\mathrm{Ra}}\langle\theta\delta(\nabla_1 . \underset{\sim}{u})\rangle_1 = 0 \qquad (8.26)$$

where

$$\nabla_1 . \underset{\sim}{u} = \frac{\partial u}{\partial x} + \frac{\partial v}{\partial y}$$

μ is a Lagrange multiplier which is introduced because the velocity components, linked by the continuity equation, are not independent. After that the variation has been performed, it is seen that μ is equal to the pressure disturbance divided by $2\mathrm{Ra}$. Physically, the first term inside the accolades is linked to the energy production inside the layer, while the next three terms are related to the dissipated energy.

If $\mathrm{Ma} = h = 0$, the principle (8.26) is equivalent to Palm's criterion (24). Palm must also be credited with having extended its principle to the non-linear range.

8.3. A reformulation of Serrin-Joseph's principle

Let us recall that Serrin-Joseph's principle was given the form

$$\delta\left(P - \Lambda_\lambda D + \langle\mu\nabla . \underset{\sim}{u}\rangle\right) = 0$$

with (4.25)-(4.26) as Euler-Lagrange and (4.29)-(4.31) as natural boundary equations. These relations are linear and it is therefore justified to expand their solutions in normal modes.

Let W and ϕ' be the amplitudes of the velocity and the modified temperature fields respectively; at marginal stability, they satisfy

$$(D^2 - k^2)^2 W - k^2 \frac{Ra(1 + \lambda/Ra)}{2\lambda^{1/2}} \phi' = 0 \tag{8.27}$$

$$(D^2 - k^2) \phi' + \frac{Ra(1 + \lambda/Ra)}{2\lambda^{1/2}} W = 0 \tag{8.28}$$

while at the upper surface ($z = 1$),

$$W = 0 \tag{8.29}$$

$$D^2 W + \frac{1}{2} k^2 \frac{Ma}{\lambda^{1/2}} \phi' = 0 \tag{8.30}$$

$$D\phi' + h\phi' + \frac{1}{2} \frac{Ma}{\lambda^{1/2}} DW = 0 \tag{8.31}$$

The variational principle which restitutes these equations is

$$\delta J(W, \phi') = 0$$

with

$$J(W, \phi') = \int_0^1 \Big(\frac{1}{k^2} (D^2 W - k^2 W)^2 + (D\phi')^2 + k^2 \phi'^2 - Ra \frac{1 + \lambda/Ra}{\lambda^{1/2}} W\theta' \Big) dz + \Big(h\phi'^2 + \frac{Ma}{\lambda^{1/2}} \phi' DW \Big)_{z=1} \tag{8.32}$$

At the exception of the last term, the above principle is identical to (8.25) after that λ has been set equal to Ra. But in contrast to (8.25), (8.32) is an exact principle; the price paid for it is that one does not recover the correct hydrodynamic equations.

For $Ma = 0$, both principles (8.25) and (8.32) yield the same critical Rayleigh number, from which follows that the conditions of stability are necessary and sufficient.

8.4. The numerical procedure

The critical values of the dimensionless parameters Ra and Ma corresponding to the onset of instability will be determined through the two above variational principles (8.25) and (8.32). To accomplish this task, the Rayleigh-Ritz technique is used. It consists of expanding the amplitudes W and ϕ (or ϕ') linearly in terms of independent functions $f_i(z)$, $g_j(z)$ which are continuous, differentiable and given a priori :

$$W = \sum_{i=1}^{N} a_i f_i(z) \tag{8.33}$$

$$\phi = \sum_{j=1}^{M} b_j g_j(z) \tag{8.34}$$

The constant coefficients a_i and b_j are unknown quantities to be determined by the procedure. The trial functions (8.33) and (8.34) must satisfy the essential, but not necessarily the natural boundary conditions. Indeed, the latter are automatically met by the variational principle. After insertion of the trial functions (8.33) and (8.34) in the functionals (8.25) or (8.32) and integration with respect to z , one is faced with a function I containing the unknowns a_i and b_j . They are determined by imposing that $I(a_i,b_j)$ is extremum, namely by

$$\frac{\partial I(a_i,b_j)}{\partial a_i} = 0 \qquad \frac{\partial I(a_i,b_j)}{\partial b_j} = 0 \qquad \begin{matrix} i = 1, \dots N \\ j = 1, \dots M \end{matrix} \tag{8.35}$$

The set (8.35) represents a system of $N + M$ linear homogeneous algebraic equations in the a_i's and b_j's . Non trivial solutions exist if and only if the determinant of of the coefficients vanishes. This yields a relation between the parameters Ra, h, Ma and k . By fixing two of these parameters, for instance Ra and h , one obtains the marginal stability curve for Ma versus the wave number k .

Let us take as trial functions

$$f_i(z) = z^2(1 - z)T^*_{i-1}(z) \tag{8.36}$$

$$g_i(z) = z(1 - z/2)T^*_{i-1}(z) \tag{8.37}$$

where the T^*_i's are the modified Tchebyshev functions defined in the range $0 \le z \le 1$.

The reason for choosing Tchebyshev polynomes is that they lead generally to a fast rate of convergence. Here convergence is realized with only four terms in the developments (8.33) and (8.34). It has been shown elsewhere [23,42] , that the numerical procedure developed in this section is equivalent to Galerkin's technique. Neutral stability curves (Ma versus k) are easily derived. By taking the absolute minimum of these curves, one obtains the critical Marangoni Ma^c and the critical wave number k_c .

In table 1, we have reported the numerical values of Ma^c and k_c for various values of h and for Ra = 0 . The results are compared with Nield's (43) , who used a Fourier series method. It is observed that the approximate results provided by the variational method fit remarkably well Nield's solution. Results corresponding to $Ra \neq 0$ have also been obtained (39).

Table 2 contains the results of our analysis of Serrin-Joseph's theory. The basic variational equation is now (8.32) and the trial functions are the same as for the linear problem. The new parameter λ introduced by Joseph must be determined in order that Ma^c be a maximum at fixed Ra . It is found that the critical Marangoni numbers are smaller than in the linear theory. This means that for the problem of a fluid layer heated from below and submitted to temperature dependent surface tensions, there exists a region of subcritical instability. A comparison with the results of Davis (44) shows an excellent agreement.

TABLE 1

		Variational solution		Nield's solution	
Ra	h	Ma^c	k_c	Ma^c	k_c
0	0	79.6	1.98	79.607	1.993
0	0.1	83.4	2.02	83.427	2.028
0	0.5	98.2	2.14	98.256	2.142
0	1	116.1	2.24	116.127	2.246
0	10	413.5	2.75	413.440	2.743
0	1000	32170	3.01	32170.1	3.010

TABLE 2

		Linear theory	The variational solution of the energy method		Davis results
	R_a	M_a^c	M_a^c	λ	M_a^c
h=0	0	79.61	57.46	920	56.77
	100	68.48	51.01	870	50.59
	300	45.49	36.58	770	36.51
	500	21.39	19.07	690	19.07
	669	0	0		0
h=10	0	413.46	183.9	1860	180.7
	100	378.75	174.4	1740	171.3
	300	305.00	152.4	1520	150.2
	500	225.11	125.2	1325	124.3
	989.5	0	0		0

REFERENCES

(1) O. Reynolds, Phil. Trans. Roy. Soc. A186, 123 (1895).

(2) W. Orr, Proc. Roy. Irish Acad., A27, 69 (1907).

(3) J. Serrin, Arch. Rat. Mech. Anal., 3, 1 (1959).

(4) D.D. Joseph, Stability of Fluid Motions, Springer, Berlin, (1976).

(5) A. Lindsted, Mém. Acad., St-Petersbourg, (1882).

(6) H. Poincaré, Les Méthodes Nouvelles de la Mécanique Céleste, 1892, Dover, New York, (1957).

(7) L. Landau and E. Lifschitz, Fluid Mechanics, Pergamon, New York, (1959).

(8) E. Hopf, Comm. Pure Appl. Math., 1, 303 (1948).

(9) L. Gorkov, J.E.T.P., 6, 311 (1958).

(10) W. Malkus and G. Veronis, J. Fluid Mech., 4, 225 (1957).

(11) J.T. Stuart, J. Fluid Mech., 4, 1 (1958).

(12) E. Hopf, Bericht. Math. Phys. Akad. Wiss. Leipzig, XCIV, 1 (1942).

(13) D.H. Sattinger, Topics in Stability and Bifurcation Theory, Lecture Notes in Mathematics, vol. 309, Springer, Berlin, (1973).

(14) G. Iooss and D.D. Joseph, Elementary Stability and Bifurcations Theory, Springer Berlin, (1980).

(15) S. Chandrasekhar, Hydrodynamic and Hydromagnetic Stability, Clarendon Press, Oxford, (1961).

(16) M. Lyapounov, Comm. Soc. Math. Kharkov (1893), Stability of Motion (English Translation), Acad. Press., New York, (1966).

(17) A. Movchan, Prikl. Mat. Mekh., 23, 483, (1959).

(18) V.I. Zubov, Methods of Lyapounov and their applications, P. Noordhoff, Groningen, (1964).

(19) W. Koiter, Purpose and Achievements of Research in Elastic Stability, Report n° 363, Lab. of Eng. Mech., Delft, (1968).

(20) A. Pritchard, J. Math. Anal., 4, 78 (1968).

(21) H. Jeffreys, Phil. Mag., 2, 833 (1926).

(22) L. Segel, in Non-Equilibrium Thermodynamics, Variational Techniques and Stability, p. 165, Univ. Chicago Press, Chicago, (1966).

(23) B. Finlayson, The Method of Weighted Residuals and Variational Principle, Acad. Press, New York, (1972).

(24) E. Palm, Ann. Rev. Fluid Mechanics, 7, 30 (1975).

(25) C. Normand, Y. Pomeau and M. Velarde, Rev. Modern Phys., 49, 581 (1977).

(26) F. Busse, Report Progr. in Phys., 12, 1929 (1978).

(27) J.R. Pearson, J. Fluid Mech., 4, 489 (1958).

(28) D.D. Joseph, Arch. Rat. Mech. Anal., 20, 59 (1965).

(29) P. Glansdorff and I. Prigogine, Structure, Stabilité et Fluctuations, Masson, Paris, (1971).

(30) A. Schlüter, D. Lortz and F. Busse, J. Fluid Mech., 23, 129 (1965).

(31) L. Landau, C. R. Acad. Sc. URSS, 44, 311 (1944).

(32) I. Stakgold, Soc. Ind. Appl. Math. Rev., 13, 289 (1971).

(33) J. Keller and S. Antman, eds, Bifurcation Theory and Nonlinear Eigenvalue Problems, Benjamin, New York, (1969).

(34) H. Haken, ed., Synergetics, a Workshop, Springer, Berlin, (1977).

(35) H. Swinney and J.P. Gollub, eds, Hydrodynamic Instabilities and Transition to Turbulence, Springer, Berlin, (1981).

(36) C. Bardos and D. Bessis, eds, Bifurcation Phenomena in Mathematical Physics and Related Topics, Nato Advanced Study Institute Series, Reidl, Dordrecht, (1980).

(37) M. Vainberg, Variational Methods for the Study of Non-linear Operators, Holden-Day, San Francisco, (1964).

(38) G. Lebon and G. Perez-Garcia, Bull. Cl. Sciences Acad. Roy., Belgique, 66, 520 (1980).

(39) M. Biot, Variational Principles in Heat Transfer, Oxford Math. Mono., Oxford, (1970).

(40) G. Lebon and J. Lambermont, J. Chem. Phys., 59, 2929 (1973).

(41) G. Lebon, Variational Principles in Thermomechanics, in C.I.S.M. Courses and Lectures Notes n° 262, Springer, Wien, (1980).

(42) D. Nield, J. Fluid Mech., 19, 341 (1964).

(43) S.H. Davis, J. Fluid Mech., 39, 347 (1969).

SOME PHYSICAL MECHANISMS OF HYDRODYNAMICAL INSTABILITIES

Carlos Pérez-García

Departament de Termologia
Universitat Autònoma de Barcelona
Bellaterra (Barcelona) Spain

1. INTRODUCTION

It is known that the study of systems far from equilibrium has adquired an increasing interest in the last ten years. After the success of the renormalization group techniques in the interpretation of static and dynamic critical phenomena, physicists who deal on macroscopic grounds have turned their attention to instability phenomena arising in a wide variety of physical systems. The laser in optics, the Esaki diode, the Gunn instability and the ballast resistor in electrical systems, the Rayleigh-Bénard and Taylor-Couette instabilities, the appearance of the von Kármán vortex street in fluid systems and the Belousov-Zhabotinsky reaction in chemical systems are some examples of instabilities appearing in many branches of physics |1|.

Although there are some attempts to provide a general understanding of instability phenomena, no general explanation or complete classification of them exists at present. Often these attempts rest on analogies with equilibrium phase transitions and, for this reason, instabilities are interpreted as non-equilibrium phase transitions. These analogies have been very fruitful, but a general theory for far from equilibrium systems remains to be built.

The aim of this lecture is not to provide such a general theory, but to make a comparative study of different instability phenomena in hydrodynamical systems, since they serve as models in understanding the mechanisms which lead the system to new states from instable ones. Among several hydrodynamical problems, we have selected three kind of instabilities in simple fluids, which are quite representative and can be studied by simple mathematical methods. These are the convective instability in fluids heated from below, the instability in a fluid between two rotating cylinders and the instability of flows in ducts. In the second section, by means of the linear theory, we review the chief features of the first instability in the systems mentioned above. In the third section we outline the principal characteristics of hydrodynamical systems when the first instability has arisen, and the different mechanisms that cause succesive instabilities in those systems. Finally we discuss only slightly the transition to turbulence because it is the specific subject of other lectures in this School.

II. HYDRODYNAMIC INSTABILITIES. LINEAR THEORY

II.1 Thermal convection in normal liquids

We have learned in elementary physics courses that heat can be carried by three different mechanisms: conduction, convection and radiation. Although convection is the most common mechanism for heat transport in nature, we are more familiar with the two other phenomena. Thus, convective instabilities are in some way responsible for atmospheric changes, sea currents, continental drifts, changes in the Earth's magnetic field, etc. Furthermore, it is very important to control convective instabilities in some industrial processes such as crystal growth, heating systems -specially solar devices-, drying paint films, etc.

Amongst these problems the simplest and best defined one is the so called Rayleigh-Bénard convection problem |2-3|. It deals with a shallow horizontal fluid layer of thickness d, subject to a positive temperature difference between lower and upper surfaces. H. Bénard was the first in descibing accurately the structures appearing in such a layer|4|. His work opened the systematic study on thermal convection instabilities. Later on, lord Rayleigh attempted a theoretical explanation of these phenomena |5|, after which only a few experimental or theoretical works were done until the sixties. From the sixties till now the interest for this problems has increased and more than a thousand papers on this field have been published.

When the temperature across the layer is small, no motion is observed in the fluid, and the heat is carried by conduction. Nevertheless, due to the positive character of thermal expansion coefficient in normal liquids, the situation is unstable since the fluid at the top is denser than at the bottom. Therefore, in this situation the fluid at the bottom tends to go up while the heavier fluid at the top tends to go down. At first, this unstable situation is prevented by viscous forces until a critical temperature gradient is reached. The instability occurs at the minimum temperature gradient at which a balance can be steadily maintained between kinetic energy dissipated by viscosity and internal energy released by bouyancy forces |2|. When the critical temperature

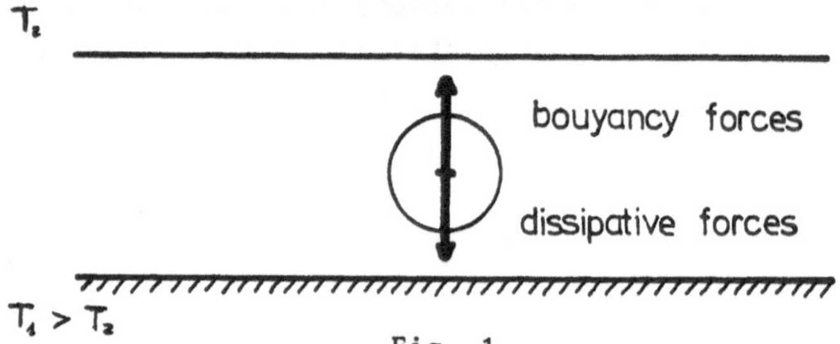

Fig. 1

gradient is reached, regular convective motions can be observed inside the fluid which lead to the formation of the so called convective cells that will be dealt with in the next section.

This explanation of the instability mechanism is due to lord Rayleigh and allows us to account for some features of convective phenomena. However we shall see later on that this mechanism is not unique and that there exist some other ones which cause different forms of convective cells.

In order to study quantitatively that mechanism an ensemble of simplifications must be assumed. This receives the name of Oberbeck-Boussinesq approximation |6| and are the following: i) the fluid is incompressible except for the external force term in the linear momentum equation. This is the simplest form to account for the bouyancy effects which are always present in thermal convection; ii) the velocity gradients are not too high and then, the viscous dissipative term in the energy balance equation can be neglected in comparison with the conductive contribution |2|. For theoretical simplicity the layer is assumed to have an infinite extent and both surfaces to have fixed properties which will be specified below.

With these approximations, the balance equation take the form

$$\nabla\cdot\underline{v} = 0 \qquad \text{(mass)} \qquad (1)$$

$$\rho_0(\partial\underline{v}/\partial t + \underline{v}\cdot\nabla\underline{v}) = -\nabla p + \rho\beta g\underline{e}_z + \mu\nabla^2\underline{v} \qquad \text{(linear momentum)} \qquad (2)$$

$$\partial T/\partial t + \underline{v}\cdot\nabla T = \kappa\nabla^2 T \qquad \text{(energy)} \qquad (3)$$

where $\underline{v}$, T, ρ and p are velocity, temperature, density and pressure fields, ρ_0 a reference density, μ the kinematic viscosity, g the acceleration of gravity, $\underline{e}_z$ the unit vector in the vertical direction and κ the thermal diffusivity. The simbol β stands for the adverse temperature gradient $\beta = \Delta T/d$. In writting these equations we have also assumed that the coefficients μ and κ are constant, though in the following section we review some interesting features when this hypothesis is not fulfilled. According to the Oberbeck-Boussinesq approximation, density ρ in the gravity term can be written as

$$\rho = \rho_0 |1 - \alpha(T - T_0)| \tag{4}$$

where α is the thermal expansion coefficient which is also assumed to be constant. It is usual to write the balance equations in a dimensionless form, scaling length, time and temperature by d, $\rho_0 d^2/\mu$ and βd respectively. As we are interested on the stability of the rest state, the dimensionless perturbed fields can be written

$$\underline{v} = \underline{u} \quad , \quad T = T_1 - \beta z + \theta \quad , \quad p = p_0 + \tilde{\pi} \tag{5}$$

where $\underline{u}$, θ and $\tilde{\pi}$ stand for the perturbations of velocity, temperature and pressure acting on the fluid. The evolution equations for these perturbations in dimensionless form read as

$$\partial \underline{u}/\partial t + \underline{u}\cdot\nabla\underline{u} = -\nabla\tilde{\pi} + Ra\theta\underline{e}_z + \nabla^2\underline{u} \tag{6}$$

$$Pr(\partial\theta/\partial t + \underline{u}\cdot\nabla\theta) = \underline{u}\cdot\underline{e}_z + \nabla^2\theta \tag{7}$$

Ra and Pr are dimensionless parameters defined as

$$Ra = \beta g\alpha d^3 \Delta T/\kappa\mu \qquad Pr = \mu/\rho_0\kappa \tag{8}$$

The first is known as Rayleigh number and accounts for the relative importance of the bouyancy effects in comparison with the viscous and thermal ones. The second is called Prandtl number and shows the ratio between viscous and thermal conduction effects.

The continuity equation (1) imposes $u_z = 0$ on the boundaries (subscript z denotes the z component) and usually it is assumed that surfaces are rigid or stress free and non deformed, leading to

$$\begin{aligned} &a)\ u_z = 0 \quad \text{at} \quad z = 0, 1 \\ &b)\ (\text{rigid})\ \partial u_z/\partial z = 0 \\ &c)\ (\text{free})\ \partial^2 u_z/\partial z^2 = 0 \end{aligned} \tag{9}$$

The Fourier law and continuity conditions require that at the boundaries

$$T = T_e \qquad \lambda \partial T/\partial z = \lambda_e \partial T_e/\partial z \tag{10}$$

The subscript e denotes values of the variables in the exterior material and λ is the heat conductivity. It is often assumed that solutions in the exterior material may obey a semiempirical relation in the form

$$\lambda_e \partial T_e/\partial z = -h(T_e - T_r)$$

where h is a heat-transfer coefficient known as Biot number and T_r a reference temperature in the exterior material. The last relation combined with (10) yields

$$\lambda \partial T/\partial z + hT = \text{prescribed function}$$

which leads to a boundary condition for temperature perturbations

$$\lambda \partial \theta/\partial z = -h\theta \tag{11}$$

On a perfectly conducting boundary,temperature perturbations have a fast relaxation, which implies that $\theta = 0$ on this boundary. This case coresponds to $h = \infty$. In a bad conductor, a slight change in the fluid temperature at the surface will not appreciably affect the rate of heat conduction and then $\partial\theta/\partial z = 0$, as obtained by making $h = 0$.

In the linear theory it is assumed that perturbations are infinitesimal so that non-linear terms in equations (6) and (7), namely $\underline{u}\cdot\nabla\underline{u}$ and $\underline{u}\cdot\nabla\theta$, can be neglected. Then, a normal mode analysis is made as it has been described in the lecture of Prof. Lebon. Let us recall that only in the unrealistic case of two free and perfectly conducting surfaces we can find analytic solutions. In this case the minimum of the marginal stability curve determines the critical values of the parameters which are

$$Ra_c = 657{,}5 \qquad \text{and} \qquad k_c = 2{,}22$$

where k is the wavenumber of the instable mode. For other cases, numerical calculations are needed, leading to the results

rigid-rigid (conducting) $Ra_c = 1707{,}7$ $k_c = 3{,}12$

rigid-free (conducting) $Ra_c = 1100{,}6$ $k_c = 2{,}68$

which agree with experimental results |2|. In these classical cases the system of equations is selfadjoint and the linear theory provides a neccesary and sufficient condition for instability.

II.2 Other mechanisms in thermal convection. Surface-tension-driven convection and convection in complex fluids.

Although Bénard suggested that the hexagonal convective cells he observed were due to surface tension effects, he did not make any quantitative study based on this assumption. After the masterwork of lord Rayleigh, people thougth that the thermal instability problem was theoretically resolved, and provided an explanation of the cells observed by Bénard. However Block |7| made some accurate experiments with thin fluid films which showed the mistake of the last assumption. In films of hidrocarbon in which hexagonal cells were present, he covered the free upper surface with a skin of silicon -insoluble in hidrocarbon- observing the dissapearance of those cells. He also observed that this thin film cooled from below, which should be bouyancy stable, showed convective cells. When heated from below, motions are observed for Ra values lower than the critical one predicted by the Rayleigh mechanism. So, Block concluded that surface tension provides a different physical mechanism for the thermal instabilities.

The first theoretical work regarding surface tension as an agent for instability was that of Pearson |8|. At that time, Pearson was working in the Imperial Chemical Industries and was shocked by the motions displayed in drying paint films, even when heated from the top. He showed that the surface tension forces are sufficient to cause instability, analyzing the case of a thin fluid layer with a free surface in absence of gravity effects. This homogeneous fluid is heated from below and its free surface is assumed to be undeformable. If a temperature perturbation takes place at the free surface, it can provoke some tractions on the fluid surface. These tractions drive away the fluid from the hotter to cooler domains, and in this way, the fluid becomes cooler because it transfers heat to outside. This causes a stationary motion at the surface which leads to a motion in the bulk. In Fig. 2, a scheme of the streamlines in a fluid whose surface tension decreases with temperature -which is the normal case- is shown.

Now we proceed to a little more quantitatively understanding of this mechanism. In a free surface subject to a surface tension, instead of (9c), the following boundary condition must be satisfied

$$\overset{\circ}{\underline{P}}{}^{V} \cdot \underline{n} = -\nabla_s \tilde{\sigma} \tag{12}$$

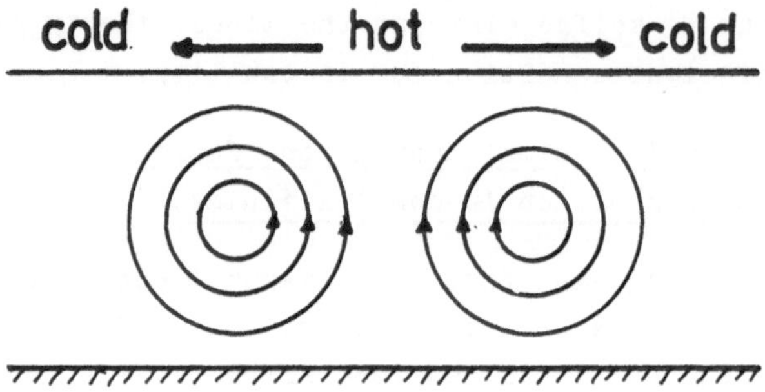

Fig. 2

where $\overset{\circ}{\underline{\underline{P}}}{}^{V}$ stands for traceless part of the viscous pressure tensor, $\underline{n}$ for the normal vector outwards the fluid, ∇_s is the gradient in the direction tangent to the surface and $\tilde{\sigma}$ denotes surface tension. If we suppose that the surface is not deformed

$$\overset{\circ}{\underline{\underline{P}}}{}^{V}\cdot\underline{n} = -\nabla_1\tilde{\sigma} \tag{13}$$

with ∇_1 the bidimensional gradient on the horizontal free surface $\nabla_1 = (\partial /\partial x)\underline{e}_x + (\partial /\partial y)\underline{e}_y$. Applying this operator to equation (13) we obtain for a Newtonian fluid

$$\mu(\partial^2 v_z/\partial x^2 + \partial^2 v_z/\partial y^2 + \partial|\partial v_x/\partial x + \partial v_y/\partial y|/\partial y) = \nabla_1^2\tilde{\sigma} \tag{14}$$

∇_1^2 denoting the horizontal Laplacian operator. Having in mind the continuity equation (1) we rewrite (14) as

$$-\mu(\partial^2 v_z/\partial z^2) = \nabla_1^2\tilde{\sigma} \tag{15}$$

The terms $\partial^2 v_z/\partial x^2$, $\partial^2 v_z/\partial y^2$ vanish because $v_z = 0$ at the surface (9a). Then, we assume that surface tension is a monotonically decreasing function of the temperature and may be expressed as follow

$$\tilde{\sigma}(T_r+\theta) = \tilde{\sigma}(T_r) - \tilde{\psi}\nabla_1^2\theta$$

where $\tilde{\psi} = -\partial\tilde{\sigma}/\partial T$. In some fluids, as those described in some recent works|9| surface tension increases with temperature and new features are expected, but for the sake of simplicity we restrict the analysis to the normal case. Equation (15) may be rewritten for velocity perturbations as

$$\mu(\partial^2 u_z/\partial z^2) = \tilde{\psi}\nabla_1^2\theta \tag{16}$$

and using the scaling described in the preceding paragraph we arrive to

$$\partial^2 u_z / \partial z^2 = \mathrm{Ma}\theta \tag{17}$$

where we have another dimensionless parameter Ma defined as

$$\mathrm{Ma} = \tilde{\psi} \Delta T d / \kappa \mu \tag{18}$$

known as Marangoni number, which accounts for the ratio between traction effects due to surface tension and dissipative effects. By means of the linear theory, Pearson obtained the critical values

$$\mathrm{Ma}_c = 80 \qquad \text{and} \qquad k_c = 2{,}0$$

for an upper insulating free surface and a rigid and conducting lower surface. This work has been of great importance in order to understand the role of surface tension in thermal convection. This theoretical result has shown to be in agreement with experiments carried out at very low gravity in spatial flights of Apollo XIV and XVII |10|. Recently, some studies has been devoted to study the influence of this instability on crystal growth in low gravity environments |11|.

Nield |12| extended Pearson's analysis to account for both gravity and surface tension effects. By a numerical method he obtained a relation between Ma, Ra, h and k which leads to the scheme drawn in Fig. 3 for the case of a free and insulating upper surface and a lower conducting and rigid. In these conditions when the surface tension effects are absent (Ma = 0) the corresponding critical values are

$$\mathrm{Ra}_c = 669 \qquad \text{and} \qquad k_c = 2{,}0$$

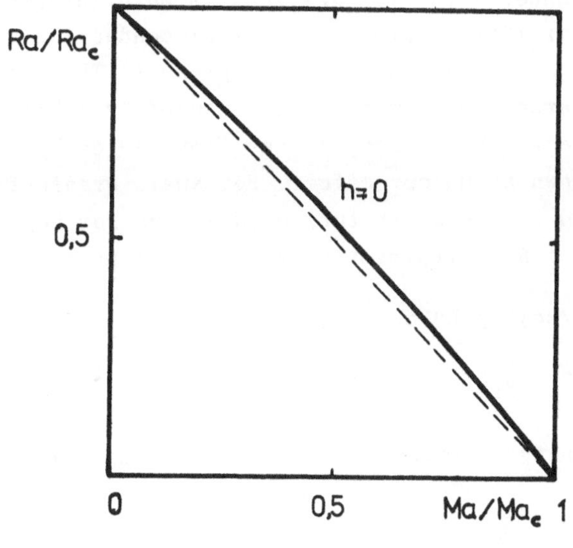

Fig. 3

In Fig. 3 the solid line corresponds to numerical calculations and the dotted one represents the relation

$$Ra/Ra_c + Ma/Ma_c = 1$$

which is very close to the numerical curve. Therefore we expect that the two effects, bouyancy and surface tractions, collaborate together on driving the system out of the stable state. Nevertheless, both effects act on different grounds as we can see in (8) and (18). Ra depends on d^3 while Ma depends on d, so we expect that for very thin layers Marangoni effects dominate, while for deeeper layers convection is driving mainly by bouyancy effects.

As it has been quoted in the lecture of Prof. Lebon, there exist some studies of the so-called Rayleigh-Marangoni convection, including bouyancy and surface tension effects, by the energy method. The first result was obtained by Davis |13| who established an upper limit for the stability of the system studied by Pearson for Ma = 58. This limit does not coincide with that obtained by the linear theory. This fact is related to the possibility of oscillatory motions inside the fluid |14| as we will see in the following paragraph.

We now summarize some of the results obtained in more complex systems as two-component fluids and liquid crystals. In both systems it is neccesary to add an extra equation corresponding to concentration or orientation of the molecules respectively. In binary liquid mixtures, concentration gradients can provoke convective motions although the temperature of the mixture is homogeneous. This case is analogous to the thermal one replacing thermal conduction by diffusion. However, the case of a binary mixture under a thermal gradient is more interesting because new features appear |15|. Apart from thermal conduction and diffusion, two cross effects, Soret and Dufour, are present. The Soret effect consists on the appearance of a matter flow caused by a temperature gradient. The reciprocal phenomenon, a heat flow caused by a concentration gradient is known as Dufour effect. For small gradients, these effects are included in the constitutive equations for heat $\underline{q}$ and mass $\underline{j}$ fluxes which reads |16| ($\tilde{\mu}$ represents the chemical potential)

$$\begin{aligned} \underline{q} &= -\lambda\nabla T - \partial\tilde{\mu}/\partial c)_{T,p} TD''\nabla c \\ \underline{j} &= -\rho_t cc'D'\nabla T - \rho_t D\nabla c \end{aligned} \tag{19}$$

where ρ_t is the total density, c stands for concentration of the heavier component and c' for the lighter one. The coefficients D, D' and D'' are respectively diffusion, thermal diffusion and Dufour coefficients. They are not independent because the two latter ones are linkend by means of an Onsager relation |16| D' = D''. In liquids, thermal conductivity is much greater than the thermal diffusion coefficient ($D/\lambda \sim 0{,}001$). Then we can neglect the Dufour contribution in comparison with thermal diffusion term. This approximation is not valid in gases since $D/\lambda \sim 1$. Introducing the constitutive equations (19) in balance equations we arrive to

$$\partial T/\partial t + \underline{v}\cdot\nabla T = \kappa\nabla^2 T$$
$$\partial c/\partial t + \underline{v}\cdot\nabla c = cc'D'\nabla^2 T + D\nabla^2 c \qquad (20)$$

When dealing with the Rayleigh-Bénard problem in binary mixtures, the expansion (4) must be modified to account for the concentration contributions

$$\rho_t = \rho_{0t}|1 - \alpha(T-T_r) + \mathcal{B}(c-c_r)| \qquad (21)$$

where $\mathcal{B} = \rho_{0t}^{-1}(\partial\rho/\partial c)_{T,p}$ describes the dependence of density on concentration of the heavier component, $\beta>0$. The motionless state of reference is characterized by

$$\underline{v} = 0$$
$$T = T_1 - \beta z \qquad \beta = (T_1 - T_2)/d$$
$$c = c_1 - \psi z \qquad \psi = (c_1 - c_2)/d$$

If we assume that some perturbations act on the system, as in (5) but adding $c = c_r + \tilde{c}$ and taking into account (20) and (21), we find

$$\nabla\cdot\underline{u} = 0$$
$$\rho_t(\partial\underline{u}/\partial t + \underline{u}\cdot\nabla\underline{u}) = -\nabla\pi + \mu\nabla^2\underline{u} + (\alpha\theta - \mathcal{B}\tilde{c})g\underline{e}_z$$
$$\partial\theta/\partial t + \underline{u}\cdot\nabla\theta = \kappa\nabla^2\theta + \beta u_z \qquad (22)$$
$$\partial\tilde{c}/\partial t + \underline{u}\cdot\nabla\tilde{c} = D'c_r c_r'\nabla^2\theta + D\nabla^2\tilde{c} + \psi u_z$$

When concentration and temperature gradients are independent, the convective currents are driven by small differences in solute concentrations and receive the name of thermohaline convection, which is very important in oceanographic studies. However, it is experimentally more interesting the case of a fluid enclosed by boundaries impermeable to mass. In this configuration, the flux of mass vanishes at the steady state, and the thermal diffusion term is balanced by the diffusion one, leading to

$$D'cc'\Delta T = -D\Delta c \qquad D'cc'\beta = -D\psi \tag{23}$$

Hence, equation (22d) now reads as

$$\partial\tilde{c}/\partial t + \underline{u}\cdot\nabla\tilde{c} = -D\psi\nabla^2\theta/\beta + D\nabla^2\tilde{c} + \psi u_z \tag{24}$$

To scale these equations we use the same adimensionalization as in the preceding paragraph but adding the quantity ψd for concentration. From the dimensionless and linearized form of equations (22) and (24) and using the normal mode technique one may arrive to a marginal stability curve. This curve relates the intrinsic parameter k to the external parameters, that in this case, are the Rayleigh number and a parameter which expresses the relative importance of diffusion and thermal diffusion effects. This parameter is known as Soret coefficient, defined as

$$S_T = D'/D$$

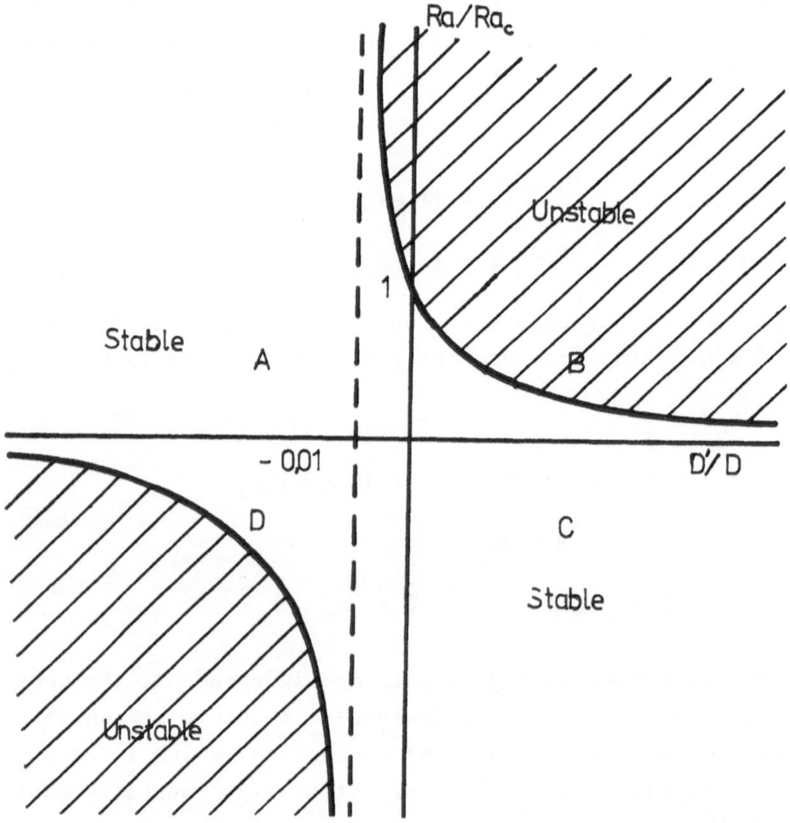

Fig. 4

Schechter et al |15| calculated the marginal stability curve for different values of the Soret coefficient, obtaining the instability scheme depicted in Fig.4 which is summarized in the following table

	D'<0	D'>0	
$\Delta T > 0$	cold hot	cold hot	density gradient
$\Delta T < 0$	hot cold	hot cold	concentration gradient

Table 1

The key to understand this table is the relation (23).

In region B (Ra>0, D'>0), the adverse density gradient is reinforced by thermal diffusion and then, the critical Rayleigh number is reduced. In region C (Ra<0, D'>0), no instability is expected because the two effects are stabilizing. In region D (Ra<0, D'<0), the system may be unstable although it is heated from above. In fact, instability appears when the thermodiffusion contribution is greater than the viscous forces and the adverse density gradient. In region A (Ra>0, D'<0), for Soret numbers sufficiently small, the destabilizing thermal expansion effect overcomes the stabilizing one due to thermal diffusion.

Liquid crystals are anisotropic fluids made by elongated molecules, which present many original phenomena. In this work we only mention some features of thermal convection in the simplest mesophase, the nematic one. The isotropic phase is characterized by a random molecular orientation, while in the nematic phase the molecules are aligned following a preferred orientation, which is described by means of the director field |17|. In a typical liquid crystal as methoxy-p-n benzildine butyl aniline (MBBA), the nematic phase appears below 47°C and above this temperature the disordered phase is present.

These systems have many anisotropic properties, but for convection problems the most interesting one is the anisotropy of the thermal diffusivity κ that may be split as

$$\kappa_a = \kappa_\parallel - \kappa_\perp \qquad (25)$$

where $\kappa_\parallel$ and $\kappa_\perp$ refer to the components parallel and perpendicular to the director |17|. This anisotropy induces a coupling between temperature and orientation fields, that together with the coupling between flow and orientation fields gives rise to some convective effects not found in isotropic fluids. For the sake of simplicity only two director orientations are considered. When the director is perpendicular to horizontal surfaces we have an homeotropic orientation. If the director has the same direction as the horizontal limiting walls we are in the presence of a planar orientation.

The equations of motion are more complex that those of binary fluids, but in order to have some idea of the phenomena taking place on this kind of fluids we recall a simplified planar model in which the linearized perturbations equations write |18|

$$\rho \partial u_z/\partial t = \alpha\rho g\theta + \mu\partial^2 u_z/\partial x^2 + \alpha_2 \partial(\partial n_z/\partial x)/\partial t$$

$$\partial\theta/\partial t - \beta u_z - \kappa\partial^2\theta/\partial x^2 + \kappa_a \beta\partial n_z/\partial x = 0 \qquad (26)$$

$$\gamma\partial n_z/\partial t + \alpha_2 \partial u_z/\partial x - \tilde{K}\partial^2 u_z/\partial x^2 = 0$$

Here n_z is the z-component of the director field, α_2 is the viscosity coefficient which takes account of the coupling between flow and orientation, γ is the rotational viscosity of the director and $\tilde{K}$ is an elastic constant limiting the director perturbations. An upwards flow of a locally warmer portion of the fluid gives rise to a destabilizing torque which tends to increase the director perturbations, but if κ_a is non-vanishing the curvature of the director causes a "heat focusing" effect. In a region where this curvature is concave downwards ($\partial n_z/\partial x < 0$) and with a positive thermal perturbation acting on, heat is "accumulated" yielding a destabilizing effect. The heat focusing term is $\kappa_a \beta\partial n_z/\partial x$ in equation (26b) and the coupling between torque and orientation is quoted in the terms multiplied by α_2 in equations (26a) and (26c).

In both homeotropic and planar cases instability may set up when the fluid is heated from below or from above. Velarde and Zúñiga | 19| reported the following critical values in MBBA layers enclosed between rigid and conducting walls

	Planar configuration		Homeotropic configuration	
	Ra_c	k_c	Ra_c	k_c
$\Delta T \uparrow$	291,9	3,0	13×10^{4}	4
$\Delta T \downarrow$	$3,89\times10^{4}$	0,28	4547	2,85

Table 2

(It may be noted that the Rayleigh number has been redefined to account for the nematic properties, as one can see in |19| and that the values of the homeotropic case correspond to overstable motions). The main difference between the isotropic and nematic Rayleigh-Bénard problem is that instability can take place when the fluid is heated from above. This situation has also been reported from fluids with internal structure |20|.

II. 3 Taylor-Couette instability

In this paragraph we outline the stability of a viscous incompressible fluid rotating between two concentric cylinders whose angular velocity can be controlled from outside. When the angular velocities are small, the flow is laminar with an angular velocity distribution given by

$$\Omega(r) = A + B/r^2 \tag{27}$$

where r stands for the radius. This distribution is obtained from the Navier-Stokes equation for an incompressible fluid confined between two rigid infinite cylindrical boundaries, on which the non-slip condition must be satisfied. In Fig. 5 the geometrical situation is sketched.

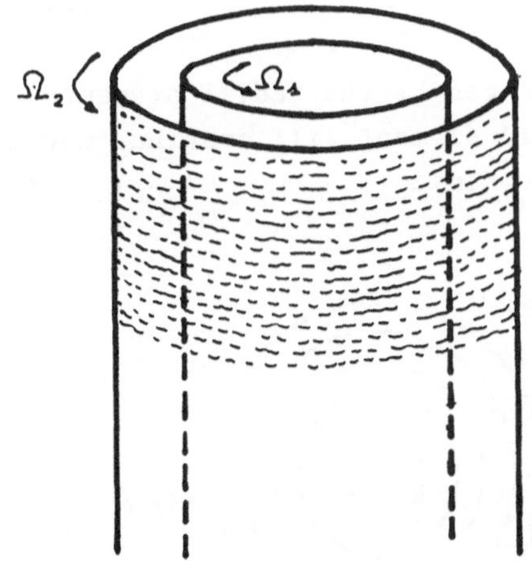

Fig. 5

Cylindrical coordinates are expressed by means of the standard notation (r, θ , z). The non-slip condition imposes on the flow

$$\begin{aligned} v_\theta &= r_1\Omega_1 \quad \text{at} \quad r = r_1 \\ v_\theta &= R_2\Omega_2 \quad \text{at} \quad r = R_2 \end{aligned} \tag{28}$$

The simplest non-trivial and time independent solution of the Navier-Stokes equation with these boundary conditions is |21|

$$v_r = v_z = 0 \quad \text{throughout the fluid}$$

$$v_\theta = r\,\Omega(r) = V(r) \tag{29}$$

where constants A and B in (27) are determined by the boundary conditions (28) giving

$$A = -\Omega_1(1-\eta)/(1-\eta^2) \qquad B = \Omega_1\, r_1^2(1-\xi)/(1-\eta^2)$$

with

$$\eta = r_1/R_2 \qquad \text{and} \qquad \xi = \Omega_2/\Omega_1 \tag{30}$$

These two parameters together with an adimensional parameter, the Taylor number defined later on, characterize the stability of those flows.

The flow described by equation (29) is known as Couette flow whose stability will be studied below.First of all we may ask for causes which can provoke this flow to be unstable. For an inviscid fluid the instability mechanism was discovered and explained in a simple manner by Rayleigh |22| and von Kármán |23|. In the steady state the centrifugal force is balanced by the radial pressure gradient $\partial p/\partial r = \rho V^2/r$. The Kelvin circulation theorem leads to a conservation of the angular momentum (per unit mass) along the ring followed by a fluid element. If such an element at a ring of radius a and velocity V_a moves radially to a more external ring of radius b, adquires a velocity $\tilde{V} = aV_a/b$. The pressure gradient at b, $(\partial p/\partial r)_b$ is balanced in each point by the centrifugal force at this ring $\rho V_b^2/b$. But when the fluid element flowing from a reaches b its centrifugal force is $\rho\tilde{V}^2/b$ which must be different from $\rho V_b^2/b$. Then, if

$$(\partial p/\partial r)_b = \rho V_b^2/b < \rho\tilde{V}^2/b \tag{31}$$

i.e., if the centrifugal force is not balanced by the radial pressure gradient, as depicted in Fig. 6b, the fluid element will continue moving outwards provoking an instability over the fluid

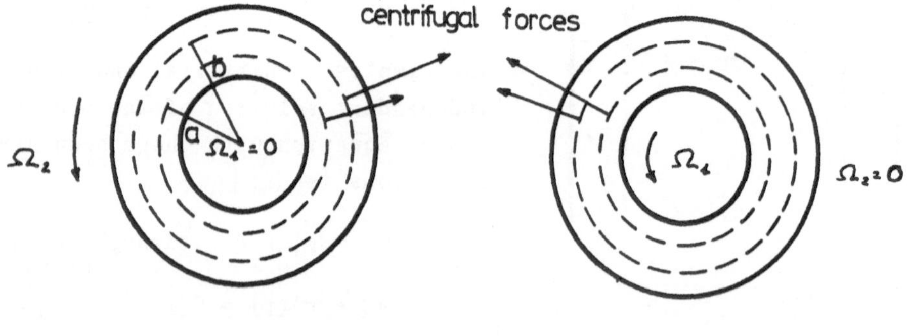

Fig. 6

The condition (31) is equivalent to $(r_aV_a)^2 > (r_bV_b)^2$. This intuitive approach leads to a criterion for stability known as Rayleigh criterion, that can be stated as follows: the necessary and sufficient condition for stability of an inviscid rotating fluid is that the square of angular momentum distribution increases outwards, i.e., $d(rV^2)^2/dr>0$. Using the solution (29) it may be shown that this criterion is equivalent to

$$\Omega_1 > 0 \qquad \text{and} \qquad \xi > \eta^2 \tag{32}$$

These conditions means that the inner cylinder must rotate in the same direction that the outer one and that the angular velocity ratio ξ must exceed the square of the radii ratio η. In particular, if the outer cylinder is kept at rest, this criterion is violated and the system is completely unstable. The stability region is drawn in Fig. 7

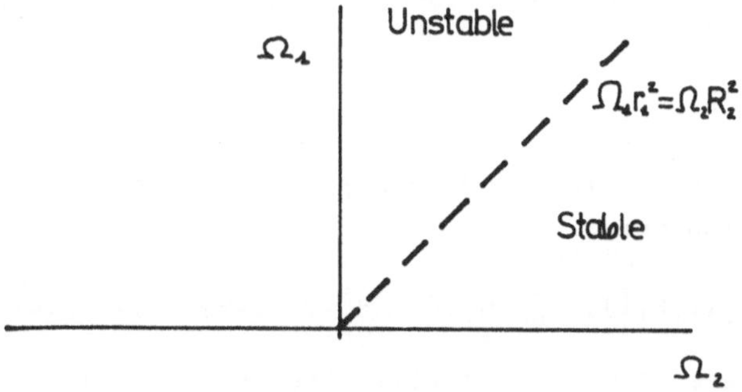

Fig. 7

Nevertheless in real viscous fluids the Rayleigh criterion provides only a neccesary condition for stability, because the viscosity acts as a stabilizing factor. The first work accounting for the viscous effects was due to Taylor |24|, who made a stability analysis based on the linear theory. Because this problem has not been studied in preceding lectures, it may be useful to discuss in some detail the linear stability analysis of the Couette flow. In order to do that one must study the evolution of velocity and pressure fluctuations u_r, u_θ, u_z and π

$$\begin{aligned} v_r' &= u_r & v_z' &= u_z \\ v_\theta' &= V(r) + u_\theta & p' &= p + \pi \end{aligned} \tag{33}$$

Introducing these expression into the Navier-Stokes equation, we arrive to the following set of equations

$$\partial u_r/\partial t + V\partial u_r/r\partial\theta - 2Vu_\theta/r + \partial\pi/\rho\partial r - \mu(\nabla^2 u_r - 2\partial u_\theta/r\partial\theta - u_r/r^2)/ =$$
$$= -(u_r\partial u_r/\partial r + u_\theta\partial u_r/r\partial\theta + u_z\partial u_r/\partial\theta + u_\theta^2/r)$$

$$\partial u_\theta/\partial t+u_r dV/dr+V\partial u_\theta/r\partial\theta +Vu_r/r+\partial\pi/\rho r\partial\theta+\mu(\nabla^2 u_r+2\partial u_r/r^2\partial\theta-u_\theta/r^2)/\rho =$$
$$= -(u_r\partial u_\theta/\partial r + u_\theta\partial u_\theta/r\partial\theta + u_z\partial u_\theta/\partial z - u_r u_\theta/r)$$

$$\partial u_z/\partial t+V\partial u_z/r\partial\theta+\partial\pi/\rho\partial z - \mu\nabla^2 u_z/\rho = -(u_r\partial u_z/\partial r+u_\theta\partial u_z/r\partial\theta+u_z\partial u_z/\partial z) \quad (34)$$

($\nabla^2 = \partial^2/\partial r^2 + \partial/r\partial r + \partial^2/r^2\partial\theta^2 + \partial^2/\partial z^2$) together with the continuity equation

$$\partial u_r/\partial r + u_r/r + \partial u_\theta/r\partial\theta + \partial u_z/\partial z = 0 \quad (35)$$

The fluid occupies the region $r_1<r<R_2$, $0<\theta<2\pi$, $-\infty<z<+\infty$, and obeys the boundary conditions

$$u_r = u_\theta = u_z = 0 \qquad \text{at} \quad r = r_1,\ R_2 \quad (36)$$

In the simplest normal mode analysis axisymmetrical disturbances are sought in the form

$$\{u_r(r,z,t),\ u_\theta,\ u_z,\pi\} = \{u(r),v,w,p_1\}\exp(\sigma t + ikz) \quad (37)$$

Introducing these solutions in (34) after some simplifications we arrive to

$$\mu(d^2/dr^2-d/rdr-1/r^2-k^2-\sigma/\mu)(d^2/dr^2+d/rdr-1/r^2-k^2)u(r) = 2\rho k^2\Omega(r)v$$
$$\mu(d^2/dr^2+d/rdr-1/r^2-k^2-\sigma/\mu)v(r) = 2A\rho u \quad (38)$$

In order to have adimensional equations we scale length by R_2, time by $\rho R_2^2/\mu$ and u by $\mu/2A\rho R_2^2$. If we denote adimensional variables by the same notation, we can rewrite equations (38) in dimensionless form as

$$(DD^* - k^2 - \sigma)(DD^* - k^2)u = - Tk^2(1/r^2 + AR_2^2/B)v$$
$$(DD^* - k^2 - \sigma)v = u \quad (39)$$

where $D = \partial/\partial r$ and $D^* = \partial/\partial r + 1/r$ and T is an dimensionless parameter called <u>Taylor number</u>, defined as

$$T = - 4\rho^2 AB/\mu^2 R_2^2 = 4\rho^2\Omega_1^2 R_2^4(1-\xi)(1-\xi/\eta^2)/\mu^2(1-\eta^2)^2 \quad (40)$$

which is the ratio between centrifugal forces and dissipative effects. The boundary conditions (36) lead to

$$u = v = Du = 0 \qquad \text{at} \qquad r = 1,\ \eta$$

In the stationary case the system (39) with the corresponding boundary conditions may be solved numerically or by variational methods.

We have quoted here the system of equations in the simplest form, but it is costumary to write them with a different adimensionalization |25|. Instead of scaling distances by R_2, the quantity $d = R_2 - r_1$ may be used. The scaling for velocity is $r_1\Omega_1$ and for u, $\mu r_1\Omega_1/2\rho A d^2$. Then perturbation equations take the same form as in (40), but with Taylor number defined now as

$$T = 4(\rho\Omega_1 r_1 d/\mu)^2 (d/r_1)\{\eta(1-\xi/\eta^2)\}/(1+\eta) \tag{41}$$

The Taylor number is positive only when $\xi<\eta^2$, which is the instability condition in an unviscid fluid.Note that this number has the form of the square of the Reynolds number defined as

$$Re = \rho\Omega_1 r_1 d/\mu \tag{42}$$

multiplied by a factor, i.e.,

$$T = Re^2 (d/r_1)\{\eta(1-\xi/\eta^2)\}/(1+\eta) \tag{43}$$

The eigenvalue problem (39) determines a marginal stability curve whose minimum give us the critical values T_c and k_c that depend on the values of ξ and η. In the case of a small gap between the cylinders and with the outer cylinder at rest one obtains

$T_c = 1708$ and $k_c = 3,12$

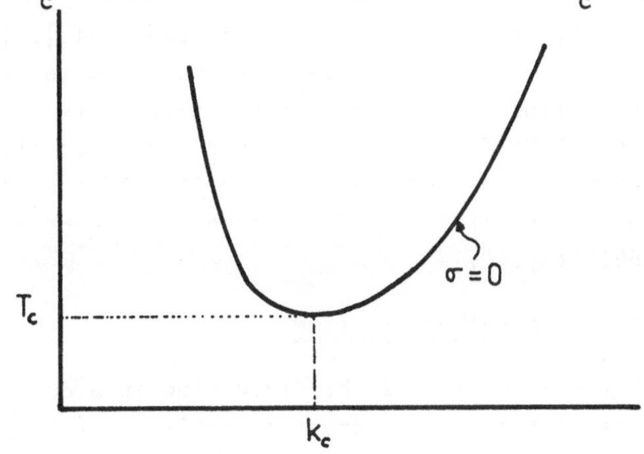

Fig. 8. Marginal stability curve.

In the stationary case the following region of stability is obtained |25|

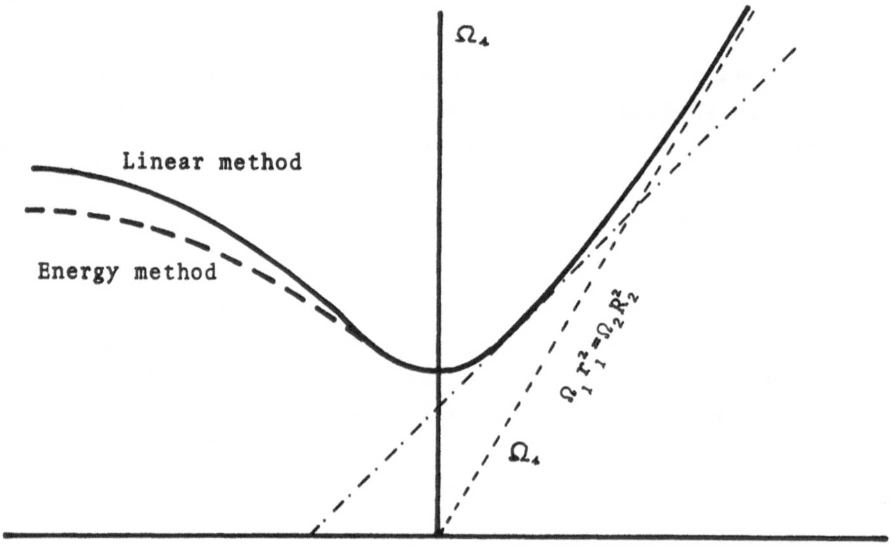

Fig. 9

In this figure one can remark that the results provided by the energy method do not coincide with the linear theory results when the two cylinders rotate in opposite directions. This fact will be analized in more detail in the following paragraph. It may also be noted that the marginal stability curve tends asymptotically to the unviscid stability line $\Omega_1 r_1^2 = \Omega_2 R_2^2$.

Experiments done by Taylor|24|,and Donnelly and Fultz|26| agree very well with the linear theory calculations and seems to confirm the assumption that the perturbations which cause instability are axysimmetric. In the case $\xi>0$,Di Prima |27| has shown theoretically that axisymmetric disturbances causes the lost of stability in the Couette flow. For $\xi<0$ it appears that nonaxisymmetric perturbations contribute notably to the lost of stability|28|.

II. 4 Instability of the Poiseuille flow.

We now ask for the stability of the fluid flow in a duct or between two fixed horizontal plates under a pressure gradient, so-called Poiseuille flow. This instability was the first to be studied, since Reynolds began to study it in 1883 |29|. For many years, it has been the prototype in studying transition to turbulence, because this transition occurs directly from laminar regime. Nowadays, after the study of the Raylegh-Bénard and Taylor-Couette problems, we can state that the transition to turbulence in that flow has a more complex nature than in the problems just mentioned |30|.

This difference is originated because a perturbation rising in a point of the fluid is carried downstream. In a confined fluid, i.e., Rayleigh-Bénard or Taylor-Couette, when Ra> Ra_c (or $T>T_c$)the small perturbations are amplified and attain a finite value, corresponding to a new structure (as we will see in section III). However,in Poiseuille flow, a perturbation that is being amplified is, at the same time, dragged away, and then, at a fixed point we observe the perturbation to decay in time, but the perturbation which travels downstream, amplifies itself in time. Therefore, when dealing with finite tubes it is possible that the perturbation leaves the tube before rising sufficiently to provoke turbulence |31|.

We start here studying the linear theory,i.e., the stability under small perturbations. The complete mathematical analysis is very complex and generally, a two-dimensional flow is taken as model. This simplification assumes that the perturbations acting on the fluid are bidimensional. In fact this approximation has proved to be valid, since in the linear range, the more general three-dimensional perturbations lead to instability for higher Reynolds numbers than two-dimensional ones. This result was proved by Squire |32|, by writing the three-dimensional perturbations as superposition of plane-wave components. So, for the linear theory we must only consider two-dimensional perturbations

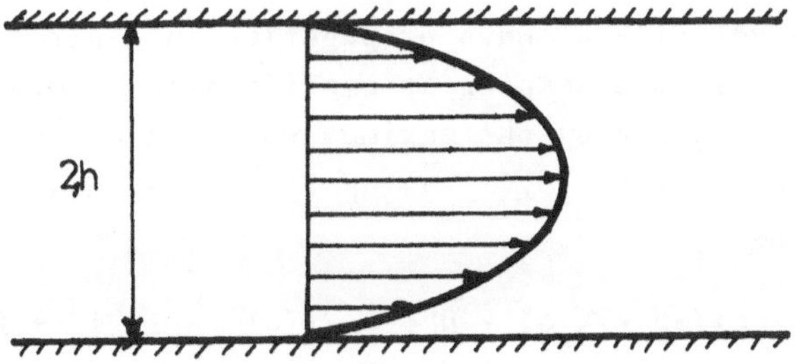

Fig. 10

The basic flow has a parabolic profile, as in Fig. 10, corresponding to a stationary solution of the Navier-Stokes equation

$$(\partial \underline{v}/\partial t + \underline{v}\cdot\nabla\underline{v}) = -\nabla p + \mu\nabla^2\underline{v} \qquad (44)$$

with the continuity equation (1). In a fluid under a constant pressure gradient $dp/dx = C$, this solution is

$$U = v_x = (-C)(h^2-y^2)/2\mu$$

Then one assumes that some two-dimensional disturbances analogues to (33) act on the fluid, which must satisfy

$$\begin{aligned} &\partial u_x/\partial t + U\partial u_x/\partial x + u_y \partial U/\partial y = -\partial\pi/\rho\partial x + \mu\nabla^2 u_x/\rho \\ &\partial u_y/\partial t + U\partial u_y/\partial x = -\partial\pi/\rho\partial y + \mu\nabla^2 u_y/\rho \\ &\partial u_x/\partial x + \partial u_y/\partial y = 0 \end{aligned} \tag{45}$$

In this two-dimensional case we can define a stream function ψ for perturbations in the form

$$u_x = -\partial\psi/\partial y \qquad u_y = -\partial\psi/\partial x \tag{46}$$

and then equations (45) read as

$$(\partial/\partial t + U\partial/\partial x)\nabla^2\psi - (\partial^2 U/\partial y^2)(\partial\psi/\partial x) = \mu\nabla^2\psi/\rho \tag{47}$$

Now, it is assumed that the solutions of this equation can be expanded in series of normal modes

$$\psi(x, y, t) = \phi(y)\exp\{i(k_x x - \sigma t)\} \tag{48}$$

The wavenumber k_x is real while σ is complex, and it is useful to define

$$c = \sigma/k_x = c_r + ic_i \tag{49}$$

where c_r is the velocity of wave propagation in x-direction and c_i, according to its sign, a damping or amplification factor. After adimensionalizing the velocity by the maximum velocity of the basic flow $U_m = U(y=h)$ and the length by h, using the normal mode decomposition (48) one obtains

$$(U - c)(\phi'' - k_x\phi) - U''\phi = -i(\phi^{IV} - 2k_x^2\phi'' + k_x^4\phi)/k_x Re \tag{50}$$

This equation is known as the Orr-Sommerfeld equation, where primes stand for the x-derivatives and Re is the Reynolds number, which is the dimensionless parameter determining the stability and defined as

$$Re = \rho U_m h/\mu \tag{51}$$

The Reynolds number accounts for the relative importance of the inertia forces and viscosity forces. The boundary conditions for this equation are that ϕ and ϕ' vanish on the boundaries. The marginal stability curve for this eigenvalue problem is obtained by setting $c_i=0$. In this problem the principle of exchange of stabilities is not accomplished and this fact implies that the perturbations can propagate along the x-direction. The determination of the marginal stability curve is a difficult task and was carried out by Lin |33| who obtained the following curve (Fig. 11)

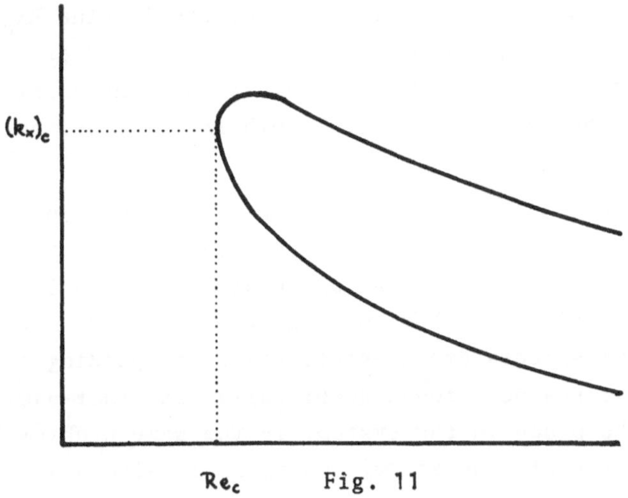

Fig. 11

where the corresponding critical values are

$$Re_c = 5772{,}22 \qquad\qquad (k_x)_c = 1{,}02$$

When $Re \to \infty$ the two branches of this curve tend to zero.

However, this theoretical result does not agree with the observed transition to the turbulence which appears for $Re \simeq 1000$. We will turn over this question in Section III.

II. 5 Overstability

When one applies the normal mode analysis generally it is assumed that the principle of exchange of stabilities holds |2|. This principle establish that the eigenvalue σ is real at marginal stability, i.e., that $\mathrm{Re}\ \sigma = 0$ also implies $\mathrm{Im}\ \sigma = 0$. When a physical system may be described by a linear set of equations, a sufficient condition for this principle to hold is that this set be selfadjoint, because this property ensures the eigenvalue to be real. (After Lebon's lecture we know that this property is also connected with the ability to construct an exact variational principle). Thus, when the linear set of equations is not selfadjoint, there exists the possibility to have oscillatory motions at marginal stability (overstability) |34|.

At marginal stability $\sigma = i\omega$ and one must split the set of equations in its real and imaginary parts. The imaginary part allows to find the frequency of oscillations ω as a function of k, while the real part gives the marginal stability curve. The minimum of this curve determines the values k_{os}, ω_{os} and Ra_{os} at which oscillatory motions may start. In general, Ra_{os} is lesser than the critical value Ra_c for stationary convection. From experimental point of view and especially in industrial processes, it is very important to know the possibility of these motions and the range of Rayleigh numbers (or other parameters) at which they may set up.

Now we study the possibility to have oscillatory motions in the physical systems we are dealing with in this section. In the Rayleigh-Bénard problem for an isotropic monocomponent fluid, the linear set of equations is selfadjoint and these motions are excluded. However, when the surface tension effects are present, the corresponding set is not selfadjoint, but it has been found numerically |35| that oscillatory motions cannot take place in the system. In the same problem for binary fluid or liquid crystals the set of equations is also nonselfadjoint. For certain values of the parameters in both problems, Lekkerkerker |36||37| and Velarde et al |15||19| have found the possibility of an overstability. In liquid crystals this oscillatory motions has been observed by Guyon et al |38|.

In the Taylor-Couette instability, the linear set of equations is not selfadjoint, but it has been proved that oscillatory motions are excluded at least for $\xi > 0$ |39|. In the Poiseuille flow we have seen, that instead of oscillatory motions, the breakdown of the principle of exchange of stabilities is associated to the appearance of a group velocity corresponding to a downstream perturbation.

III STABILITY BEYOND THE LINEAR TRANSITION THRESHOLD

Linear theory provides a sufficient condition for instability, but it does not allow to predict the form of dissipative structures appearing beyond the threshold. For example, in the Rayleigh-Bénard problem with rigid and conducting plates, cylindrical rolls are proved to be stable, while in fluids with variable surface tension hexagonal patterns are observed near the threshold. In the Taylor-Couette problem the situation is analogous, since for Taylor numbers above the threshold the Couette flow loses to be stable and Taylor vortices appear. Linear theory cannot predict the form of these structures because above the threshold it predicts an exponential growth of the perturbations,but near the transition point nonlinear terms become important, and that theory ceases to be valid.

Several methods, mainly perturbative and numerical, have been developed to account for nonlinear terms. Perturbative methods allow to know the behaviour of the system above,but near the transition , while numerical ones may be applied in a wide range beyond the critical point. Perturbative schemes are the main tool we use in order to discuss theoretically the nonlinear regimes. Some experimental findings related to those calculations are also quoted in this section. Although, we do not describe in detail the many routes to turbulence for our hydrodynamical problems because it will be the specific subject of two lectures in this School. In Rayleigh-Bénard and Taylor-Couette problems we recall only the nonlinear instabilities whose mechanism have received some theoretical explanation. The laminar Poiseuille flow comes to turbulence without any intermediate instability. This direct transition can be studied by several methods, but here we only discuss the weakly nonlinear theory because it is based on perturbative methods similar to those used in the other two problems we are concerned with.

III. 1 Perturbative methods in the Rayleigh-Bénard instability

a) Boussinesquian fluids

We start this paragraph describing the perturbation method due to Gorkov |40|, Malkus and Veronis |41|. In this method, the variables of the system are expanded in series of a small parameter ε

$$u_z = (u_z)_0 + \varepsilon(u_z)_1 + \varepsilon^2(u_z)_2 + \ldots$$
$$Ra = Ra_c + \varepsilon Ra_1 + \varepsilon^2 Ra_2 + \ldots \tag{52}$$

where $(u_z)_0 = 0$ because we develop around the rest state, and Ra_c is the critical Rayleigh number found by the linear theory. If it is assumed that nonlinear solutions must be rolls in y-direction, one arrives to the following velocity dependence (for details see Lebon's lectures)

$$u_z \sim \{Ra - Ra_c\}^{1/2}$$

This dependence suggests an an analogy between the transition to instability far from equilibrium and mean field theories in equilibrium phase transitions |42|. Experimentally, Bergé |43| obtained exponent 1/2 in the case of a Bénard cell with conducting walls.

An important result using this perturbation method is due to Schlüter et al |44|, who study the form and stability of the dissipative cells in the Rayleigh-Bénard problem. They assume the perturbations to have the form

$$(u_z)_1 = \sum_{n=-N}^{+N} \exp\, i(\underline{k}_n \cdot \underline{r}) f(z) \tag{53}$$

whose solutions are rolls for N=1, and hexagons for N=3 (see Fig. 12)

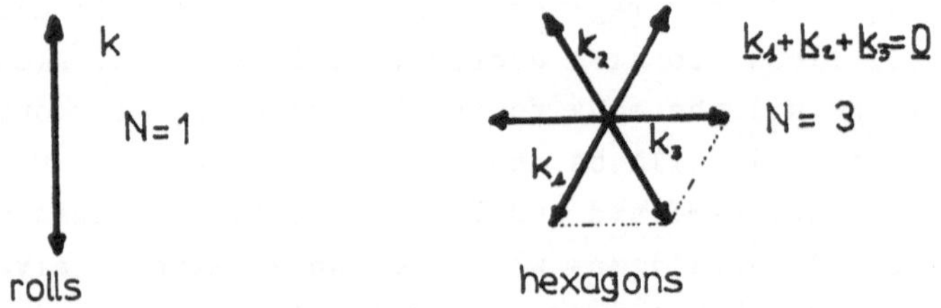

Fig. 12

In order to determine which is the new stable state it is neccesary to make a stability analysis of these flows. To do that, Schlüter et al |44| assumed that some small perturbations act on the flows and have the form

$$\begin{aligned} \tilde{u}_z &= (\tilde{u}_z)_0 + \varepsilon(\tilde{u}_z)_1 + \varepsilon^2(\tilde{u}_z)_2 + \dots \\ \sigma &= \sigma_0 + \varepsilon\sigma_1 + \varepsilon^2\sigma_2 + \dots \end{aligned} \tag{54}$$

The coefficients σ_1 and Ra_1 vanish for symmetry reasons (the system is selfadjoint) and the coefficient σ may be written as

$$\sigma = \varepsilon^2\sigma = -2k^2(Ra - Ra_c) \tag{55}$$

For general perturbations these authors find that the case $N \neq 1$ correspond to unstable states. i.e., the rolls are the only possible stable solutions. These solutions are unstable for perturbation with $k < k_c$, but a region of stability exists for $k > k_c$, as depicted in Fig. 13

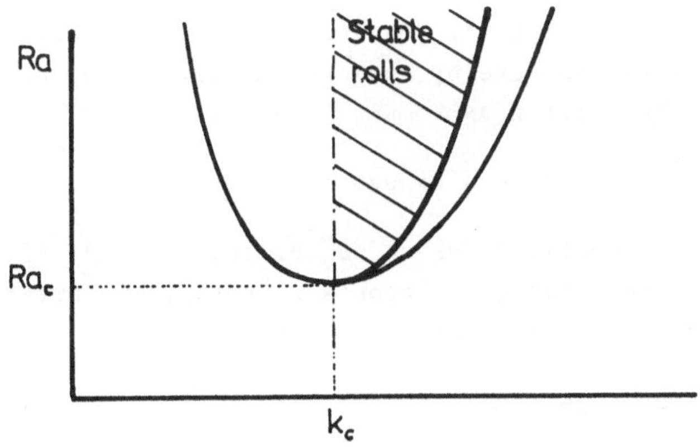

Fig. 13

It must be pointed out the importance of the selfadjointness property in the derivation of these results. The parabolic form of the dependence of u_z on $(Ra-Ra_c)$ is due to this property. Because in the case we are studying that property holds, the results obtained by the linear or energy methods coincide and then, the <u>bifurcation</u> to new states <u>is normal</u> as drawn in Fig. 14

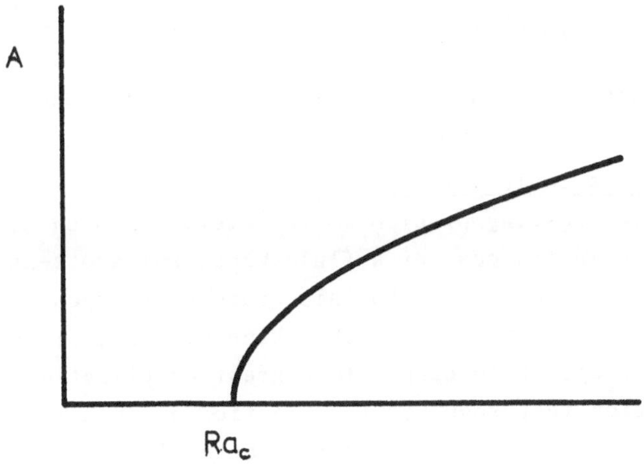

Fig. 14

A slightly different method is due to Stuart and Watson |45|. In their method, an expansion in the following form is assumed

$$\theta = \theta_0 + \theta_1 + \theta_2 + \dots$$
$$u_z = (u_z)_0 + (u_z)_1 + (u_z)_2 + \dots \tag{56}$$

Then, instead of dealing with expansions in ε, one assumes that in the case where the system develop motions in form of two-dimensional rolls, solutions can be written as

$$(u_z) = A(t) \cos kx \operatorname{sen} \pi z \tag{57}$$

where A(t) is an amplitude which can evolve in time. After aplying the integrability condition (see Lebon's lecture), one arrives to an equation for this undetermined amplitude in the form

$$dA/dt = \sigma A - \gamma A^3 + \dots \tag{58}$$

Restricted to second order in the integrability conditions, i.e., in order three in A^3,one obtains a stationary solution **as**

$$A_s = (\sigma/\gamma)^{1/2} = k\gamma^{-1/2}(Ra - Ra_c)^{1/2} \tag{59}$$

which agrees with the development of Gorkov-Malkus-Veronis. To establish the stability of these solutions (two-dimensional rolls) one assumes a small perturbation $\tilde{A}$ acting on the stationary solutions A_s. These perturbations must obey

$$d\tilde{A}/dt = -2k^2(Ra - Ra_c)\tilde{A} \tag{60}$$

The solution of this equation decays for $Ra > Ra_c$ showing the stability of the rolls.

b) Fluids with variable surface tension

The first interesting problem is to determine which is the form of the stable cells in the case of a fluid whose surface tension depends on temperature. This seems to be the main cause of hexagonal pattern formation observed by Bénard |3|. In this kind of fluids, the linear problem is not selfadjoint leading to a great complication in the solvability condition that requires the solution of both linear and linear adjoint set of equations in order to be applied. Scanlon and Segel |46| solved the problem in the unrealistic case of a semiinfinite layer of a fluid without bouyancy effects. In that work the authors made a nonlinear analysis à la Stuart-Watson, assuming as reference solutions

$$(u_z)_1 = (A_{02}\cos 2ly + A_{11}\cos kx \cos ly) \text{ sen } z \qquad (61)$$

which allows us to recover the two-dimensional rolls when $A_{11}=0$, and the hexagonal pattern when $k^2 = 3l^2$ and $A_{11} = 2A_{02}$. After using the solvability conditions, one arrives to the following system of equations for the amplitudes

$$\begin{aligned} dA_{11}/dt &= \sigma A_{11} - \underline{\gamma A_{11}A_{02}} - RA_{11}^3 - PA_{11}A_{02}^2 \\ dA_{02}/dt &= \sigma A_{02} - \underline{\tfrac{1}{4}\gamma A_{02}^2} - R_1A_{02}^3 - \tfrac{1}{2} PA_{11}^2A_{02} \end{aligned} \qquad (62)$$

with $P = 4R - R_1$ and , R and R_1 parameters determined by those conditions. The symmetry $A \to -A$ is broken in these equations, because some terms, as those underlined, are even in the amplitudes. The analysis of these equations leads to the conclusion that hexagons are stable solutions for $\Delta Ma = (Ma - Ma_c)/Ma_c$ between $-0{,}023<\Delta Ma<64$. this nonlinear analysis confirms the possibility of subcritical instabilities, which was predicted by the discrepancy between linear and energy methods. Althougth the use of the Stuart-Watson method is not justified for so great supercritical values as $64Ma_c$, it ensures the stability of hexagonal patterns in the supercritical region near the threshold. These two features indicate that the bifurcation of this problem is inverted. This fact can be found by making $A_{11} = 2A_{02}$ in the stationary solution of (62) leading to

$$\sigma A_{11} - (\gamma/2)A_{11}^2 - RA_{11}^3 = 0 \qquad (63)$$

which corresponds to an inverted bifurcation as drawn is Fig. 15

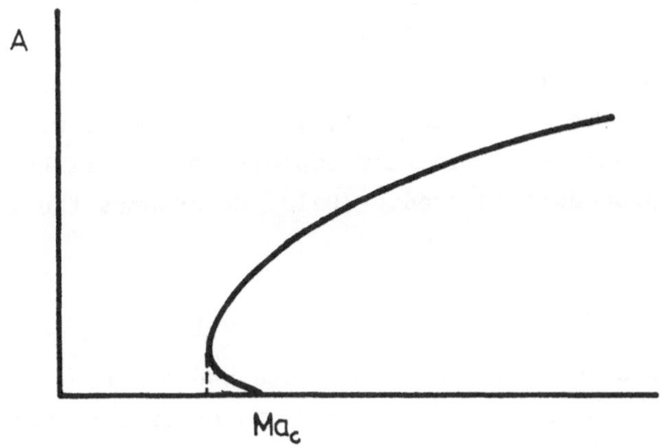

Fig. 15

c) <u>Fluids with variable viscosity</u>

Another interesting problem is the instability of a fluid whose viscosity is variable with temperature. This non-Boussinesquian effect introduces new features in the nonlinear problem which we study now. Let we start with the experimental findings obtained by Graham |47|. He observed that the preferred structure in the Rayleigh-Bénard problem for gases is the hexagonal pattern, even with the fluid enclosed between two rigid plates, but with a dextrorsum circulation sense, i.e., the gas goes downward at the center of the cell. In normal liquids, the motion is in the sinistrorsum sense, since the liquid moves upward at the center of the cell (see Fig. 16)

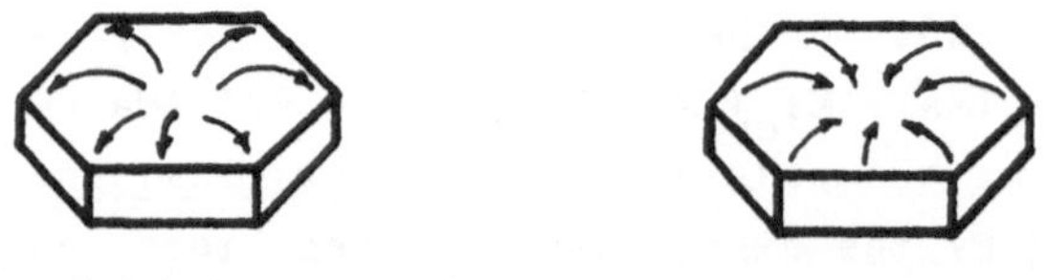

Fig. 16

Graham associates the change in circulation to the fact that the viscosity decreases with the temperature in normal liquids, while it increases in normal gases. The work of Tippelskirch |48| with liquid sulfur confirmed this hypothesis, since this fluid has a viscosity decreasing with the temperature $d\mu/dT<0$ in the temperature interval $120°C<T<153°C$ and increasing $d\mu/dT>0$ for $153°C<T<180°C$. He observed that the hexagonal pattern in the fluid was sinistrorsum in the range with $d\mu/dT<0$ and dextrorsum in the range with $d\mu/dT>0$.

Among the theoretical works on this problem, those due to Busse |49| and Palm and coworkers |50| stand out. The first deals with the Gorkov-Malkus-Veronis technique, while Palm employs the Stuart-Watson one. We follow here the procedure adopted by Palm. He assumes the viscosity to vary as

$$\mu = \mu_r + \Delta\mu \cos \tilde{\alpha}(T-T_r) \tag{64}$$

where subscript r denotes a reference state and $\tilde{\alpha}$ a constant. This is a very particular dependence, but it allows to calculate more easily the nonlinear equations in the case $\Delta\mu/\mu_r \ll 1$ and it has shown to be qualitatively correct. As solutions of the nonlinear equations expression (61) may be taken, leading to the following equations for the amplitudes

$$dA_{11}/dt = \sigma A_{11} - \underline{b_1 \Delta\mu A_{11} A_{02}} - b_2 A_{11}^3 - b_3 A_{11} A_{02}^2$$

$$dA_{02}/dt = \sigma A_{02} - \underline{\frac{1}{4} b_1 \Delta\mu A_{11}^2} - b A_{02}^3 - \frac{1}{2} b_3 A_{11}^2 A_{02} \qquad (65)$$

with $b_3 = 4b_2 - b$. Note that these equations have the same structure as those for the surface tension problem (62). In the present case, the tems corresponding to a symmetry breaking are multiplied by $\Delta\mu$, and therefore one expects that when $\Delta\mu \ll \sigma$ the stable structure corresponds to roll pattern, while if $\Delta\mu \simeq \sigma$ the stable pattern has hexagonal form. This theoretical result have been confirmed, even qualitatively in the experiments made by Hoard et al |51|.

d) Other properties variables with the temperature

Dubois et al |52| studied experimentally the stable structure in the Rayleigh-Bénard problem in water near 4°C, where the thermal expansion coefficient is anomalous. In this region the density cannot be considered to vary linearly with temperature, but to vary as

$$\rho = \rho_r\{1 - \tilde{\gamma}(T - T_{max})^2\}$$

where $\tilde{\gamma}$ determined experimentally is $8{,}11\times10^{-6}K^{-2}$ and $T_{max} = 3{,}98$°C. These author observed that the hexagonal pattern was stable near the linear threshold, with a jump in velocity which varies discontinously from zero to a definite value, corresponding to an inverted bifurcation, as drawn in Fig.17. When the Rayleigh number is increased transition from hexagonal to roll patterns is observed which shows a hysteresis loop as depicted in Fig. 17

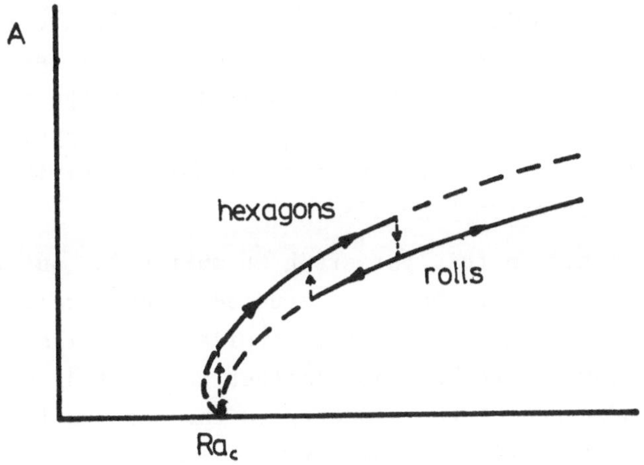

Fig. 17

Busse |53|, in a very important theoretical work, studied the influence of several non-Boussinesquian effects on pattern formation in Rayleigh-Bénard problem. In the simple case of a fluid whose density does not vary linearly with temperature

$$\rho = \rho_r\{1 - \alpha(T-T_r) + \tilde{\eta}(T-T_r)^2\}$$

(where $\tilde{\eta}$ accounts for the nonlinear effect) he arrived by using the G-M-V perturbative technique to the stability diagram quoted in Fig. 18

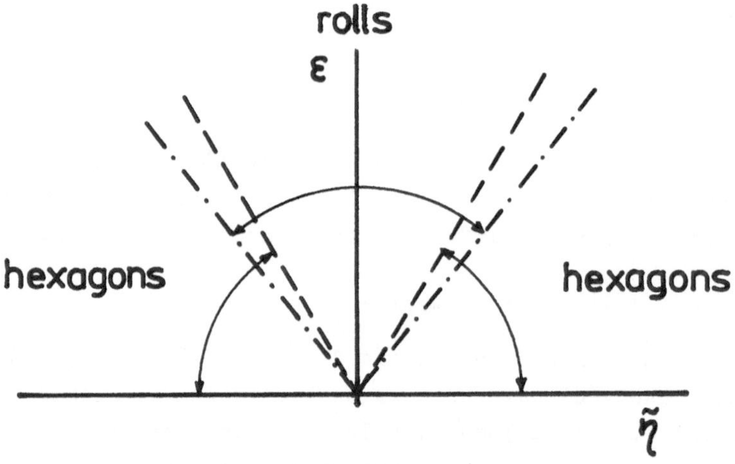

Fig. 18

For a particular value of $\tilde{\eta}$ when the Rayleigh number increases, first an hexagonal pattern is obtained; for higher Ra both rolls or hexagons can develop, but for sufficiently high Ra only rolls are stable cells. The same behaviour has been obtained for other properties variable with temperature. These results agree qualitatively with the features observed by Dubois et al |52| and also with the findings of Tippelskirch |48|, since according to the sign of non-Boussinesquian contribution, one has dextrorsum or sinistrorsum hexagons.

As final comments on this paragraph we must point out that although the transition hexagon-roll has been studied theoretically and observed experimentally, an explanation of the physical mechanism which causes this transition remains to be done. Another interesting problem is to determine the influence of the geometry on the pattern formation. In theoretical works an infinite geometry is assumed, while in experiments one deals with a finite geometry. The consideration of lateral boundaries makes calculations more complex even for the linear problem, but recent theoretical works |54-56| show that these boundaries are determinant for the wavelength selection. An interesting feature observed experimentally |43| but not yet explained theoretically is the tendency of rolls to orient themselves perpendicular to the lateral walls also.

III. 2 Instabilities at higher Rayleigh numbers

In this paragraph we describe in some detail the main experimental findings in the Rayleigh-Bénard problem when Rayleigh number is increased. Here we only consider the case of Boussinesquian fluids enclosed in a cell with a high aspect ratio, i.e., lateral lengths are higher than d, the gap between the two plates.(This quantity is the ratio between these lengths $\Gamma_x = L_x/d$ or $\Gamma_y = L_y/d$ and it has been shown as an important parameter in the nonlinear problems). For this system, the theoretical analysis of Schlüter et al |44| predicted that only rolls are structurally stable near the threshold. However, the random orientation of possible initial disturbances induces a randomness on the supercritical pattern, which consists on rolls coupled at random (see Fig. 8 of Bergé's lecture). This introduces a spatial chaos known as phase turbulence |57|.

The degeneracy in the initial orientation of perturbations may be avoided if the box enclosing the fluid has a small aspect ratio (typically $\Gamma \simeq 2$). The side walls exert a stabilizing effect upon the spatial modes which prevents the appearance of the phase turbulence. This features are described in Dubois' seminar.

Another way to prevent initial random orientation consists on the imposition of a initial orientation. This may be made by an experimental technique largely used in last years and due to Chen and Whitehead |58|. When the fluid is near but above the threshold, a grid with slits spaced out with a desired wavelength is placed upon the upper plate limiting the fluid. Then, by shining with a lamp, one can induce supercritical thermal gradients in some places which cause the appearance of rolls into the fluid with a well defined form. Increasing the ΔT between the plates one can maintain these rolls and study their evolution in time. Different evolutions and instabilities |49|, |59-63| are obtained depending on the values of Rayleigh and Prandtl numbers, and those of the aspect ratio. Busse and Clever obtained numerically the stability diagram drawn in Fig. 19 which has a surprising similarity with the form of a duck head |49|.

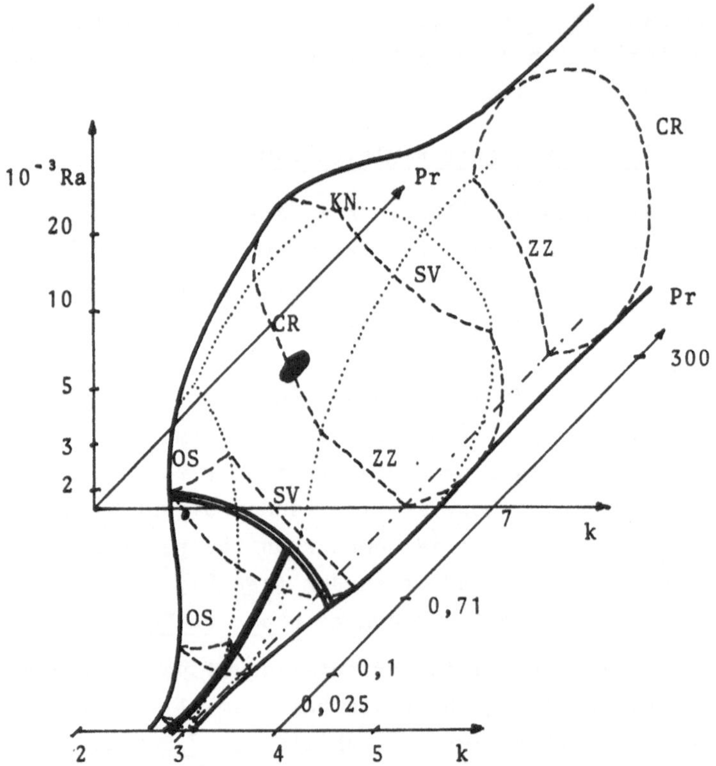

Fig. 19: "Duck head" stability region from Busse and Clever.

The stability region of rolls is limited by different instabilities. At low Pr an instability appears in the form of an oscillation (OS) of the rolls, which induces a time dependence on the convective flow. These oscillations propagate along the rolls and are characterized by a strong component of vertical vorticity. Another kind of instability that appears at low and intermediate Pr is the skew varicose instability (SV) which tends to increase the wavelength of the original rolls when Ra is increased.

At intermediate and high Pr, the instability may develop in the form of cross-rolls (CR), where the original rolls are replaced by rolls in the perpendicular direction and with a smaller wavenumber. A modification of this instability seems to be the knot (K) instability present at intermediate Pr and for high Ra. In it, no perpendicular rolls raise, but the original rolls tend to concentrate up and down motions in spoke patterns. This structure seems to be more efficient for heat transport. Another instability raising at intermediate and high Pr is known as zig-zag instability (ZZ) which induces wavy distortions in the rolls. This instability appears for wavenumbers smaller than the critical one and tends to decrease the wavelength of the original rolls by lengthening the boundary between rolls. Experimentally, this instability has been observed to induce new rolls with smaller wavenumber and oriented with an angle of 45° respect to the original ones. The physical mechanism responsible for these instabilities are not completely understood, but it seems that they emerge from instabilities in thermal boundary layers.

When the Rayleigh number is increased still more, a new configuration with smaller rolls that superpose to the original ones appears. This pattern of motion is three-dimensional and it is known as bimodal convection. We must point however that the spectral analysis of this pattern |64| reveals not to be a simple superposition of perpendicular rolls, since a coupling between the modes is present. For fluids with high Pr, bimodal convection emerges for Ra>26.000 |43||65| and its mechanism lies on the efficiency to transport the heat througth the boundary layers. In the convection pattern of rolls, the temperature distribution is approximately as sketched in Fig. 20

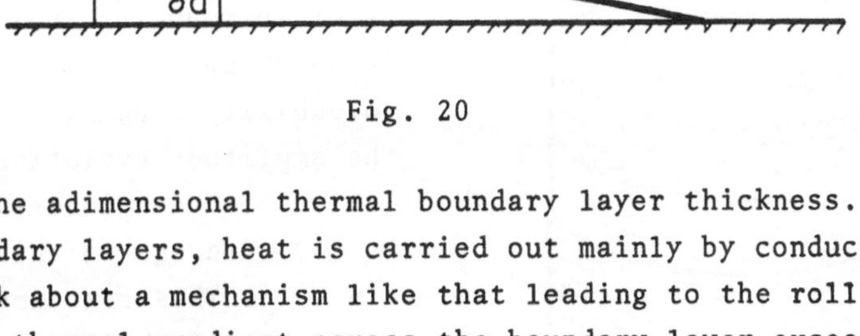

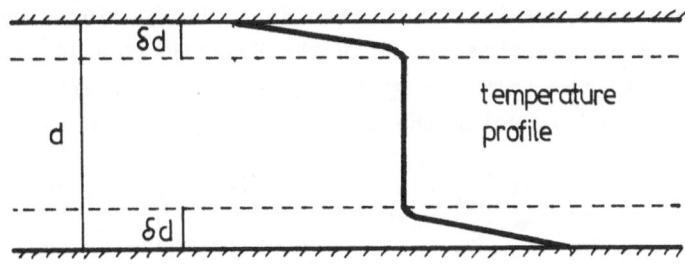

Fig. 20

where δ is the adimensional thermal boundary layer thickness. In the thermal boundary layers, heat is carried out mainly by conduction. Then one can think about a mechanism like that leading to the roll pattern. So, when the thermal gradient across the boundary layer exceeds a critical value, an instability may develop in these layers. A criterion for instability ougth to be established by means of the heat flux that writes as

$$\underline{q} \sim \nabla T \sim Ra/2\delta$$

and recalling that $Ra_c \sim d^3$ one can assume that the value of Rayleigh number Ra_b for the raise of bimodal convection must satisfy $Ra\delta^3/2 > Ra_c$. A criterion for instability in these layers is then

$$q < \frac{1}{2} Ra (Ra/2Ra_c)^{1/3}$$

So, that instability appears when heat flux grows less than the 4/3 power of Rayleigh number. The predominance of the perpendicular rolls in the thermal boundary layers seems to be in agreement with this interpretation |65|.

II. 3 Taylor-Couette instability. Perturbative approaches.

a) Taylor vortices.

The linear theory allow us to determine the threshold T_c at which the instability sets up but, in order to obtain the form of the stable pattern developing on the system, a nonlinear analysis must be done. The full nonlinear equations have the form of (40), adding nonlinear contributions

$$(DD^* + \partial^2/\partial z^2 - \partial/\partial t)(DD^* + \partial^2/\partial z^2)u - T\Omega(r)\partial^2 v/\partial z^2 = N_r(u,v)$$

$$(DD^* + \partial^2/\partial z^2 - \partial/\partial t)v - u = N_\theta(u,v) \tag{68}$$

where N_r and N_θ are quadratic polynomials in u, v and their derivatives. These equations near the linear threshold can be developed by means of perturbative schemes as those used in the Rayleigh-Bénard problem. In order to determine the form and stability of Taylor vortices, we use the Stuart-Watson method assuming the nonlinear solution to have the form

$$v = V(r) + A(t)v_1(r)\cos kz + A^2\{v_{20}(r) + v_{22}\cos 2kz\} + \dots \tag{69}$$

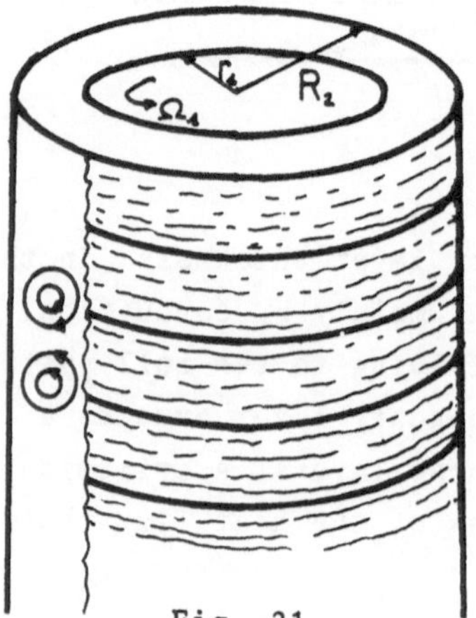

Fig. 21

After introducing this solution in equations (68) and using the solvability conditions, one obtains the amplitude evolution equation (58), whose stationary solution is (59), though in this case $\sigma = T - T_c$. This solution is valid near the threshold and gives toroidal motions as sketched in Fig. 21, known as Taylor vortices. Then, the analogy between equilibrium phase transitions and the

Rayleigh-Bénard problem may be extended to the present problem. Solution (59) has been shown in this case in agreement with the local velocity measurements made by Gollub and Freilich |66|.

Kogelman and Di Prima |67| using a method developed by Eckhaus |68|, determined the range of stable wavenumbers k for the Taylor vortices, which lies between

$$k - k_c < \frac{1}{3}(k_i - k_c) \quad \text{where } k_i = k_1 \text{ or } k_2$$

as schematized in Fig. 22

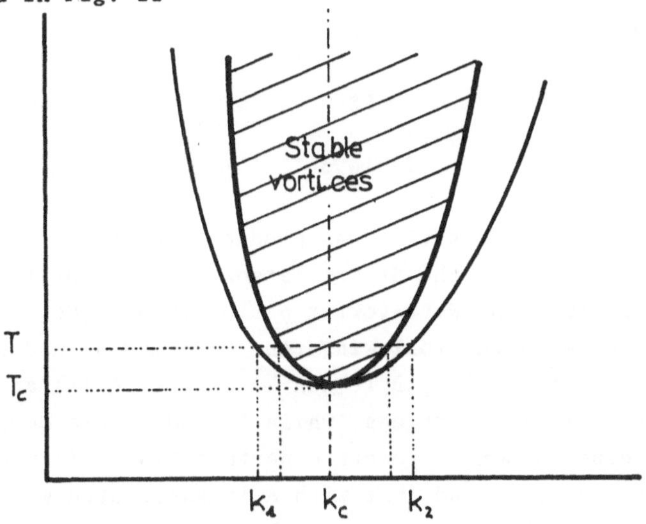

Fig. 22

This result was confirmed experimentally by Snyder |69| and Burkhalter and Koschmieder |70|. These authors obtained for different experimental situations the observed wavenumber lying between the range calculated by Kogelman and Di Prima.

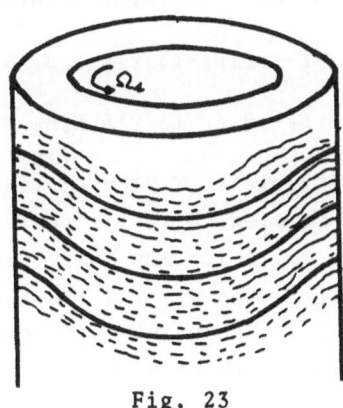

Fig. 23

b) Wavy vortices

The stability of Taylor vortices was studied experimentally by Coles |71|, who observed azimuthal motions of the vortices for Taylor numbers higher than the critical T_c (see Fig. 23). This new instability leads to a wavy-vortex flow with a definite azimuthal wavenumber. The corresponding Taylor number Re_w for which this instability appears are very sensitive to the ratio η and to the aspect ratio L/d. He observed for η = 0,874 and L/d = 27,9 the sequence quoted in table 3

Re_w/Re_c	vortices	waves
1	28	0
1,25	28	4
2,17	24	5
3,21	22	5
3,57	22	6
6,66	22	5
8,50	22	4

Table 3

Theoretically this transition was predicted by Davey et al |72| for the case $\xi=0$ and $\eta\to 1$, with the Stuart-Watson method, but taking into account the possibility of wavy-vortex perturbations. So, they analized the interaction between the following four fundamental modes: $A_0 \cos kz$, $B_0 \cos kz$, $A_1(t) \cos kz \exp im\theta$, $B_1(t) \operatorname{sen} kz \exp im\theta$. $A_0(t)$ and $B_0(t)$ are real amplitudes of Taylor vortices, while A_1 and B_1 are complex amplitudes corresponding to nonaxisymmetric perturbations. The mode with A_0 describes Taylor vortices and that with A_1 is associated with a wavy vortex with m waves, while the modes with B's amplitudes are added in order to have a mathematically complete system of equations. After using the solvability conditions at each order and truncating the equations at cubic order one obtains (~ denotes the complex conjugate)

$$dA_0/dt = \sigma A_0 - A_0^3 - A_0 B_0^2 - 6A_0|A_1|^2 - 2A_0|B_1|^2 - 2B_0(A_1\tilde{B}_1 - \tilde{A}_1 B_1)$$

$$dB_0/dt = \sigma B_0 - B_0^3 - B_0 A_0^2 - 6B_0|B_1|^2 - 2B_0|A_1|^2 - 2A_0(A_1\tilde{B}_1 - \tilde{A}_1 B_1)$$

$$dA_1/dt = \sigma_1 A_1 - 3A_1|A_1|^2 - 2A_1|B_1|^2 - 3A_1A_0^2 - \gamma A_1 B_0 - (3-\gamma)B_1A_0B_0 - \tilde{A}_1 B_1^2$$

$$dB_1/dt = \sigma_1 B_1 - 3B_1|B_1|^2 - 2B_1|A_1|^2 - 3B_1B_0^2 - \gamma B_1 A_0 - (3-\gamma)A_1A_0B_0 - \tilde{B}_1 A_1^2 \tag{72}$$

The two growth rates σ and σ_1 correspond respectively to Taylor vortices and to wave amplitudes. Then σ is real and σ_1 complex, with its real part representing the growing of the wavy mode while its imaginary part is associated with the wave speed. The system (72) has several stationary solutions : when all the amplitudes vanish, the solution corresponds to the Couette flow, for $A_0\neq 0$ and $A_1=0$ (for simplicity we assume $B_0=B_1=0$) we have Taylor vortices, for $A_0=0$ and $A_1\neq 0$ we have a pure nonaxisymmetric mode and finally, when A ≠0 and A ≠0 the wavy vortex is present.

The wavy-vortex transition observed experimentally depends strongly on η and L/d, and one may have different number of cells and waves for the same value of T, depending upon the manner in which the final speed is reached |71|. Another feature not yet understood is the growth of the wavy modes appearing in a configuration with η= 0,876 and L/d = 80 described in a work by Donnelly et al |73|. In this experiment it is observed that for Re/Re_c = 1,2 the number of waves is 2, for Re/Re_c = 1,35 is 1, and for a final state with Re/Re_c = 2,1 the number of waves is 6. Between the extreme values a dislocation activity seems to appear. It may perhaps be associated to a phase diffusion as that described by Pomeau and Manneville |54| in the Rayleigh-Bénard problem.

c) Transition to turbulence in the Taylor-Couette instability.

Apart from that dislocation activity, evolution to turbulence observed by Fenstermacher et al |74| in Taylor-Couette flow for η=0,877 and L/d = 20 shows many steps, as quoted in table 4

Re/Re_c	Spectrum	
1	-	Couette flow
1-1,2	-	Steady Taylor flow
1,2- ∿10,1	f_1	Wavy vortex with a characteristic frequency f_1
10,1-∿12	f_1, f_2	Appearance of a quasiperiodic regime with two frequencies
12-∿19	f_1,f_2, B	A broad band, corresponding to a weak turbulence appears
19-∿21,9	f_1, B	f_2 disappears
>21,9	B	f_1 also dissapears and only the broad band remains

Table 4

In a recent work, Gorman and Swinney |75| have observed that the frequency f_2 corresponds to a modulation of the azimuthal waves. The emergence of a broad band in the spectrum is associated to a weak turbulence. However, even when the flow may be considered as turbulent, the Taylor vortex pattern remains observable as depicted in Fig. 24. Moreover, in a more recent experimental work Walden and Donnelly |76| reported the existence of a reemergent order characterized by a reappearance of f_1 in a configuration similar to that of Fenstermacher et al, for values $28Re_c$- $36Re_c$.

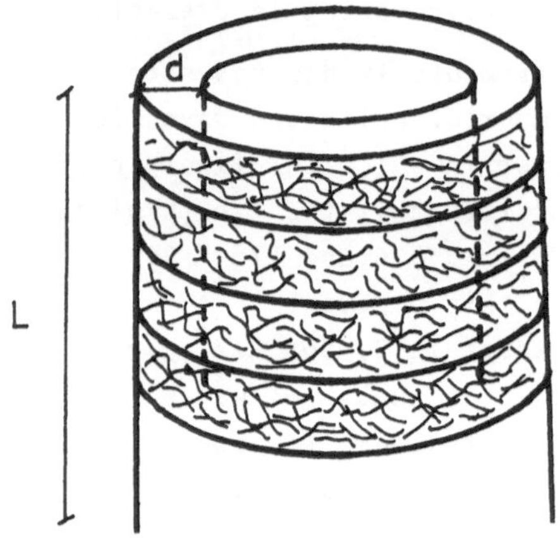

Fig. 24

As in the Rayleigh-Bénard problem, theorethical analysis deals with cylinders with infinite length. The results obtained agree with the experiments made in large aspect ratio cells. In some recent works, the boundary conditions corresponding to a finite length have been taken into account theoretically, leading to a different structure for the linear threshold. The transition in this case, comes from an inverted bifurcation |77|. Experimental and theoretical results in cells with small aspect ratio reveal also the appearance of anomalous modes |78|. All these features manifest the riches of this kind of phenomena, which have not yet received a complete theoretical explanation. Most of the theoretical works are done by numerical simulation, but besides this kind of methods it is very important to understand the physical mechanisms which provoke the different transitions.

III. 4. Transition to turbulence in Poiseuille flow. Perturbative approach.

It was pointed out in II.4 that linear theory cannot describe correctly the instability of planar Poiseuille flow. Linear theory predicts that this transition takes place when Re > 5772, while experimentally this transition is observed for Re > 1000. In that paragraph we explained the difficulties inherent to Poiseuille flow in order to study the evolution of perturbations acting on. Here we only analyse the results obtained by some perturbative and numerical methods close to those used in Rayleigh-Bénard and Taylor-Couette problems. A more complete analysis is quoted in a recent work by Maslowe |79|.

Instead of an exponential growth of fluctuations, as predicted by linear theory, one arrives, after application of the Stuart-Watson method to an amplitude equation in the form

$$d|A|^2/dt = 2|A|^2\{k_x c_i + a_2|A^2| + a_4|A|^4 + \dots\} \qquad (73)$$

where the coefficients a_{even} are real numbers given by the corresponding solvability conditions.

Considering only terms below the quartic order stationary solutions of this equation are obtainable only if $k_x c_i/a_2 < 0$. In such a case we have two possibilities: i) $c_i>0$ implies $a_2>0$ and therefore, a linerly unstable perturbation will evolve towards a steady finite amplitude $A_{st}= -k_x c_i/a_2$, that corresponds to a supercritical transition; ii) $c_i<0$ implies $a_2>0$ which means that a perturbation that would be damped in linear theory can grow, provided the initial amplitude is greater than a value $|A|^2> -k_x c_i/a_2$, leading to a subcritical instability. Reynolds and Potter |80| obtained numerically a value of a_2 that leads to such an instability, in accord with experimental results. However tipical experimental values for transition to turbulence are in the range $1000 < Re < 1500$, which are very far from the value of $Re_c=5700$, and therefore a perturbative analysis around this value ceases to be valid.

Recently, Orszag and Kells |81| have done an interesting work based on numerical analysis whose main conclusions are the following: i) for two-dimensional perturbations the transition to turbulence occurs at Re=2800 that is very close to the result obtained by the Stuart-Watson method and it has the subcritical character expected; ii) three-dimensional finite-amplitude effects are the mean cause of transition to turbulence. This numerical result agrees with the conjecture of Klebanoff et al |82| that three-dimensional disturbaces control the nonlinear development; iii) all disturbances decay either for Poiseuille or for plane Couette flows for Re<500.

IV. FINAL COMMENTS

The main conclusion one may reach in view of theoretical and experimental works concerning hydrodynamic instabilities is that these still constitute open problem for research. In the present work, we have outlined some of the mechanism which cause instabilities, but it is evident that in most cases the explanation is just intuitive.

In the linear case these mechanisms are well understood. For supercritical situations, perturbative methods seem to provide satisfactory results but their validity is restricted to a narrow region near the threshold. However, these methods cannot explain physically the cause of supercritical instabilities, which finally lead the system to a turbulent state. Therefore, the understanding of transition to turbulence depends on the physical interpretation of these intermediate instabilities. This transition seems not to be an universal phenomenon, but to have different features for different systems and, even in each system, to depend on some parameters (aspect ratio, Prandtl number, boundary conditions, etc.).

This general understanding will not be raised without an adequate thermodynamic theory. In the present developments the local equilibrium hypothesis is always assumed. Although this frameworks is only valid for near equilibrium states, it may explain some of these intermediate mechanisms. But a complete interpretation and classification of these far-from-equilibrium phenomena, like that obtained by the scaling hypothesis in critical phenomena, depends on the development of a general thermodynamic formalism valid for systems far from equilibrium.

ACKNOWLEDGEMENTS

We are specially greateful to Dr. D. Jou for many interesting remarks and suggestions. This work was partially financed by the Comisión Asesora de Investigación Científica y Técnica of Spanish Government.

REFERENCES

|1| H. Haken, Synergetics: An Introduction , 2nd edition (Springer-Verlag, Berlin, 1978)

|2| S. Chandrasekhar, Hydrodynamic and Hydromagnetic Stability (Oxford Univ. Press, London, 1961)

|3| Ch. Normand, Y. Pomeau and M.G. Velarde, Rev. Mod. Phys., 49, 581 (1977)

|4| H. Bénard, Rév. Gén. Sci. pures appl., 11, 1261, 1309 (1900)

|5| Lord Rayleigh, Scientific Papers, 6, 432 (Cambridge Univ. Press, London, 1916)

|6| D.D. Joseph, Stability of Fluid Motions, 2 vol. (Springer-Verlag, Berlin, 1976)

|7| M.J. Block, Nature, 178, 650 (1956)

|8| J.R.A. Pearson, J. Fluid Mech., 4, 489 (1958)

|9| E. Guyon et J. Pantaloni, Comptes Rendus Acad. Sci. Paris, 290B, 301 (1980)

|10| P.G. Grozka and T.C. Bannister, Science, 176, 506 (1972); 187, 165 (1975)

|11| Y. Malméjac, ed., Microgravity Research in Space, Brochure ESA BR-05 (1981)

|12| D.A. Nield, J. Fluid Mech., 19, 341 (1964)

|13| S.H. Davis, J. Fluid Mech., 39, 347 (1969)

|14| G. Lebon and C. Pérez-García, Bull. Acad. Roy. Belg., Classe Sci., 66, 520 (1980)

|15| R.S. Schechter, M.G. Velarde and J.K. Platten, Adv. Chem. Phys., 26, 256 (1974)

|16| S.R. de Groot and P. Mazur, Non-Equilibrium Thermodynamics (North-Holland, Amsterdam, 1962)

|17| E. Guyon and P. Pieranski, Physica, 73, 184 (1974)

|18| E. Guyon et M.G. Velarde, J. Physique, 39, L-205 (1978)

|19| M.G. Velarde and I. Zúñiga, J. Physique, 40, 725 (1979)

|20| G. Lebon and C. Pérez-García, Int. J. Engng. Sci., 19, 1321 (1981)

|21| L. Landau et E. Lifchitz, Mécanique des Fluides (Mir, Moscou, 1971)

|22| Lord Rayleigh in ref |4|, p. 447

|23| T. von Kármán, Proc. 4th. Congr. Appl. Mech., Cambridge (1934), p. 54

|24| G.I. Taylor, Philos. Trans. Roy. Soc. London, A223, 289 (1923)

|25| R.C. Di Prima and H.L. Swinney in Hydrodynamic Instabilities and the Transition to Turbulence, H.L. Swinney and J.P. Gollub, eds. (Springer-Verlag, Berlin 1981)

|26| R.J. Donnelly and D. Fulz, Proc. Roy. Soc. London, A258, 101 (1960)

|27| R.C. Di Prima, Phys. Fluids, 4, 751 (1961)

|28| E.R. Krueger, A. Gross and R.C. Di Prima, J. Fluid Mech., 24, 521 (1966)

|29| O. Reynolds, Philos. Trans. Roy. Soc. London, 186, 123 (1883)

|30| H.A. Rose and P.L. Sulem, J. Physique, 39, 441 (1978)

|31| A.S. Monin and A.M. Yaglom, Statistical Fluid Mechanics (MIT Press, Cambridge, 1971)

|32| H.B. Squire, Proc. Roy. Soc. London, A142, 621 (1933)

|33| C.C. Lin, The Theory of Hydrodynamic Stability (Cambridge Univ. Press, London 1961)

|34| C. Pérez-García and J.M. Rubí, Int. J. Engng. Sci (in press)

|35| M. Takashima, J. Phys. Soc. Japan, 20, 810 (1970)

|36| H.N.W. Lekkerkerker, Physica, 93A, 307 (1978)

|37| H.N.W. Lekkerkerker, J. Physique, 38, L-277 (1977)

|38| E. Guyon, P. Pieranski and J. Salan, J. Fluid Mech., 93, 65 (1979)

|39| C.S. Yih, Arch. Rat. Mech. Anal., 47, 288 (1972)

|40| L.P. Gorkov, Sov. Phys.-JETP, 6, 311 (1957)

|41| W.V.R. Malkus and G. Veronis, J. Fluid Mech., 4, 225 (1958)

|42| J.P. Boon dans Les Inestabilités Hydrodynamiques en Convection Libre, Forcée et Mixte, J.C. Legros et J.K. Platten, eds., Lectures Notes in Physics, 72, p. 1 (Springer-Verlag, Berlin, 1978)

|43| P. Bergé in Fluctuations, Instabilities and Phase Transitions, T. Riste, ed. (Plenum Press, New York, 1975)

|44| A. Schlüter, D. Lortz and F.H. Busse, J. Fluid Mech., 23, 129 (1965)

|45| J.T. Stuart, Ann. Rev. Fluid Mech., 3, 347 (1971)
|46| J.W. Scanlon and L.A. Segel, J. Fluid Mech., 30, 149 (1967)
|47| A. Graham, Philos. Trans. Roy. Soc. London, A232, 285 (1933)
|48| H. Tippelskirch. Beitr. Phys. Atmos., 29, 37 (1956)
|49| F.H. Busse, Rep. Progr. Physics, 41, 1929 (1978)
|50| E. Palm, Ann. Rev. Fluid Mech., 7, 39 (1975)
|51| C.O. Hoard, C.R. Robertson and A. Acrivos, Int. J. Heat & Mass Transfer, 13, 849 (1970)
|52| M. Dubois, P. Bergé and J.E. Wesfreid, J. Physique, 39, 1253 (1978)
|53| F.H. Busse, J. Fluid Mech., 28, 223 (1967)
|54| Y. Pomeau and P. Manneville, J. Physique, 40, L-609 (1979)
|55| M.C. Cross, P.G. Daniels, P.C. Hohenberg and E.D. Siggia, Phys. Rev. Lett., 45, 898 (1980)
|56| G. Dewell, D. Walgraef and P. Borckmans, J. Physique, 42, L-361 (1981)
|57| P. Bergé and M. Dubois in Dynamical Critical Phenomena, C.P. Enz, ed., Lectures Notes in Physics, 104, p. 288 (Springer-Verlag, Berlin, 1979)
|58| M.M. Chen and J.A. Whitehead, J. Fluid Mech., 31, 1 (1968)
|59| F.H. Busse and J.A. Whitehead, J. Fluid Mech., 47, 305 (1971)
|60| F.H. Busse and J.A. Whitehead, J. Fluid Mech., 66, 67 (1974)
|61| J.A. Whitehead and G.L. Chan, Dynamics of Atmospheres and Oceans, 1, 33 (1974)
|62| F.H. Busse and R.M. Clever, J. Fluid Mech., 91, 319 (1979)
|63| R.M. Clever and F.H. Busse, J. Fluid Mech., 102, 61 (1981)
|64| V. Croquette, Rapport de stage, CEA Saclay, unpublished (1978)
|65| F.H. Busse, J. Math. Phys., 46, 140 (1967)
|66| J.P. Gollub and M.H. Freilich in Fluctuations, Instabilities and Phase Transitions, T. Riste, ed., p. 195 (Plenum Press, New York, 1975)
|67| S. Kogelman and R.C. Di Prima, Phys. Fluids, 13, 1 (1970)
|68| W. Eckhaus, Studies in Non-Linear Stability Theory (Springer-Verlag, Berlin, 1975)
|69| H.A. Snyder, J. Fluid Mech., 35, 273, 337 (1969)
|70| J.E. Burkhalter and E.L. Koschmieder, Phys. Fluids, 17, 1929 (1974)
|71| D. Coles, J. Fluid Mech., 21, 385 (1965)
|72| A. Davey, R.C. Di Prima and J.T. Stuart, J. Fluid Mech., 31, 17 (1968)
|73| R.J. Donnelly, K. Park, R. Shaw and R.W. Walden, Phys. Rev. Lett., 44, 987 (1980)

|74| P.R. Fenstermacher, H.L. Swinney and J.P. Gollub, J. Fluid Mech., 94, 103 (1979)

|75| M. Gorman and H.L. Swinney, J. Fluid Mech. (to appear)

|76| R.W. Walden and R.J. Donnelly, Phys. Rev. Lett., 42, 301 (1979)

|77| J.T. Stuart and R.C. Di Prima, Proc. Roy. Soc. London, A372, 357 (1980)

|78| T.B. Benjamin and T. Mullin, Proc. Roy. Soc. London, A377, 221(1981)

|79| S.A. Maslowe in Hydrodynamic Instabilities and the Transition to Turbulence, H.L. Swinney and J.P. Gollub, eds., p. 181(Springer-Verlag, Berlin, 1981)

|80| W.C. Reynolds and M.C. Potter, J. Fluid Mech., 27, 465 (1967)

|81| S.A. Orszag and L.C. Kells, J. Fluid Mech., 96, 159 (1980)

|82| P.S. Klebanoff, K.D. Tidsdrom and L.M. Sargent, J. Fluid Mech., 12, 1 (1962)

HYDRODYNAMIC FLUCTUATIONS NEAR THE RAYLEIGH-BENARD INSTABILITY

DAVID JOU

Departament de Termologia
Universitat Autònoma de Barcelona
Bellaterra (Barcelona) Spain

1. INTRODUCTION

In the last years, a great attention has been payed both to second -order equilibrium phase transitions and to non-equilibrium instabilities in a wide variety of physical systems. The similarities between both phenomena have been often emphasized in the literature [1-7] .Some lecturers in this course [8-9] have insisted on this analogy, which provides a fruitful unifying framework and allows to interrelate progress in both fields. The purpose of this lecture is to deepen in some aspects of this analogy, with special emphasis on the thermodynamic point of view.

First of all we recall some notions of second-order phase transitions. Let us assume, for instance, a ferromagnetic material near its critical point, in absence of any external magnetic field. When the critical temperature is approached from below, the spontaneous magnetization M, the so-called order parameter in this problem, tends to zero. Furthermore, the fluctuations δM of the order parameter, which are proportional to the magnetic susceptibility, become highly enhanced near this point. Also, the correlation length ξ and the relaxation time τ of these fluctuations increase extraordinarily near the critical point. It has been observed that the behaviour of these quantities near the critical point can be described by means of power laws such as

$$M \sim (T-T_c)^{\beta} \quad ; \quad \langle \delta M \, \delta M \rangle \sim (T-T_c)^{-\gamma}; \quad \tau \sim (T-T_c)^{-z} \quad ; \quad \xi \sim (T-T_c)^{-\nu} \qquad (1)$$

The quantities β, γ, z and ν are the so-called critical exponents of the problem. In the classical mean-field theory (in the Weiss model or in the more general Landau theory of second-order phase transitions) these critical exponents have the values

$$\beta = 1/2 \quad ; \qquad \gamma = 1 \quad ; \qquad z = 1 \quad ; \qquad \nu = 1/2 \tag{2}$$

The same values are obtained for analogous critical exponents in the mean-field theory of critical points in the gas-liquid transition of fluids (Van der Waals theory) or in many other critical points of different physical systems. These values of the exponents, however, are not observed in practice since they are changed by the enhanced fluctuations.

As some previous lecturers have commented on [8-9], when one takes the convective velocity as the order parameter of the convective phase in the Rayleigh-Bénard instability, one can show theoretically, and it can be verified experimentally, that the behaviour of the instability may be described by the following power laws

$$V \sim \varepsilon^{1/2} \quad ; \qquad \tau \sim \varepsilon^{-1} \quad ; \qquad \xi \sim \varepsilon^{-1/2} \tag{3}$$

where $\varepsilon = (R-R_c)/R_c$, R being the dimensionless Rayleigh number which will be defined below and R_c the critical Rayleigh number. Therefore, three of the "critical exponents" on the hydrodynamic instability coincide with the corresponding ones in the mean-field theory of second-order equilibrium phase transitions. The first purpose of this lecture is to explore the behaviour of the velocity fluctuations near the Rayleigh-Bénard instability in order to check whether the critical exponent γ of the instability coincides with that of the second-order phase transition.

The features outlined in (3) may be summarized in a phenomenological potential of the Ginzburg-Landau form. Such a potential, as proposed by Wesfreid et al. [6] is given by

$$\Phi = -\tau_0^{-1} \int [(\varepsilon/2)V^2 - (1/4)V^4/V_0^2 + (\xi_0^2/2)(\partial V/\partial x)^2]\, dx \tag{4}$$

where V is the velocity; ξ_0 a correlation length and τ_0 a relaxation time whose experimental values, according to [6], are $\xi_0^2 = 0{,}148\ell^2$, $\tau_0 = 0{,}063\ell^2/\chi$ and $V_0 = 10{,}5\chi/\ell$ with χ the thermal diffusivity and ℓ the thickness of the fluid layer. It can be shown that the minima of Φ determine the stationary states and that the evolution equation for the velocity is given by

$$\partial V/\partial t = -\partial\Phi/\partial V \qquad (5)$$

If the non-linear terms are neglected, results summarized in (3) are directly derived from (4) and (5). The potential (4) is very useful to accumulate the experimental information in a very compact and elegant form. However, it would be desirable to have a theoretical deduction of this potential starting from the hydrodynamic equations. This is the second aim of this lecture.

The minima of the non-equilibrium potential (4) determine the steady states in the same way as the minima of the usual thermodynamic potentials specify the equilibrium states. The equilibrium thermodynamic potentials have yet some other properties. Indeed, they are related to the probability of the fluctuations around equilibrium states through a relation of the form $Pr \sim \exp(\Delta F/kT)$, with F the free energy, if the system is hold at a fixed temperature, and so on for the different physical conditions and the corresponding thermodynamic potentials. On the other hand, they can be used as Lyapunov functions for the analysis of the stability of the corresponding equilibrium states. The third aim of this lecture is to show that the non-equilibrium potential (4) is also related to the probability of fluctuations and to the stability of the corresponding steady states in the same way as the equilibrium thermodynamic potentials.

Finally, we ask for the influence of the non-linear terms of the hydrodynamic equations on the behaviour of the convective instability. There are three main possibilities. Non-linear terms may change the critical exponents (Fig.1), as it happens in the critical points of many systems, due to enhanced fluctuations. A second possibility is that non-linear terms cut down the divergence of the fluctuations, the relaxation time and the correlation length (Fig.2). A third possibility is that fluctuations change the order of the transition from a continuous second-order phase transition to a discontinuous first-order one (Fig.3). The fourth aim of this lecture is to examine the different possibilities that arise when non-linear terms are considered.

This lecture is restricted to amplitude fluctuations of velocity and we do not consider the problem of phase fluctuations of velocity [9-11]. It must be kept in mind, however, that the latter may be of more physical interest than the former ones, which in general are negli

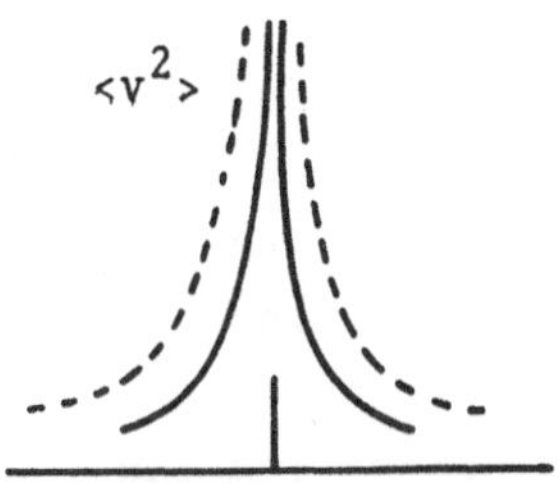

Fig.1

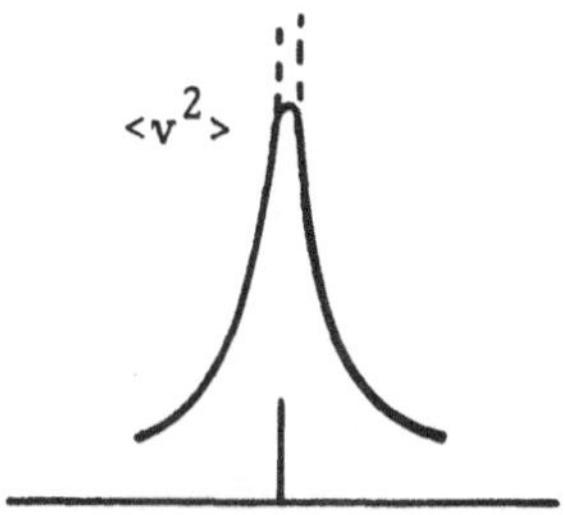

Fig.2

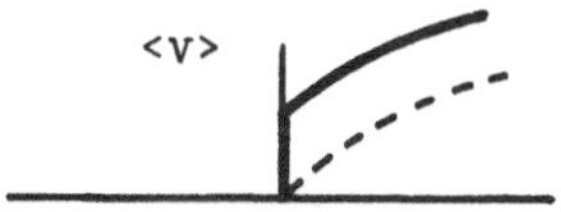

Fig.3

The Figures represent the different possibilities which may arise when the non-linear terms of the hydrodynamic equations are considered. Each possibility is commented in the text.

gible. Nevertheless, the analysis of phase fluctuations is quite recent and we simply omit it here. Let us finally mention that the purposes of this lecture are mainly introductory for a non specialized reader, and that we have aimed at a synthesis of some fundamental points rather than at a detailed analysis of technical specific points.

In Section 2 we discuss the amplitude fluctuations of velocity in the linear hydrodynamic theory and we compare their behaviour with that of the order parameter fluctuations in continuum phase transitions. In Section 3, we obtain from the hydrodynamic equations a Ginzburg-Landau potential and we show its connection to the probability of the fluctuations around the corresponding stationary state, and to the stability of the steady state. In Section 4 we discuss the influence of the non-linear terms on the behaviour of the hydrodynamic instability, starting from the non-equilibrium potential of the previous Section. This analysis indicates that the non-linear terms cut down the divergence of the fluctuations in the convective threshold. In Section 5 we present some other features of the hydrodynamic instability, namely, that the presence of fluctuations makes it a discontinuous transition, or the influence of the boundary surfaces on the critical exponents. In the final remarks, we ask for the validity of a local-equilibrium theory for the non-equilibrium thermodynamic noise.

2. HYDRODYNAMIC CRITICAL EXPONENTS IN THE LINEAR THEORY

The problem of hydrodynamic fluctuations near the convection threshold was considered for the first time by Zaitsev and Shliomis in 1970 [1], in the context of a linearized hydrodynamic theory with fluctuating terms described by the well-known Landau-Lifshitz formulae. We summarize in this Section their main results.

We consider the problem of the stability of a plane horizontal fluid layer at rest, heated from below. For sufficiently strong heating, the buoyance force due to the volume expansion of the liquid near the bottom may overcome the viscous shear forces and the fluid starts moving. The liquid undergoes a transition from a state where heat is transported by pure heat conduction to another state of combined heat conduction and heat convection. While the purely conducting state is homogeneous, the convective state may have some different structures, depending on the boundary conditions. We will analyse the simple case of a plane container of rectangular basis with free boundaries at fixed temperature at the top and at the bottom of the fluid, and we will assume that, after the transition, the motion of the fluid has the structure of parallel two-dimensional rolls along the shortest dimension of the rectangular basis. Though physically unrealistic, this case contains all qualitative features of the transition and allows for an analytic solution.

The basic hydrodynamic equations describing the state of the fluid are [1,3,12]

$$\nabla.\underline{v} = 0 \tag{6}$$

$$\partial\underline{v}/\partial t + (\underline{v}.\nabla)\underline{v} = -\nabla p + \nabla^2\underline{v} + R^{1/2}T\underline{e}_z + \nabla.\underline{\underline{S}} \tag{7}$$

$$P(\partial T/\partial t + \underline{v}.\nabla T) = \nabla^2 T + R^{1/2}\underline{v}.\underline{e}_z - \nabla.\underline{q} \tag{8}$$

where $\underline{v}$, T and p are the corresponding deviations of fluid velocity absolute temperature and pressure from their respective values in the purely heat-conducting state. The unit vector $\underline{e}_z$ is in the vertical upward direction. We scale lengths by ℓ (the thickness of the layer), times by ℓ^2/ν and temperatures by $\nu\ell^{-1}(\nu\Delta T/g\beta\chi\ell)^{1/2}$, with χ the thermal diffusivity, ν the kinematic viscosity, ΔT the difference of temperatures between the bottom and the top of the layer, β the volume expansion coefficient of the fluid and R and P the Rayleigh number and the Prandtl number, respectively, defined by

$$R = g\beta\Delta T\ell^3/\nu\chi \quad ; \quad P = \nu/\chi \tag{9}$$

The fluctuating stress tensor $\underline{\underline{S}}$ and heat flux $\underline{q}$ are delta-correlated Gaussian fields in space and time, described by the Landau-Lifshitz relations which in the present notation take the form |1|

$$\langle\underline{\underline{S}}\rangle = \langle\underline{q}\rangle = 0$$

$$\langle q_i(\underline{r},t)q_j(\underline{r}',t')\rangle = 2Q_1\delta_{ij}\delta(\underline{r}-\underline{r}')\delta(t-t')$$

$$\langle S_{ij}(\underline{r},t)S_{kl}(\underline{r}',t')\rangle = 2Q_2(\delta_{ik}\delta_{jm} + \delta_{il}\delta_{jk})\delta(\underline{r}-\underline{r}')\delta(t-t') \qquad (10)$$

$$\langle S_{ij}(\underline{r},t)q_l(\underline{r}',t')\rangle = 0$$

with

$$Q_1 = g\beta kT^2/\rho C_v\Delta T\nu^2 \quad ; \qquad Q_2 = kT/\rho\ell\nu^2 \qquad (11)$$

where k is the Boltzmann constant.

The boundary conditions corresponding to free surfaces at $x_3 = 0$ and $x_3 = 1$ are

$$v_3 = T = 0 \quad ; \; \partial_3^2 v_3 = \partial_3 v_1 = \partial_3 v_2 = 0 \qquad (12)$$

Boundary conditions in the x_1,x_2 plane are also needed. For the moment we assume a sufficiently extended horizontal fluid layer wherea number of rolls can be accomodated.

In order to analyse the stability of the solutions of the equations (6)-(8), we linearize these equations in $\underline{v}$ and T, and for the moment we neglect the random terms $\underline{\underline{S}}$ and $\underline{q}$. If we assume a time dependence of the form exp$(-\lambda t)$, as it is usual in the normal modes method , the corresponding linearized forms of equations (7) and (8) are respectively

$$-\lambda\underline{v} = -\nabla p + \nabla^2\underline{v} + R^{1/2}T\underline{e}_z \qquad (13)$$

$$-\lambda PT = \nabla^2 T + R^{1/2}\underline{v}.\underline{e}_z \qquad (14)$$

With the boundary conditions (12), these equations have real eigenvalues and their complete set of eigenfunctions $u_n = (\underline{v}_n, T_n, p_n)$ can be normalized as [1]

$$\int \{v_n(\underline{r})v_m^*(\underline{r}) + PT_n(\underline{r})T_m^*(\underline{r})\}\, dV = \delta_{nm} \qquad (15)$$

where the index n denotes $(\underline{k},n)$, where $\underline{k}$ is a wavenumber in the layer plane and n a discrete number associated with the vertical wavenumber. The linearized equations (13) and (14) with boundary conditions (12) have the following set of eigenvalues

$$\lambda_{n,k} = (2P)^{-1}\{(P+1)(\underline{k}^2+n^2\pi^2) - |(\underline{k}^2+n^2\pi^2)(P-1)^2 + 4PR(\underline{k}^2+n^2\pi^2)^{-1}|^{1/2}\}$$

whose corresponding eigenfunctions are |14|

$$v_{1n,k}(\underline{r}) = A_{n,k}\underline{k}^{-2}(-ik_x n\pi)\exp(-ikx)\cos n\pi z$$

$$v_{2n,k}(\underline{r}) = 0$$

$$v_{3n,k}(\underline{r}) = A_{n,k}\exp(-ikx)\sin n\pi z \qquad (17)$$

$$T_{n,k}(\underline{r}) = A_{n,k}R^{1/2}(\underline{k}^2+n^2\pi^2-\lambda_{n,k}P)^{-1}\exp(-ikx)\sin n\pi z$$

$$p_{n,k}(\underline{r}) = A_{n,k}\underline{k}^{-2}\, n\pi(\underline{k}^2+n^2\pi^2 - \lambda_{n,k}P)^{-1}\exp(-ikx)\cos n\pi z$$

where the coefficient $A_{n,k}$ can be found from the normalization condition (15).

It is seen that $\lambda_{1,k}$ vanishes for the smallest critical Rayleigh number $R_c = 27\pi^4/4$ at the critical wavenumber $k_c = \pi/\sqrt{2}$. Near the critical Rayleigh number, $\lambda_{1,k}$ is represented as

$$\lambda_{1,k} = 3\pi^2\varepsilon/2(1+P) \qquad (18)$$

From now on, we focus our attention mainly on this slow mode. The expression (18) shows clearly the phenomenon of critical slowing down. Indeed, the relaxation time of this mode, $\tau_c = \lambda_{1,k}^{-1}$ diverges when R tends to R_c as

$$\tau_c = 2(1+P)(3\pi^2)^{-1}\varepsilon^{-1} \qquad (19)$$

which shows clearly that the corresponding "critical exponent" z for the Rayleigh-Bénard instability is z = 1, in accordance with (3).The investigation of the stability of the motion pattern obtained after the transition is analogous to that in the case of the stability of the rest state. In this case, the relaxation time of the fluctuation which attenuates most slowly goes as $\tau_c(R>R_c)=(1/2)\tau_c(R<R_c)$. We do not study this result here.

In order to understand the behaviour of the velocity fluctuations near the critical point, we have to introduce into the set of equations (13-14) an independent term coming from the random forces $\underline{\underline{S}}$ and $\underline{q}$ of (7) and (8). We expand the solution of the inhomogeneous linearized set in the complete systems of the eigenfunctions of the homogeneous problem [1], i.e.

$$(\underline{v},T,p) = \Sigma_n c_n(t)(\underline{v}_n,T_n,p_n) \quad (20)$$

where the $c_n(t)$ are time-dependent coefficients which are to be obtained and the functions $(\underline{v}_n,T_n,p_n)$ are given by (17). When this expansion is introduced into the following equations

$$\partial\underline{v}/\partial t = -\nabla p + \nabla^2\underline{v} + R^{1/2}T\underline{e}_z + \nabla.\underline{\underline{S}} \quad (21)$$

$$P\partial T/\partial t = \nabla^2 T + R^{1/2}T\underline{v}.\underline{e}_z - \nabla.\underline{q} \quad (22)$$

and taking into account that $(\underline{v}_n,T_n,p_n)$ satisfy (13) and (14), we get

$$\Sigma_n(\dot{c}_n + \lambda_n c_n)v_{n\alpha} = \partial S_{\alpha\beta}/\partial x_\beta \; ; \quad \Sigma_n P(\dot{c}_n + \lambda_n c_n)T_n = -\partial q_\beta/\partial x_\beta \quad (23)$$

where $\alpha = 1,2,3$ and the repeated subindices stand for the corresponding sum. We multiply the first equation times v^*_m, the second one times T^*_m, we add them and integrate over the volume. Having in mind the orthonormality conditions (15) we obtain

$$\dot{c}_n + \lambda_n c_n = \int (v_{n\alpha}\partial S_{\alpha\beta}/\partial x_\beta - T_n\partial q_\beta/\partial x_\beta)\, dV \quad (24)$$

The integral in the right-hand side can be transformed into

$$\dot{c}_n + \lambda_n c_n = \int(\underline{q}.\nabla T_n - \underline{\underline{S}}:\nabla\underline{v}_n)\, dV \quad (25)$$

with the double dot between two tensors standing for their double contraction. To obtain (25) we have assumed that fluctuations vanish at the boundaries.

From (25) we obtain for the components of the Fourier transforms of $c_n(t)$ the following equation

$$(-i\omega+\lambda_n)c_n(\omega) = \int\{q_\alpha(\omega)\partial T_n/\partial x_\alpha - S_{\alpha\beta}(\omega)\partial v_{n\alpha}/\partial x_\beta\}dV \quad (26)$$

From this expression, and using the correlation function of the components $\underline{q}(\omega)$ and $\underline{\underline{S}}(\omega)$, we get

$$\langle c_m(\omega)c^*_n(\omega')\rangle = 4\pi(Q_1K_{mn} + Q_2J_{mn})(-i\omega+\lambda_m)^{-1}(i\omega+\lambda_n)^{-1}\delta(\omega-\omega') \quad (27)$$

where $J_{mn} = \int \mathrm{rot}\underline{v}_m.\mathrm{rot}\underline{v}_n\, dV$ and $K_{mn} = \int\nabla T_m.\nabla T_n\, dV$. From here, the following correlation functions are obtained for the coefficients $c_n(t)$

$$<c_m(t)c_n(t+t')> = 2(Q_1K_{mn}+Q_2J_{mn})(\lambda_m+\lambda_n)^{-1} \exp(-\lambda_n t') \tag{28}$$

From (28), the correlation functions of the temperature fluctuations and of the velocity components can be calculated; for instance, we have for the velocity

$$<v_\alpha(\underline{r},t)v_\beta(\underline{r},t+t')> = \Sigma_{m,n}<c_m(t)c_n(t+t')>v_{n\alpha}(\underline{r})v_{m\beta}(\underline{r}) \tag{29}$$

We will limit our analysis to the case of a discrete spectrum for the eigenvalues λ_n. This is the case of a box with finite lateral dimensions L. Analysis of the continuous spectrum (infinite plane box) would lead to the result that correlation length diverges as $\varepsilon^{-1/2}$. Since this result has been already commented by other lecturers [9] and since it can be recovered in the linear approximation from the potential (4), which we will show later to be valid, we omit here the case of continuous spectrum. In the discrete spectrum, since λ_1 and λ_2 are separated by a finite interval, the main contribution to the sum (29) comes from the term with m=n=1, for which

$$<c_1(t)c_1(t+t')> = (Q_1K_{11}+Q_2J_{11})\lambda_1^{-1} \exp(-\lambda_1 t') \tag{30}$$

From (13) and (14) one can obtain |1| that $J_{11}=K_{11}= \lambda_1\varepsilon^{-1}$, and consequently the second moments of the velocity fluctuations are given by

$$<v_\alpha(\underline{r},t)v_\beta(\underline{r}',t+t')> = (Q_1+Q_2)\ \varepsilon^{-1} \exp(-t'/\tau_c)v_{1\alpha}(\underline{r})v_{1\beta}(\underline{r}') \tag{31}$$

with τ_c given by (19). From (31) it is seen that the order parameter fluctuations,i.e. the convective velocity fluctuations, diverge near the critical point according to ε^{-1}, and therefore we have shown that the fourth critical exponent γ of the Rayleigh-Bénard instability coincides with the mean-field value $\gamma = 1$ of equilibrium critical points.

Ligth-scattering is one of the possible methods to probe the behaviour of the fluid fluctuations. However, to observe large fluctuations in the order parameter, scattering angles as small as 10^{-3}, 10^{-4} radians are needed, so that the experimental problem is far from trivial [7] . Another possible experimental method for testing these results is the analysis of the Brownian motion of small particles as the system approaches the critical Rayleigh number . The corresponding theoretical analysis of this problem has been carried out by Lastovka and Boon

[14]. The Langevin equation of motion for a Brownian particle of mass m in a laboratory fixed frame is

$$m d\underline{u}/dt + \zeta\{\underline{u} - \underline{v}(\underline{r})\} = \underline{f}(t) \tag{32}$$

where $\underline{u}$ is the velocity of the particle, $\underline{v}(\underline{r})$ the fluid velocity at the particle position $\underline{r}$, ζ the Stokes drag coefficient and $\underline{f}$ the usual random noise whose second moments are

$$\langle \underline{f}(t)\underline{f}(t')\rangle = 4kT\zeta\, \underline{\underline{U}}\delta(t-t') \tag{33}$$

Taking into account both the equilibrium effects described by (33) and the singular effects described by the anomalous part of the velocity fluctuations one arrives at the following expression for the total mean square displacement of the particles

$$\langle \Delta x^2(t-t_0)\rangle = 4(D+D^*)(t-t_0) \tag{34}$$

where D is the usual diffusion coefficient and D^*, related to the anomalous fluctuations of the velocity, is given according to the Zaitsev-Shliomis theory by [14]

$$D^* = (1/2)(kT/\rho\ell^3)(\nu+\chi)(3\nu\chi k_c^2)^{-1}\varepsilon^{-2} \tag{35}$$

with k_c the critical wavenumber. For a spherical particle of radius a= 0,3cm, ℓ=0,1cm and $\nu/\chi = 10$, D^* becomes equal to D at $\varepsilon \sim 0,03$.This would provide a possibility of testing the value of the critical exponent γ.

3. A NONEQUILIBRIUM THERMODYNAMIC POTENTIAL FOR STATIONARY STATES

The first aim of this Section is to obtain a nonequilibrium thermodynamic potential of the form (4) from the hydrodynamic equations.The second aim is to show that this potential is related to the probability of the fluctuations and to the stability of the stationary states in the same way as the usual equilibrium thermodynamic potentials are related to the probability of the fluctuations and to the stability of the equilibrium states.

In order to obtain the required non-equilibrium potential we follow the procedure proposed by Graham[2-4]. We look for an equation for the complex amplitude of the slow mode, which was designed in our previous analysis as c_1 and which we will denote as $w(x_1,x_2,t)$ in the following.

We start from the equations of motion (6)-(8), where we explicitly distinguish between the length scale of the convection cells, wich is set by the layer thickness ℓ, and the length and time scales in which the amplitude w changes. This goal is achieved by introducing

$$x_1 = \xi\varepsilon^{-1/2} \quad ; \quad x_2 = \zeta\varepsilon^{-1/4} \quad ; \quad t = \tau\varepsilon^{-1} \quad ; \quad u = \varepsilon^{1/2} w(\xi,\zeta,\tau) \qquad (36)$$

with $\varepsilon = (R-R_c)/R_c$. This choice of scales is well known|15|. For instance, the choice of the time scale is based on the development (18); the choice of the prefactor $\varepsilon^{1/2}$ in u in (36) has its origin in (3), since it is known that the amplitude of the velocity behaves as $\varepsilon^{1/2}$ for small amplitudes. Since R(k) has a minimum for $k_1=k_c$ and $R=R_c$, the quantities $|k_1-k_c|^2 \sim R(k_1)-R_c \sim\varepsilon$ are both of the same order. Then, one substitutes into the equations (6)-(8)

$$\partial_1 \to \partial_1 + \varepsilon^{1/2}\partial_\xi \; ; \quad \partial_2 \to \partial_2 + \varepsilon^{1/4}\partial_\zeta \; ; \quad \partial_t \to \partial_t + \varepsilon\partial_\tau \qquad (37)$$

Collecting equal powers of ε, we may write the evolution operator (6)-(8) in powers of $\varepsilon^{1/4}$

$$L = L_0 + \varepsilon^{1/4}L_{1/4} + \varepsilon^{1/2}L_{1/2} + \dots \qquad (38)$$

and we may look for a solution in powers of $\varepsilon^{1/4}$ as

$$u = {}^{1/2}(u_0 + \varepsilon^{1/4}u_{1/4} + \varepsilon^{1/2}u_{1/2} + \dots) \qquad (39)$$

The result of this iterative scheme, for whose details the reader is referred to |3|, in the order $\varepsilon^{3/2}$ is the following Langevin equation for the complex amplitude w[2-4]

$$(1+P)\partial w/\partial t = \{(3\pi^2/2)\varepsilon - (1/2)P^2|w|^2\}w + 4\{\partial_1 - (i\pi/\sqrt{2})\partial_2^2\}w + \Gamma \qquad (40)$$

with the noise Γ specified by the corresponding properties deduced from the Landau-Lifshitz hydrodynamic noise (10)-(11)

$$\langle\Gamma\rangle = \langle\Gamma^{*}\rangle = \langle\Gamma\Gamma\rangle = 0$$

$$\langle\Gamma^{*}(x_1,x_2,t)\,\Gamma(x_1',x_2',t')\rangle = \pi^2(Q_1+Q_2)\delta(x_1-x_1')\delta(x_2-x_2')\delta(t-t') \qquad (41)$$

For the sake of simplicity we assume from now on that the problem is two-dimensional in x_1 and x_3 , and we neglect the variation of the pa rameters with x_2. From the Langevin equation (40)-(41) one can obtain a stochastically equivalent Fokker-Planck equation for the probability distribution W of the slow mode amplitude w fluctuations. Since this amplitude is a continuous field, its probability distribution will be a functional. We formally consider w and w^* as independent fields and denote δ_w the functional derivative with respect to w(x) with $w^*(x)$ fixed This Fokker-Planck equation takes the form

$$(1+P)\partial W/\partial t = \int d^2x\{\delta_w[(-(3\pi^2/2)\varepsilon+(1/2)P^2|w|^2)w - 4\partial^2 w/\partial x_1^2+ Q\delta_{w^*}]W\}+ \text{c.c.} \tag{42}$$

with $Q = \pi^2(Q_1+Q_2)/2(1+P)$. This may be written, by taking into account the detailed balance symmetry, in the form

$$\partial W/\partial t = Q(1+P)^{-1}\int d^2x\{\delta_w[\delta_{w^*}\Phi(\{w\}) +\delta_{w^*}]W + \text{c.c.} \tag{43}$$

with a suitably chosen functional $\Phi(\{w\})$. By comparison of (42) and (43) it is found that

$$\Phi(\{w\}) = Q^{-1}\int d^2x\{-(3\pi^2/2)\varepsilon|w|^2 + (P^2/4)|w|^4 + 4|\partial w/\partial x_1|^2\} \tag{44}$$

The Fokker-Plack equation (43) obviously has the following time-independent solution

$$W(\{w\}) = N \exp\{-\Phi(\{w\})\} \tag{45}$$

where N is a normalization constant. This result shows clearly the central role of the potential Φ in this theory. The steady state is obtained by minimizing Φ with respect to w, subject to the specific boundary conditions of the problem. The extrema of Φ satisfy the condition

$$-(3\pi^2/2)\varepsilon w + (1/2)P^2|w|^2 w - 4(\partial^2 w/\partial x^2) = 0 \tag{46}$$

For $\varepsilon<0$, i.e. for $R< R_c$, the only solution of (46) satisfying $|w(x)|$ finite when $x \infty$ is $w = 0$, which describes the purely heat-conducting state. It can be verified that this state corresponds to the only minimum of Φ. For $\varepsilon>0$, i.e. for $R >R_c$, $w=0$ is a maximum of Φ, and the state which absolutely minimizes Φ is that given by $|w|^2 = (3\pi^2/P^2)\varepsilon$, which corresponds to a completely regular alignment of two-dimensional rolls of wavevector corresponding to k_c. We then recover the previous result that $w \sim\varepsilon^{1/2}$.

A second property of Φ is that it is related to the evolution of the velocity amplitude through (5). Indeed, by taking averages in equation (43) one gets [3]

$$\langle \dot{w}(x,t) \rangle = -Q(1+P)^{-1} \langle \delta_w * \Phi(\{w\}) \rangle \tag{47}$$

so that Φ acts as a potential of a generalized force $-\langle \delta_w * \Phi \rangle$ which tends to restore the steady state when it has been perturbed by a fluctuation. Dropping the averages in (47), we recover the hydrodynamic equation (40) without the random forces.

A third property of Φ, analogous to those of the thermodynamic potentials, is that it can serve as a Lyapunov function which shows the relative stability of the states which minimize it. Indeed, Φ may be chosen positive by adding to it a convenient constant. On the other hand as a consequence of (47), Φ has the property

$$\dot{\Phi} = -2Q(1+P)^{-1} \int d^2x \, |\delta_w \Phi|^2 < 0 \tag{48}$$

so that Φ has indeed the properties of a Lyapunov function.

Finally, a fourth analogy of the potential Φ with the equilibrium thermodynamic potentials is that it is related to the fluctuations near the corresponding steady state by the relation (45), in the same way as the free energy, for instance, is related to the probability of the fluctuations near equilibrium states at fixed temperature by $Pr \sim \exp(\Delta F/kT)$.

The potential Φ in (44) has the form of a Ginzburg-Landau potential, and therefore the development of this Section has given a theoretical basis to the phenomenological potential (4). Also, it has presented a deeper insigth into its analogies with the usual thermodynamic potentials. The use of Φ allows to unify the concepts of equilibrium phase transitions with the non-equilibrium transitions into a compact and common formalism. From this potential, the results stated in (3) are easily recovered when one discards the non-linear terms. Our next question is which are the effects of the non-linear terms on the behaviour of the Rayleigh-Bénard transition.

4. THE INFLUENCE OF THE NON-LINEAR TERMS ON THE CRITICAL BEHAVIOUR

First of all, we evaluate the region where the non-linear terms of (44) become relevant. The ratio of the non-linear terms to the linear ones is of the order of $P^2|w|^2/\varepsilon$. This ratio can also be written in the form $V_0^2/V_c^2\ \varepsilon$, where we have taken into account the adimensionalization of the velocities w by the factor $V_0 \sim w(\nu/\ell)$ and where V_c is consequently defined as $V_c = \chi/\ell$. A typical value of V_c is of the order of 10 μm/s [12]. The value of V_0 cannot be made arbitrarily low, since it is limited by the thermodynamic fluctuations, and therefore the minimum value experimentally available is given by $V_0 = (2kT/\rho\ell^3)^{1/2} = 3\times10^{-3}$ μm/s for $\rho = 1\ g/cm^3$, T=300K and $\ell = 1$ cm. In consequence, the non-linear terms will be comparable to the linear ones when $V_0^2/V_c^2\varepsilon \sim 1$,i.e.for $\varepsilon \sim 10^{-8}$. The critical region turns out to be impracticably small, since it would imply a perfect control on the temperature up to a millionth of degree. Therefore, in principle, the results of this section are mainly of theoretical interest.As we have commented in the introduction, in this critical region the non-linear terms may change the classical mean-field behaviour summarized in (3) according to the three main possibilities outlined in Fig.1-3.

In order to analyze the influence of the non-linear terms we follow the method used by Graham, starting from the potential (44) in a quasilinear approximation in the heat-conducting region. Indeed, for $\varepsilon < 0$, the main effect of the fourth-order term on this potential is to change slightly the size of the second-order term. This effect can be taken into account in an average fashion by replacing $|w|^4$ by the partly averaged term $4\langle|w|^2\rangle|w|^2$.Hence, we replace (44) by

$$\Phi = L_2 Q^{-1}\int dx\{(-(3\pi^2/2)\varepsilon + P^2\langle|w|^2\rangle)|w|^2 + 4|\partial w/\partial x|^2\} \tag{49}$$

where $\langle|w|^2\rangle$is to be determined consistently.

By introducing spatial Fourier transforms, we may write the transform $K_w(q)$ of the correlation function

$$K_w(q)=\int dx\ e^{iqx'}\langle w^*(x)w(x+x')\rangle \tag{50}$$

in the form

$$\langle w^*(q)w(q')\rangle = 2\pi\delta(q-q')K_w(q) \tag{51}$$

The average of (51) may be evaluated from (49) and it is obtained

$$K_w(q)=(Q/L_2)\{-(3\pi^2/2)\varepsilon + P^2\langle|w|^2\rangle + 4q^2\}^{-1} \tag{52}$$

Expression (51) implies that $\int(2\pi)^{-1}K_w(q)dq = \langle|w|^2\rangle$, which, together with (52) gives the following equation for $\langle|w|^2\rangle$

$$\langle|w|^2\rangle^3 - (3\pi^2/2P^2)\varepsilon\langle|w|^2\rangle^2 - (Q^2/16P^2L_2^2) = 0 \tag{53}$$

From here we obtain as limiting expressions for $\langle|w|^2\rangle$

$$\begin{aligned} \langle|w|^2\rangle &\sim (Q/4PL_2)^{2/3} && \text{when } \varepsilon \quad 0 \\ \langle|w|^2\rangle &\sim Q/2\pi L_2(6|\varepsilon|)^{1/2} && \text{when } \varepsilon < 0 , \quad |\varepsilon| >> 1 \end{aligned} \tag{54}$$

Since the velocity fluctuations are given by

$$\langle v_3(x_1,x_3)v_3(x_1',x_3')\rangle = 2\langle|w|^2\rangle \sin\pi x_3' \sin\pi x_3 \cos(\pi/2)(x_1-x_1')\, e^{-|x_1-x_1'|/\xi_1} \tag{55}$$

it is seen that the velocity fluctuations do not diverge at the critical point, as they do in the classical linear theory, but that they are cut off by the non-linear terms.

The function $K_w(q)$ given in (52) may also be written in the form

$$K_w(q) = 2\langle|w|^2\rangle\xi_1\{1 + (\xi_1 q)^2\}^{-1} \tag{56}$$

which makes evident that the correlation length ξ_1 behaves as

$$\xi_1 = \{-(3\pi^2/8)\varepsilon + (P^2/4)\langle|w|^2\rangle\}^{-1/2} \tag{57}$$

and therefore, at the critical point ($\varepsilon=0$), ξ_1 does not diverge, but it takes the limiting value $\xi_1 = 2(2L_2/P^2Q)^{1/3}$, while sufficiently far from the critical point it behaves asymptotically as $\xi_1 \sim 2\sqrt{2}/\pi\sqrt{3\varepsilon}$.

Finally, in order to obtain the value of the relaxation time at the critical point, we introduce the Laplace-Fourier transform

$$K_w^L(qz) = \int dt' \int dx'\, e^{iqx'-izt'} \langle w^*(x+x',t+t')w(x,t)\rangle \tag{58}$$

which may be evaluated from (49) as

$$K_w^L(qz) = K_w(q)(1+P)\{(1+P)iz + (3\pi^2/2)|\varepsilon| + P^2\langle|w|^2\rangle + 4q^2\}^{-1} \tag{59}$$

From here, one has in the space-time domain

$$\langle w^*(x,t)w(x',t')\rangle = \langle|w|^2\rangle \exp\{-|x-x'|/\xi_1 - |t-t'|/\tau_1\} \tag{60}$$

with ξ_1 given by (57) and τ_1 related to it through

$$\tau_1 = (1+P)\xi_1^2/4 \tag{61}$$

Therefore, the relaxation time does not diverge at the instability point but it has there the value

$$\tau_1 = (1+P)(4L_2/P^2Q)^{2/3} \tag{62}$$

while far from the critical point it is given by

$$\tau_1 = 2(1+P)/3\pi^2|\varepsilon| \qquad \varepsilon < 0 \quad |\varepsilon|>>1 \tag{63}$$

We do not give here the results for the heat-convecting region [3]. We limit ourselves to comment that the relation (61) remains valid in this zone, and that $\tau_1(+)$ and $\xi_1(+)$ of the convecting zone are related to $\tau_1(-)$ and $\xi_1(-)$ of the conducting zone as $\tau_1(+) = 2\tau_1(-)$, $\xi_1(+)= \sqrt{2}\xi_1(-)$, as we have already observed in the linear case[3].

These results may be summarized in the following ideas: the correlation length, the relaxation time and the velocity fluctuations become very large (in comparison with their usual values) near the critical point. The non-linear terms of the hydrodynamic equations prevent these variables from diverging at this point, contrary to the predictions of the classical linear theory. However, in all practical cases, the transition region is so small that the instability behaves as a critical point with classical exponents, as it is observed experimentally |6,16|. This is a difference with a critical point, where the non-linear terms change the mean-field critical exponents but do not prevent the divergence of the correlation length and fluctuation time.

These results can be confirmed from a more fundamental point of view without making a direct use of the potential (49), but starting directly from the hydrodynamic equations in the framework of mode-coupling theories. We expand the solutions of the non-linear hydrodynamic equations (7)-(8) in the complete set of eigenfunctions of the homogeneous linear problem, as we have done for the linear inhomogeneous case in (20). Then it is found that the amplitudes $c_n(t)$ satisfy the following set of equations

$$\dot{c}_n + \lambda_n c_n + \Sigma_{ml} V_{nml} c_m c_l = f_n(t) \tag{64}$$

where f_n is given by the second term of (25) while the mode-coupling coef ficient V_{nml} is given by[13,17]

$$V_{nml}(\underline{k},\underline{k}',\underline{k}'') = (1/2)\int_0^1 dz \int d^2x [\underline{V}_n^*(r)\{(\underline{V}_1(\underline{r}).\nabla)\underline{V}_m(\underline{r}) + (V_m(\underline{r}).\nabla)\underline{V}_1(\underline{r})\} +$$

$$+ PT_n^*(\underline{r})\{\underline{V}_1(\underline{r}).\nabla T_m(\underline{r}) + \underline{V}_m(\underline{r}).\nabla T_1(\underline{r})\}] \qquad (65)$$

These mode-coupling coefficients have several symmetries. Kawasaki|18| was the first to show that $V_{nnn}=0$, and it is also possible to see|13| that $V_{nml} = V_{nlm}$ and that

$$\left.\begin{matrix} V_{nmn}(\underline{k},\underline{k}',\underline{k}'') \\ V_{mnn}(\underline{k},\underline{k}',\underline{k}'') \end{matrix}\right\} = \left\{\begin{matrix} 0 & m \neq 2n \\ \text{non-zero} & m=2n \end{matrix}\right. \qquad (66)$$

If one assumes that the only relevant coupling is that between the first and the second modes (m=1, n=2) and with several simplifications concerning the $\underline{k}$'s , one can arrive [13] to the following equation for the amplitude of the first mode

$$\dot{c}_1 + \lambda_1 c_1 + \gamma|c_1|^2 c_1 = f_1(t) \qquad (67)$$

where the non-linear term describes the interaction between the fundamental and the second mode, with the coupling coefficient given by $\gamma = -4V_{121}(k_c,0,k_c)V_{211}(0,k_c,-k_c)\lambda_{20}^{-1}$. From (67) one can obtain the most salient features of non-linear fluctuations, as the non-divergence of the correlation length and the fluctuations [13]. However, in order to have a more detailed and quantitative description of the maximum value of the fluctuations and of the correlation time and length, one must go to more complicated schemes, starting from (64)-(65), as can be found , for instance,in the approach of Graham and Pleiner[17]. We do not deal here with these approaches, which basically confirm the features obtained from the potential Φ in the quasilinear approximation.

5. OTHER NON-CLASSICAL ASPECTS OF THE HYDRODYNAMICAL FLUCTUATIONS IN THE RAYLEIGH-BENARD INSTABILITY

The analysis of the precedent Section has shown that the non-linear terms cut down the divergences predicted by the linear theory, according to the possibility presented in Fig.2. However, this is not the only possibility, and in fact other theoretical analyses have shown that the non linear terms may affect the transition as indicated schematically in Fig 1 and Fig.3 of the Introduction. The aim of this Section is to explore briefly in which circumstances these behaviours may be expected, according to the analyses of Swift and Hohenberg [19] and of Procaccia and Goldhirsch[20,21].

a) Swift-Hohenberg's theory: possibility of a first-order transition.

Swift and Hohenberg have analysed the behaviour of fluctuations near the convective instability. They obtained a generalized thermodynamic potential very similar to Graham's one (44) , with the important difference that the rotational symmetry of the starting hydrodynamic equations is preserved in their model. From the observation of the close analogy between their thermodynamic potential and that of a model of condensation of a liquid to a non-uniform state (Brazovskii model), they can translate the main consequences of the latter model to the problem of the hydrodynamic instability. Though the absence of cubic terms indicates that the transition is of second order, when fluctuations are taken into account this feature may be changed, depending on the lateral dimension L of the plate. If we call λ the coupling constant of the non-linear terms of the potential, which is related to $kT/\nu^2 \ell\rho$, and $k_c \sim \ell$ the critical wavevector, one can distinguish three regions:

i) $Lk_c \gg \lambda^{-2/5}$: no conclusions are derived from the comparison with Brazovskii's model.

ii) $\lambda^{-2/5} \gg Lk_c \gg \lambda^{-1/3}$: Brazovskii's model predicts that the fluctuations will change the original second-order phase transition into a first-order one.

iii) $\lambda^{-1/3} \gg Lk_c$: this is the region analysed by Graham, and the results of the precedent Section apply.

The new aspect outlined by Swift and Hohenberg is the possibility, under certain conditions, of a discontinuous first-order transition. However the jump in the order parameter (the velocity) is of the order of $\Delta w = \nu k_c \lambda^{1/3}$. Since in the Bénard case $\lambda \sim 10^{-9}$, this jump will not be observable in practice.

b) Goldhirsch-Procaccia's theory: a change in the critical exponents

Recently, Goldhirsch and Procaccia[20,21] have pointed out the influence of the properties of the boundaries on the critical behaviour of the Rayleigh-Bénard instability. Their analysis is based on the renormalization-group approach, and we skip its details here and deal only with the main conclusions, which are the following ones: when the boundaries are highly conducting (copper plates) ,the thermal fluctuations at the boundaries are rapidly and efficiently wiped out, since they diffuse through the boundaries. In this case, the problem is two-dimensional, and the renormalization-group analysis has not any fixed point, so that it indicates a first-order transition, in accordance with the predictions of Swift and Hohenberg.

However, if the thermal conductivity of the boundaries is smaller than that of the fluid, thermal fluctuations at the boundaries have to be taken into account (glass plates). In this case, the problem is three-dimensional and the renormalization group analysis leads to a second-order phase transition with the following exponents

$$\beta = 0{,}583 \qquad \nu = 0{,}583 \qquad z = 2\nu \tag{68}$$

This result is in excellent agreement with the experimental values obtained by Bergé and Dubois [22]. This analysis predicts also a cross-over from these non-classical exponents to the classical exponents in a very small zone near the critical point. This feature is unexpected from the previous experience with critical phenomena, where the non-classical exponents are found precisely in the immediate vicinity of the critical point.

Let us finally mention two recent papers by Enz[23,24] in which a perturbation formalism which works far from equilibrium is proposed, in view of a possible renormalization-group application. The main result consists in an explicit construction of the unperturbed stationary state which, although it is Gaussian, is non trivial far from equilibrium.

6. CONCLUDING REMARKS

Two final remarks concerning two general hypotheses behind the line of thought examined in this paper deserve some attention . One of them is the expression of the non-equilibrium hydrodynamic noise, while the other one refers to some possible limits of the analogy between non equilibrium instabilities and equilibrium phase transitions.

With only one exception[20-21] all the authors mentioned in this paper assume that in non-equilibrium the random noise of the hydrodynamic equations, due to the fluctuations of the dissipative fluxes, is given by the local-equilibrium Landau-Lifshitz expressions. As a consequence, the effect of the fluctuations even at the critical point turns out to be very small. However, it has been pointed out on an experimental basis[21] that the noise in the transition region must be three or four orders of magnitude larger than the noise predicted by the local-equilibrium expressions.

Procaccia and Goldhirsh[21] have presented a phenomenological mechanism for noise amplification. In their analysis, the velocity and temperature fluctuations are correlated. Indeed, when a mass of fluid at a given temperature experiences a velocity fluctuation, it is transported into a region at a different temperature and it is "felt" as a temperature fluctuation. From a simple dynamical model wich takes into account viscous drag and heat diffusion, they have calculated the enhancement factor of the non-equilibrium fluctuations compared to the local-equilibrium estimates. In some circumstances, this factor is of the order of 10^4 to 10^6. This makes observable the effect of the fluctuations, as it is seen experimentally in the anomalous critical exponents (68). Therefore, it is desirable to analyse the random forces in non-equilibrium systems from first principles.

We have examined this problem from a thermodynamic point of view [25], starting from a generalized non-equilibrium entropy|26| which depends on the dissipative fluxes as well as on the classical variables. The Gibbs equation corresponding to this entropy has the form|26|

$$ds = \theta^{-1}du + \theta^{-1}\pi dv - (\tau_1 v/\lambda T^2)\underline{q}.d\underline{q} - (\tau_0 v/\zeta T)p^v dp^v - (\tau_2 v/2\mu T)\overset{\circ}{\underline{\underline{P}}}{}^v : d\overset{\circ}{\underline{\underline{P}}}{}^v \qquad (69)$$

In this expression, s is the specific entropy per unit mass, u and v the specific internal energy and the specific volume, the τ's are the relaxation times of the corresponding dissipative fluxes($\underline{q}$,heat flux,p^v scalar viscous pressure, $\overset{\circ}{\underline{\underline{P}}}{}^v$ viscous pressure tensor), while θ and π are given in terms of the absolute temperature T and the thermodynamic pressure p as

$$\begin{aligned} \theta^{-1} &= T^{-1} - (1/2)(\partial\alpha_1/\partial u)\underline{q}^2 - (1/2)(\partial\alpha_0/\partial u)p^{v2} - (1/2)(\partial\alpha_2/\partial u)\overset{\circ}{\underline{\underline{P}}}{}^v : \overset{\circ}{\underline{\underline{P}}}{}^v \\ \theta^{-1}\pi &= T^{-1}p - (1/2)(\partial\alpha_1/\partial v)\underline{q}^2 - (1/2)(\partial\alpha_0/\partial v)p^{v2} - (1/2)(\partial\alpha_2/\partial v)\overset{\circ}{\underline{\underline{P}}}{}^v : \overset{\circ}{\underline{\underline{P}}}{}^v \end{aligned} \qquad (70)$$

where $\alpha_1 = \tau_1 v/\lambda T^2$, $\alpha_0 = \tau_0 v/\zeta T$ and $\alpha_2 = \tau_2 v/2\mu T$. From this generalized entropy and from an extension of the Einstein hypothesis for the probability of fluctuations, one may get the following equations for the non-equilibrium fluctuations in the presence of a temperature gradient

$$\begin{aligned} \langle\delta u\delta q\rangle &= (kq_0/m\alpha_1\Delta)\{1+(A/\Delta)q_0^2\}^{-1}\{(\partial\alpha_1/\partial u)(\partial T^{-1}p/\partial v) - (\partial\alpha_1/\partial v)(\partial T^{-1}/\partial u)\} \\ \langle\delta q\delta q\rangle &= (k\lambda T^2/mv\Delta)\{1+(A/\Delta)q_0^2\}^{-1}\{\Delta + q_0^2|(\partial T^{-1}/\partial v)(\partial^2\alpha_1/\partial v\partial u) - \\ &\quad - (1/2)(\partial T^{-1}/\partial u)(\partial^2\alpha_1/\partial v^2) - (1/2)(\partial T^{-1}p/\partial v)(\partial^2\alpha_1/\partial u^2) \end{aligned} \qquad (71)$$

where $A = (\partial T^{-1}/\partial u)\{(1/\alpha_1)(\partial\alpha_1/\partial v)^2 - (1/2)(\partial^2\alpha_1/\partial v^2)\} + (\partial T^{-1}p/\partial v)$ $\{(1/\alpha_1)(\partial\alpha_1/\partial u)^2 - (1/2)(\partial^2\alpha_1/\partial u^2)\} - 2(\partial T^{-1}/\partial v)\{(1/\alpha_1)(\partial\alpha_1/\partial u)(\partial\alpha_1/\partial v)$ $- (1/2)(\partial^2\alpha_1/\partial u\partial v)\}$ and $\Delta = (\partial T^{-1}/\partial u)(\partial T^{-1}p/\partial v) - (\partial T^{-1}/\partial v)^2$. These expressions show the non-equilibrium corrections to the classical formulas. In particular, the Landau-Lifshitz formula is modified by the presence of a temperature gradient, and the variables of different time reversal parity become correlated. For values of ∇T near $(-\Delta/A\lambda^2)^{1/2}$ the fluctuations may become greatly enhanced. This effect is of statistical origin and is different from that considered by Procaccia and Goldhirsch. Further work on this line is in progress.

The aim of the approaches quoted in this short review is to emphasize the analogy between nonequilibrium instabilities and equilibrium second-order phase transitions. While many features are indeed common to both phenomena, the ultimate arguments of this analogy have recently been the subject of a controversy on whether the dissipative structures represent or not a broken symmetry[27,28]. The present state of this controversy is the observation that, rather than the translational symmetry, the main symmetry breaking in the Rayleigh-Bénard problem is related to a discrete symmetry under velocity reversal [29] so that the above mentioned analogy is indeed not just superficial, but has a fundamental meaning and interpretation.

REFERENCES

(1) V.M.Zaitsev and M.I.Shliomis, Sov.Phys.JETP 32 (1971) 866-870

(2) R.Graham, Phys.Rev.Lett.31 (1973) 1479-1482

(3) R.Graham, Phys.Rev.A 10 (1974) 1762-1784

(4) R.Graham in Order and Fluctuations in equilibrium and non-equilibrium Statistical Mechanics (Nicolis,Dewel and Turner, eds) Wiley, New York,1981, pp.235-288

(5) W.A.Smith, Phys.Rev.Lett.32 (1974) 1164-1168

(6) J.Wesfreid, Y.Pomeau, M.Dubois, C.Normand, P.Bergé, J.Physique Lett. 39 (1978) 725-731

(7) H.N.W.Lekkerkerker and J.P.Boon in Fluctuations, Instabilities and Phase Transitions (T.Riste,ed) Plenum Press, New York, 1975

(8) P.Bergé, this volume

(9) J.E.Wesfreid, this volume

(10) G.Dewel, D.Walgraef and P.Borckmans, J.Physique Lett.42 (1981)361

(11) J.Toner and D.Nelson, Phys.Rev.B. 23 (1981) 316

(12) C.Normand,Y.Pomeau and M.G.Velarde, Rev.Mod.Phys. 49 (1977) 581-624

(13) Y.Tsuchiya, J.Phys.Soc.Japan 43 (1977) 1823-1831
(14) J.B.Lastovka and J.P.Boon, Phys.Rev.A 14 (1976) 1583-1586
(15) A.Newell and J.Whitehead, J.Fluid Mech. 38 (1969) 279
(16) J.Wesfreid, P.Bergé and M.Dubois, Phys.Rev.A 19 (1979) 1231-1233
(17) R.Graham and H.Pleiner, Phys.Fluids 18 (1975) 130-140
(18) K.Kawasaki, J.Phys.A 6 (1973) L4-L6
(19) J.Swift and P.C.Hohenberg, Phys.Rev.A 15 (1977) 319-328
(20) I.Procaccia and I.Goldhirsch, in Systems far from Equilibrium (L. Garrido,ed) Sitges Conference on Statistical Mechanics, Springer, Berlin, 1980
(21) I.Goldhirsch and I.Procaccia, Phys.Rev.A 24 (1981) 580-597
(22) P.Bergé and M.Dubois, Phys.Rev.Lett.32 (1974) 1041-1043
(23) C.P.Enz, J.Stat.Phys. 24 (1981) 109-117
(24) C.P.Enz in Field Theory, Quantization and Statistical Physics (E. Tirapegui,ed) Reidel, New York, 1981, pp.263-275
(25) D.Jou, J.E.Llebot and J.Casas-Vázquez, Phys.Rev.A 25 (1982)
(26) D.Jou, J.M.Rubí and J.Casas-Vázquez, Physica A 101 (1980) 588-598
(27) D.L.Stein, J.Chem.Phys. 72 (1980) 2869-2874
(28) D.Walgraef, G.Dewel and P.Borckmans, J.Chem.Phys.74 (1981) 755-757
(29) L.Sneddon, Phys.Rev.A 24 (1981) 1629-1632

SOME TOPICS ABOUT THE TRANSITION TO TURBULENCE

P. Bergé
Service de Physique du Solide et de Résonance Magnétique
CEN-Saclay - 91191 Gif-sur-Yvette Cedex
France

Abstract

The basic concepts allowing to define the different states of a dynamical system are illustrated through the behaviour of the Rayleigh-Bénard convection in confined geometry. The interest of the Poincaré sections of the phase space is pointed out in parallel with that of the iterated maps. Finally one shows how the deterministic chaos can be understood, thanks to the concepts of strange attractor.

I. *UNSTEADY REGIMES AND TURBULENCE*

We deal with an unsteady regime as far as a variable representing a dynamical system is time dependent.

In order to characterize such a time dependent regime we can use 3 different methods (suppose that the relevant variable is the velocity V, as in the hydrodynamic instabilities).

First, we can record the velocity versus time. This gives at least a qualitative idea of the time dependent regime (fig.1).

Secondly, we can perform a Fourier analysis (F.F.T.) of the velocity fluctuations. Then, a power spectrum is obtained which reveals the frequencies of the different modes of fluctuations. (fig.2)

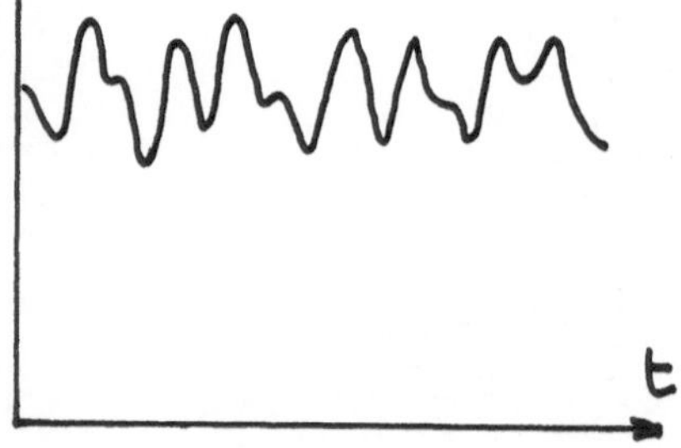

Fig.1. Temporal variation.

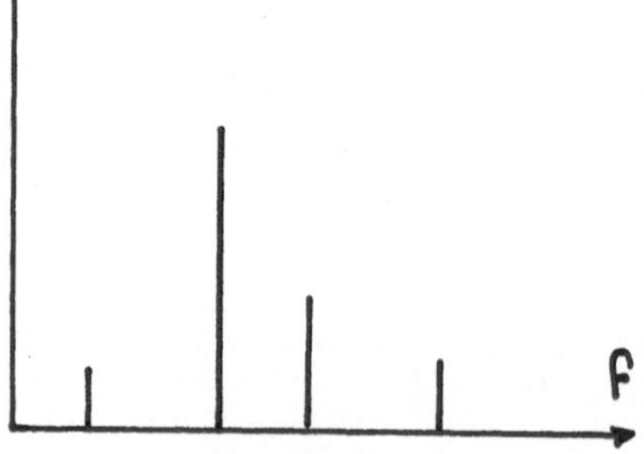

Fig.2. Fourier Spectrum.

Even more important is to consider a third type of diagram = the phase-space diagram.

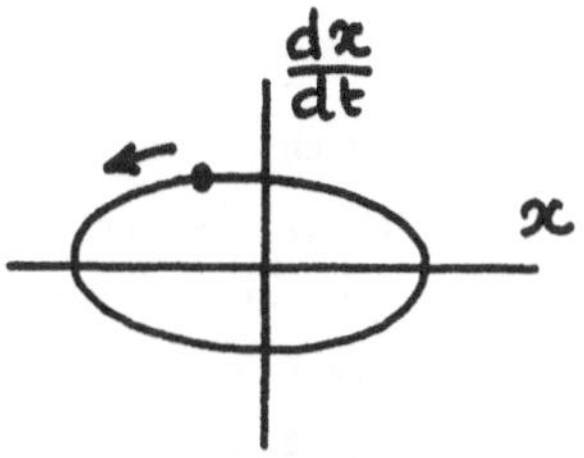

Fig.3. Phase space diagram

For example, in order to characterize the motion of a simple oscillator we can draw a closed loop by plotting the amplitude of the motion versus its time derivative (fig.3). In such a phase diagram a steady regime will be obviously represented by a single point.

A dynamical system is characterized -among others- through its number of degrees of freedom - i.e. the number of independent variables necessary to determine the state of the system.

In order to illustrate this point let us return to a simple oscillator namely a pendulum (oscillating in its linear domain). This dynamical system is completely determined if we know -at a given instant- two and only two variables for example the position x and the velocity dx/dt. The point representing the motion describes an ellipse in a phase diagram whose axis are x and dx/dt (fig.4). So, in principle, we can say that a simple oscillator has two degrees of freedom, for example x and dx/dt.

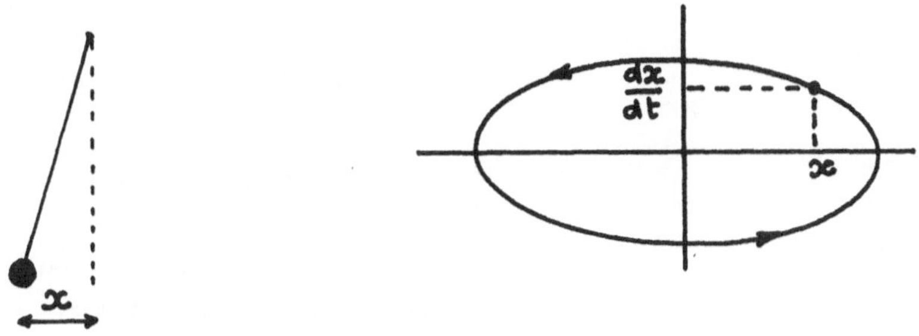

Fig.4. Phase space diagram of a pendulum

But, in practice, we will consider further oscillations (and oscillators) with almost constant energy that means that the knowledge of only one variable determines the state of the oscillator. In that sense, one says that one oscillator have only one degree of freedom which is the phase of the oscillation. But have in mind that the dimensionality of the corresponding phase space is nevertheless two.

Let us finally recall a definition of the turbulence or chaos. We emphasize here that the following definition do not consider any spatial criterium like the scale of the eddies etc...

It is said that a system is turbulent if there is a broad and continuous part in the Fourier spectrum of any variable representing the system, for example, in the velocity spectrum (fig.5).

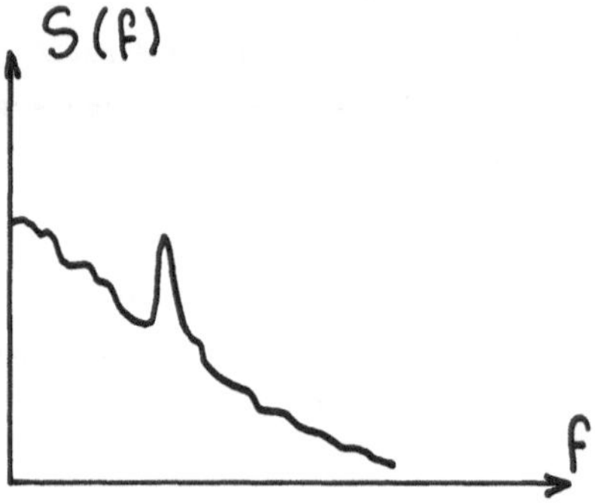

Fig.5. Broad band spectrum

Let us remind that the auto-correlation function <V(t).V(t+τ)> is the Fourier transform of the power spectrum. The existence of a broad band spectrum implies then an autocorrelation function (A.F.) which tends to zero when the time increases. Because the A.F. measures the time similarity, then the presence of turbulence implies a loss of the similarity, a loss of the memory of the initial states.

In other words, the knowledge of the time behaviour in the past does not permit to predict the behaviour in the future ; turbulence corresponds to unpredictability.

A very important consequence of this loss of similarity (or memory) of the initial states is the divergence of the trajectories in phase space : two very neighbouring (then very similar) trajectories diverge (and then loose their similitude) if the regime is turbulent (fig.6).

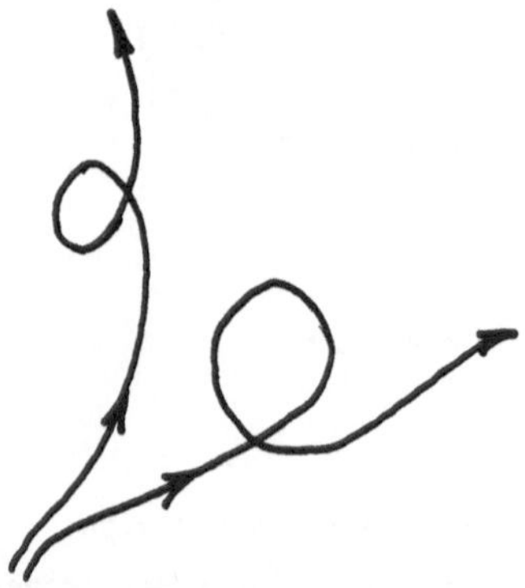

Fig.6. Divergence of the trajectories in the phase space.

At the opposite, any unsteady regime, even very complicated, but not turbulent, is characterized by :

- A Fourier spectrum composed of only sharp lines.
- An autocorrelation function which remains finite even at large times.
- The fact that two neighbouring trajectories in the phase space remain neighbouring at any time.

The problem of the transition to turbulence when the external stress applied to a system is increased has been subjected to many theoretical ideas. In order to schematize the question let us recall two opposite concepts.

Landau [1] proposed that by increasing the Reynolds number (i.e. the velocity) of a fluid, the successive appearance of many independent (incommensurate) frequencies in the Fourier spectrum of the velocity will produce a very complicated behaviour ; this regime can be considered as practically turbulent in the limit of a very large number of independent frequencies. Then, according to Landau, the time dependent velocity takes the form

$$V(t) = f(\omega_1 t, \omega_2 t, \ldots \omega_N t)$$

with N large.

Obviously the appearance of turbulence is related to a system with a large number of degrees of freedom (N). At the opposite, a challenging mechanism was proposed by Ruelle and Takens [2] . For the moment let us just say that, basically, and according to these authors, only the occurrence of 3 incommensurate frequencies can produce turbulence. (Here the system has a small number of degrees of freedom).

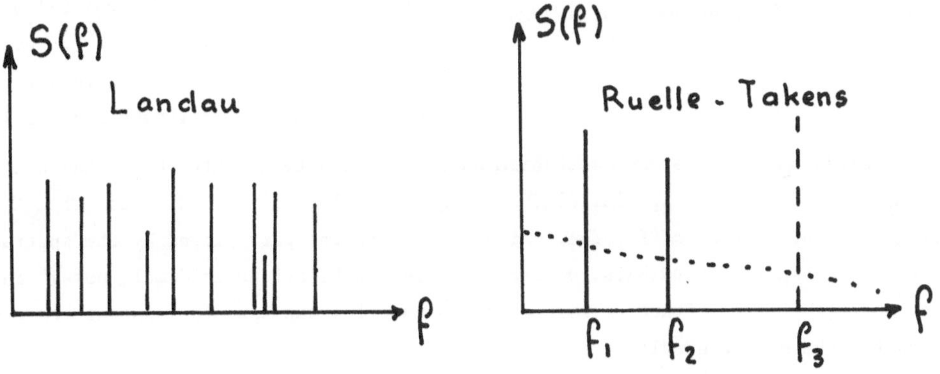

Fig.7. Two schemes for the transition to turbulence

II. SHORT DESCRIPTION OF THE BEHAVIOUR OF RAYLEIGH-BENARD INSTABILITY (*High Prandtl number case*)

a. *Large boxes*

The word "large" has to be understood as "large horizontal extend compared to the depth of the fluid layer". For this purpose the "aspect ratio" Γ of the box containing the fluid is introduced :

$$\Gamma_x = \frac{L_x}{d}$$

$$\Gamma_y = \frac{L_y}{d}$$

If $\Gamma_x, \Gamma_y >> d$,

the observed convective structure is generally at random and contains defects [3]

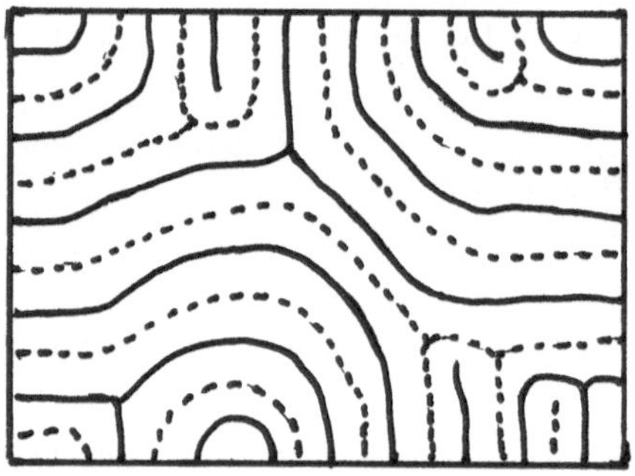

Fig.8. A roll arrangement in a large box.

(fig.8) . Nevertheless, after long transient periods, we can notice that a steady state is reached, steady state which is stable in a finite range of R_a number above R_{a_c} (R_{a_c} is the critical R_a number which corresponds to the appearance of the rolls). On the contrary, at a certain threshold R_{a_T} (whose value is not universal but ranges about 10 R_{a_c}) the steady state becomes unstable and the velocity behaves erratically : then, in that case, we observe a direct transition between the steady state and the turbulent one. Patient and careful observations [4] show that this turbulence is due to very complex erratic motions of the convective structure including defects (Phase fluctuations). We can hardly believe that this mechanism have something to do with the Landau picture but here, and due to the spatial disorder, we obviously deal with a system with a large number of degrees of freedom. Owing to the leading part played by the spatial disorder, we can expect to have completely different behaviour in "small boxes" in which the stabilizing effect of the lateral boundaries can keep the spatial order by fixing the phase of the rolls.

b) *Small boxes*

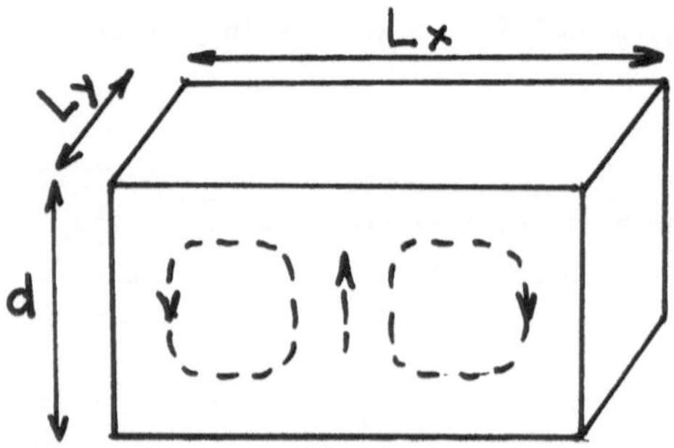

Fig.9. A scheme of a small box

The "small boxes" considered here are such as

$$\Gamma_x = \frac{L_x}{d} = 2$$

$$\Gamma_y = \frac{L_y}{d} = 1,2$$

(see fig. 9).

Then basically two rolls can be present in such a box and it is clear that the position (or phase) of each one is well stabilized by the presence of its lateral boundary. As a matter of fact this stabilizing effect is very efficient : the steady spatial order can be maintained up to 200 R_{a_c} or more (this contrasts very much with the behaviour of a large box in which the position (phase) of the rolls becomes time dependent comparatively near R_{a_c}).

We will now describe a typical route from the steady to unsteady convection. At a definite threshold R_{a_1} (>200 R_{a_c}) the steady state is replaced by an oscillating regime at a well defined frequency f_1 (including eventually higher harmonics nf_1).

This means that the velocity in any point of the box is modulated with a constant period (periodic regime). On the other hand the Fourier spectrum of the velocity contains only sharp peaks at $f_1(2f_1,3f_1,...)$; finally a phase space diagram obtained, for example, by plotting on a x,y recorder the velocity versus its time derivative $\frac{dV}{dt}$ consists in a closed loop.

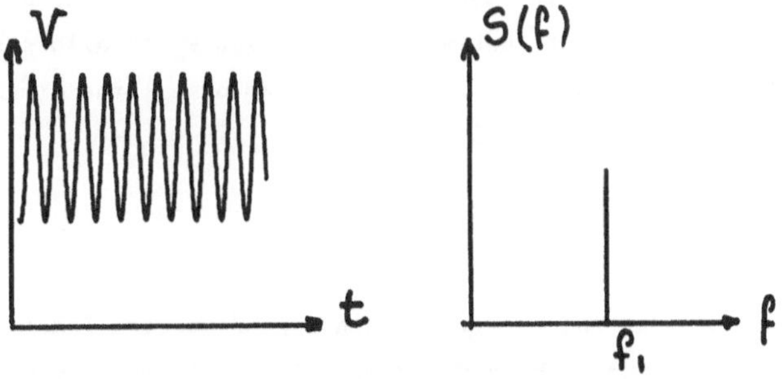

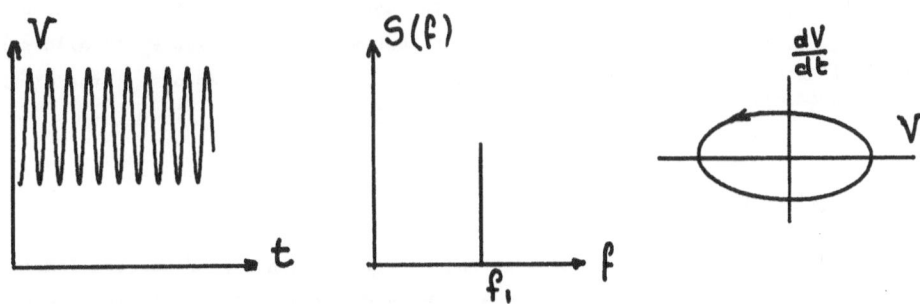

Fig.10. The 3 representations of a periodic regime

Such a closed loop is called a limit cycle and we say that, at R_{a_1}, the system bifurcates from the steady state (a fixed point) to a stable limit cycle.

The mechanism responsible for this oscillatory regime (and more generally for most of the oscillatory regimes we have found in the R.B. convection) are related to instabilities near, or in, the thermal boundaries layer. In many cases these thermal oscillators are localized in space and correspond to thermal heterogeneities like thermal droplets [5] or plumes [6] periodically growing and being advected.

The second step breaking the (mono) periodic regime corresponds to the appearance at a new threshold $R_{a_2} > R_{a_1}$ of a new frequency f_2 superimposed to the former one f_1. More generally the two frequencies f_1 and f_2 are unrelated, more precisely they are incommensurate* .

But for simplicity let us assume that only f_1 and f_2 are present and incommensurate. This new regime is called biperiodic or more precisely quasi-periodic to recall that its properties are intermediate between a periodic regime and a chaotic one.

----

*incommensurate means that one cannot find integers n and m such, as $mf_1 = nf_2$.

In order to draw the corresponding phase space trajectories, a two dimensional representation can no longer be used.

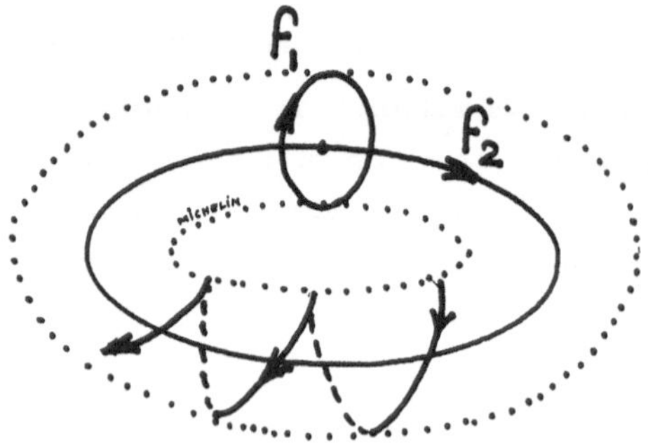

Fig.11. Phase space diagram in a quasi periodic regime

The limit cycle representing f_1 has to wind with the frequency f_2 and thus the representation, in a three dimensional phase space, takes the form of a torus (see fig.11). At R_{a_2} we say that the stable limit cycle bifurcates to an invariant torus T^2.

The fact that the phase space is no more representable in two dimensions leads us to use a transformation which allows to represent the section of a three dimensional phase space.

c) *Poincaré section and iterated maps*

The interest of the Poincaré section is not only to obtain a good bidimensional representation of a complex tridimensional phase diagram, but also to make useful comparison with very simple mathematical models labelled generically as "iterated maps".

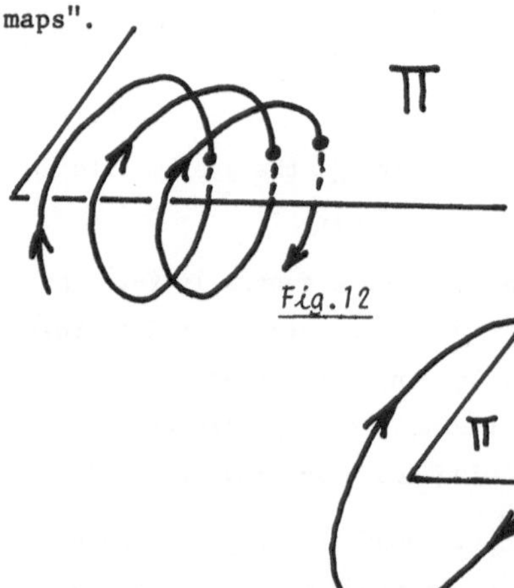

Fig.12

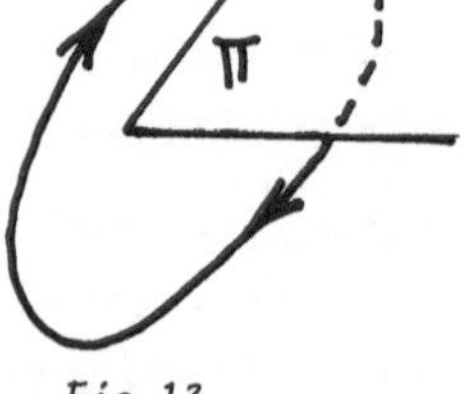

Fig.13

Very simply the Poincaré section is a cut, a section, of all the trajectories (here in 3 dimensions) by a given plane π (see fig. 12). Then, an ensemble of points are obtained which corresponds to the Poincaré section of the phase diagram. Obviously this simple transformation lowers by one the dimensionality of the representation.

A limit cycle (phase space in 2 dimensions) is tranformed into a single point (analogous to a fixed point (see fig. 13)). Much more interesting is the following : the trajectories in a quasi periodic regime (torus T^2) drawed in two dimensions show a very intricate mixture of trangled lines (the plane projection of the trajectories lying on a torus). On the contrary, the Poincaré section is a simple closed loop (analogous to a limit cycle). One can see on fig.14a some trajectories in the phase space in a quasi-periodic regime (obtained with electronic oscillators) and the corresponding Poincaré section. The presence of harmonics of the fundamental

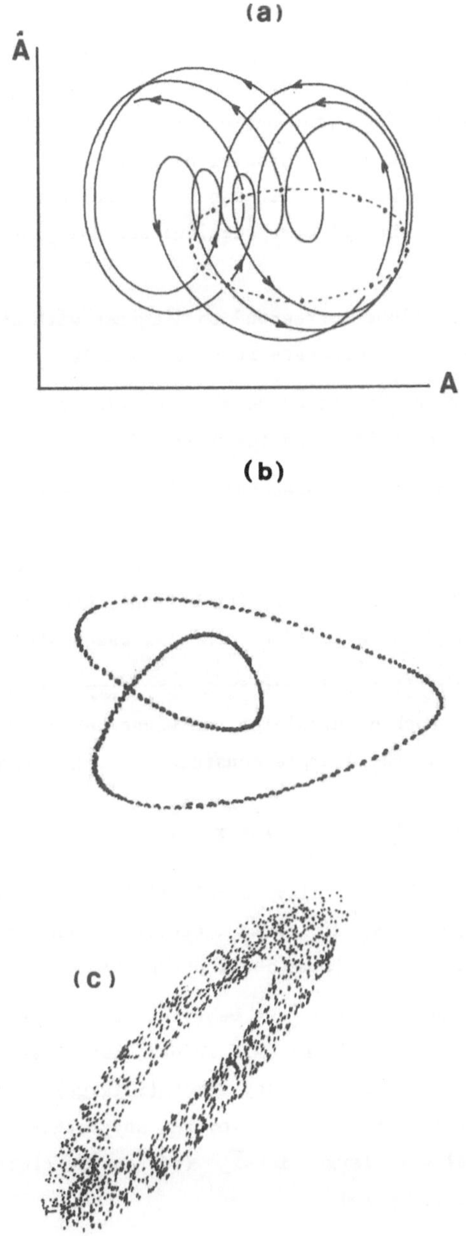

Fig.14. Some Poincaré sections
a) quasi periodic without harmonics (T^2) (full lines = 3 phase loops, dots and dashed line = Poincaré section)
b) quasi periodic regime T^2 with harmonics
c) quasi periodic regime with 3 frequencies (T^3).

frequencies may give rise to some more complicated sections than a simple ellipse (see fig. 14b), but as far as the regime remains quasi-periodic(torus T^2, phase space at 3-dimensions)the Poincaré section is a closed loop. Reciprocally if the Poincaré section is a closed loop one can affirm that the dimensionality of the corresponding phase space is 3. By opposition see on fig. 14c the section of a phase space trajectories in a space of dimension 4 (3 independent frequencies). For experimental illustrations of the Poincaré sections see the paper by M. Dubois in this volume [7].

Furthermore, the Poincaré sections correspond to diagrams with discretized time, very useful to do comparison with discrete iterative models.

What is the interest to use the "iterated maps" as models to understand some properties of bifurcation and the transition to turbulence ?

If we start with the full equations representing the properties of an hydrodynamical system like R.B. convection in the time dependent state, this corresponds to an untractable problem. Then the idea was to simplify and truncate these equations as Lorenz did [8] . This model is very interesting but the resolution of the set of the famous 3 equations still requires a top-desk computer associated with a digital plotter.

Even much more simple are the iterated maps from very simple functions which can be processed by a programmable pocked calculator and whose behaviour are very similar to those of dynamical systems. For example consider the well known iteration [9][10]

$$x_{i+1} = 1 - a\,x_i^2 \qquad , \quad -1 < x < +1 \qquad ,$$

a being the "control parameter" which plays a role similar to that of the Rayleigh number R_a. The successive points x_i, x_{i+1}, x_{i+2} etc... can be considered as the Poincaré section of a (virtual) two dimensional phase space.

Let us enter a bit more in the detail of the behaviour of this simple map : $f(x) = 1 - ax^2$ on the interval -1, +1. As said before, for a given a and initial value of x,one iterates the function f(x) many times (i large). For certain values of a $a < .75$ there is a stable fixed point, i.e. for any initial value of x in the interval -1, 1 the final value (i large) is $\bar{x}_*$ (see on the fig.15 the corresponding Poincaré map of a stable limit cycle).

For $.75<a<1.25$ there are two fixed points $\overline{x_1}$ and $\overline{x_2}$ corresponding to a Poincaré section of a periodic orbit of period 2 (sub-harmonic of the previously obtained limit cycle, see fig.16).

Let us remark furthermore that such an iterated map well describes the concept of bifurcations when a -the control parameter- is increased. (See on fig. 17 the first bifurcation at a = 0,75).

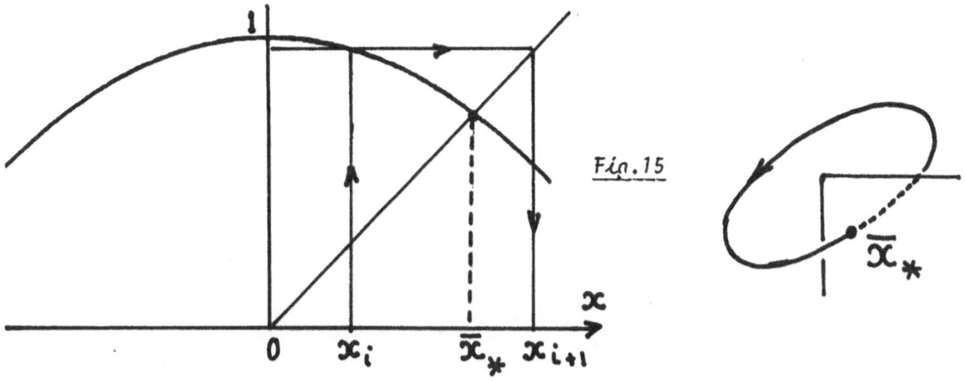

Fig.15

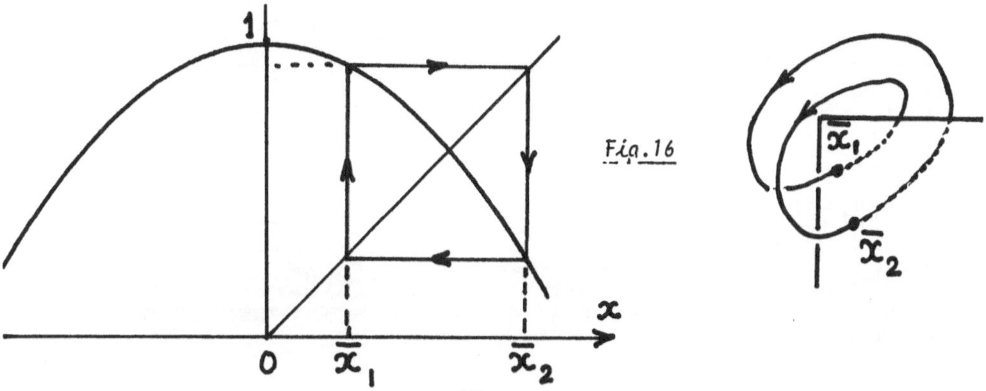

Fig.16

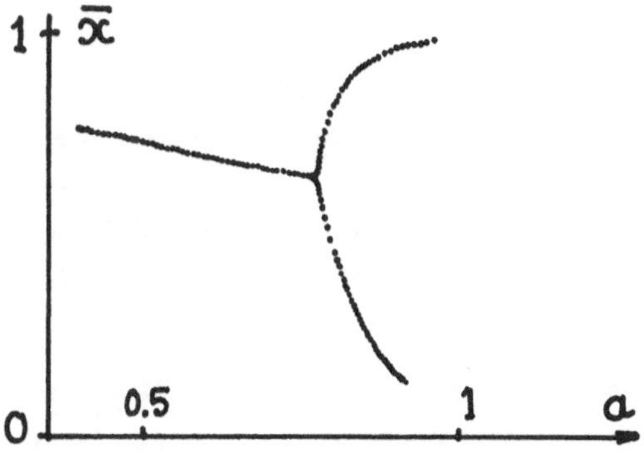

Fig.17. *The first bifurcation of* $f(x) = 1-ax^2$ *at* $a = 0.75$.

d. *Attractors*

Let us first emphasize that the existence of attractors is closely related to the existence of dissipation. For example, in steady R.B. convection, the convective amplitude (velocity) is the result of a balance between the energy supplied to the system and the dissipation of this system.

If we pull this system away from its state of equilibrium (but not too far) the system will return (due to viscous and thermal dissipation) to its former state of equilibrium, see fig.18).

In the phase space, the trajectory is attracted, converges toward the fixed point F corresponding to the equilibrium state. We say that F is an attractive fixed point.

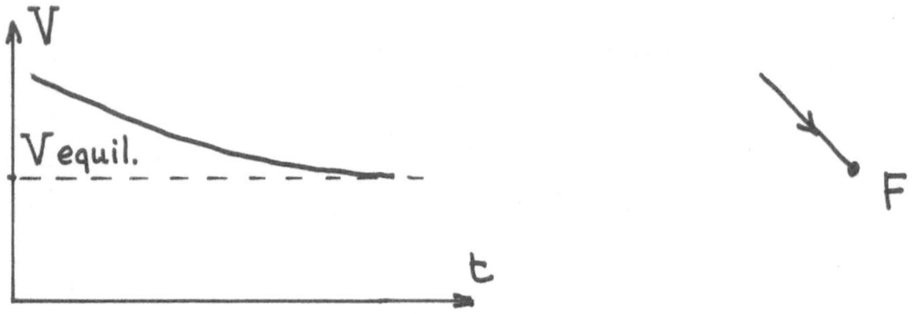

Fig.18. Attraction to a fixed point.

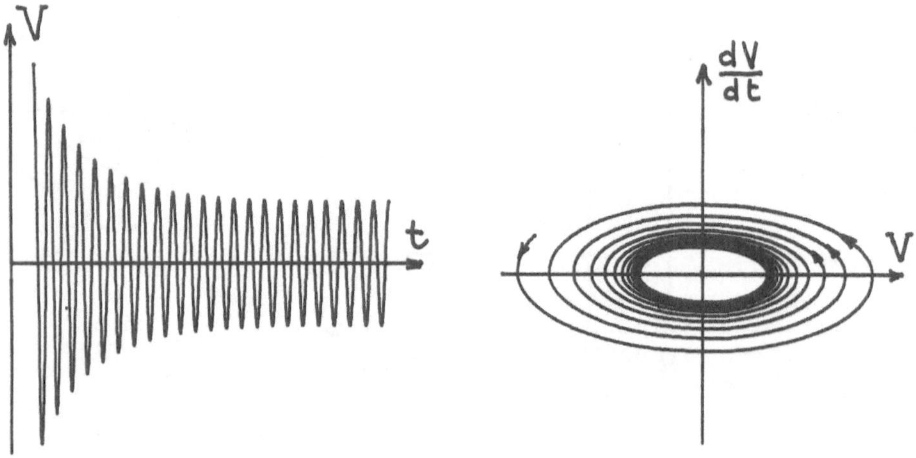

Fig.19. Attraction to a limit cycle.

On the same manner and for the same reasons if R.B. convection is oscillating according to an (equilibrium) stable limit cycle, if we perturb this state, the system will return to its equilibrium (see fig.19).

In that case we deal with an attractive limit cycle or periodic attractor (Note that dissipation contracts phase space, more precisely dissipation lowers the measure of the phase space). More generally if we consider the quasi periodic regime in a dissipative system we deal with a quasi periodic attractor (or an attractive torus).

III. MORE ABOUT A ROUTE TO TURBULENCE, STRANGE ATTRACTOR

Let us remind that in R.B. convection a route (not the unique route !) may be :

- at R_{a_1} appearance of the periodic regime
- at R_{a_2} appearance of the quasi-periodic regime.

What happens by a further increase of the R_a number ?

- at R_{a_3} turbulence or chaos appears (i.e. the spectrum is no more composed of sharp lines but also broad band noise begin to appear especially near zero frequency. For more details about this route (and others), see [11][12][13].

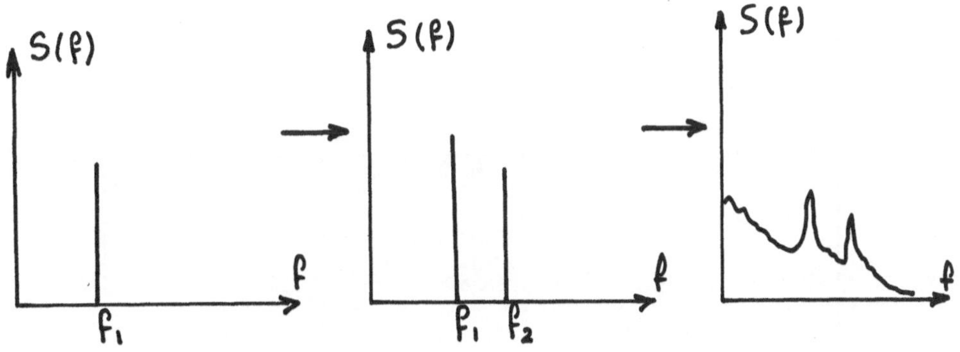

Scheme of the route through biperiodism.

This kind of route is in very good agreement with the ideas of Ruelle-Takens though it is not yet proved that at R_{a_3} it is the appearance of a third frequency f_3 which produces chaos.

Nevertheless a basic question remains : how is it possible to reconcile the apparent discrepancy between the determinism of the system (the existence of 2 may be 3 coupled oscillators) and the resulting chaos which is -in essence- a non predictible state.

In fact this point can be solved through the concept of "strange attractors".

To illustrate the question of deterministic systems versus chaos we will return to the iterated maps yet examined above, but here, we will consider a system in two dimensions, the Henon [14] map :

$$\begin{cases} x_{i+1} = y_i + 1 - a\, x_i^2 \\ y_{i+1} = b\, x_i \end{cases}$$

This simple transformation is, indeed, deterministic !

Furthermore, for b = 0,3 (the value usually adopted in this problem) this Hénon map is attractive (b is the analog of the dissipation parameter).

If a = 1.3 the successive iterations of all initial point in a certain ensemble converge to a set of 7 points ; then we have a periodic attractor of period 7.

On the contrary, if a = 1.4, all the initial points (in a certain ensemble) are attracted on a complicated structure of strange aspect (and strange properties)

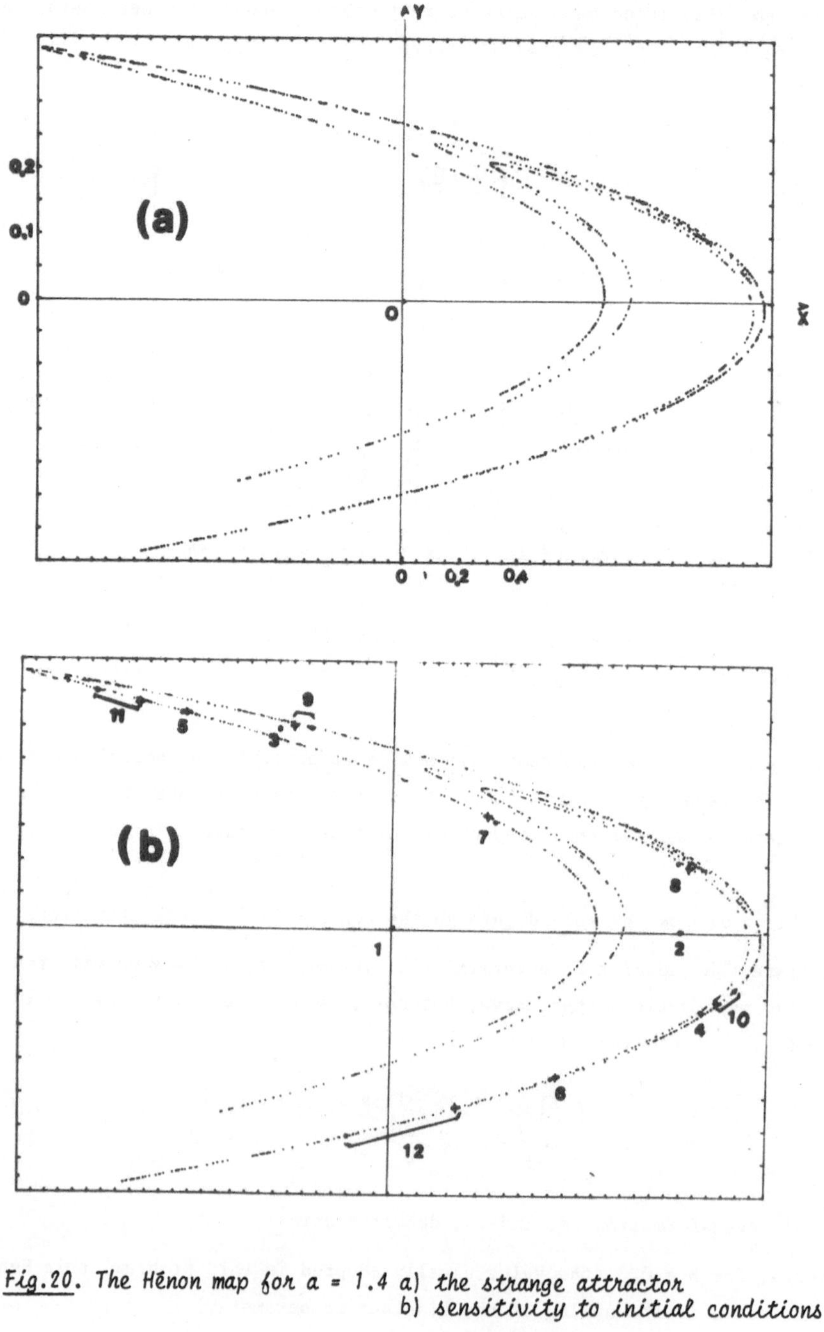

Fig.20. The Hénon map for $a = 1.4$ a) the strange attractor
b) sensitivity to initial conditions

$$\begin{cases} x_1 = y_1 = 0 & \bullet \\ x'_1 = y'_1 = 10^{-3} & + \end{cases}$$

(the numbers refer to the order i of the iteration).

which is called "strange attractor" (see fig. 20a).[21][22]

We refer the reader to specialized papers to know many of the curious properties of this strange object. But in what sense the successive points evolving on a strange attractor can represent turbulence or chaos ? Remember that an iterated map can be considered as the Poincaré section of a (fictive) phase space of higher dimension. Here the Henon map can be considered as the section of a phase space of dimension 3.

The fundamental property of such an attractor is the following : if we start with two very neighbouring points the successive iterations produce a divergence of the two corresponding trajectories (this divergence being exponential) (see fig. 20b). Now, remember that this divergence of the trajectories in a phase space is a definition of turbulence (see above). This example clearly shows that there is no discrepancy between a deterministic system and a chaotic or turbulent behaviour. This fundamental property of any strange attractor related to the divergence of very neighbouring trajectories is called "sensitivity to initial conditions" or S.C.I. It corresponds to the fact that if *in principle* the system is predictible (the same iteration performed twice gives indeed the same final state) in practice it is *not*, due to even infinitesimal errors which always affect the initial conditions and are exponentially amplified when the time runs.

Lorenz, who invented the first and most famous example of deterministic chaos through the appearance of a strange attractor in the solution of his 3 equations, gave this strinking illustration of the S.C.I. Referring to the general unpredictability for (long-term) weather forecasting, Lorenz said that even a very small change of initial conditions such as that produced by the motion of butterfly wings would have (unpredictible) large consequences for the next future behaviour of the atmosphere (considered as a dynamical system).

From the behaviour of the Henon map we can give a simplified but general definition of a strange attractor.

If the trajectories in a phase space are such as :

1) in a certain neighborhood of a given region A they are attracted into A and they remain inside A,

2) almost all pairs of trajectories initially neighbouring diverges inside A (S.C.I.)

then we deal with a strange attractor (fig. 21).

Last return to the Ruelle and Takens model

Let us remind that Ruelle and Takens predict in essence that 3 coupled oscillators with incommensurate frequencies may produce chaos through the appearance of a strange attractor.

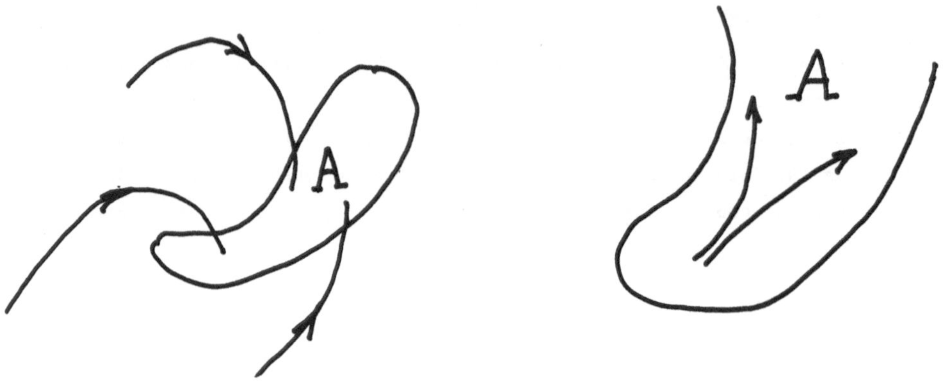

Fig.21. Attraction and S.C.I. of a strange attractor.

Initially,the two conditions of attraction and divergence of the trajectories might appear uncompatible ; indeed it is impossible to get a strange attractor if the phase space trajectories are attracted on an finite object at two dimensions : in a bounded two dimensional area the divergence of trajectories is not possible. For example, for a quasi-periodic attractor with two frequencies f_1, f_2 all the trajectories remain on the surface of a torus T^2. We can better visualize this surface by unfolding the torus into a flat area. (See Fig.22).

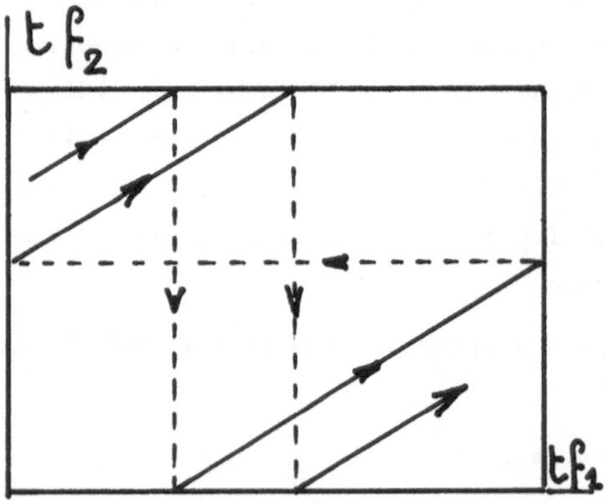

Fig.22. Unfolded torus T^2.

Consider a small perturbation on such a quasi-periodic regime tending to produce a slight divergence of the trajectories ; because these trajectories will soon have nowhere to go they have to tangle and in particular they intersect : that means that for a quasi-periodic regime the effect of an instability will be the synchronisation of the two oscillators (torus $T^2 \rightarrow$ limit cycle).

On the contrary if we want to keep the possibility to form a strange attractor we have to consider an attractive object with 3 dimensions : for example in the case of the attractor described by Rossler [15] (some kind on simplified Lorenz attractor) the trajectories can diverge on a two dimensional spiral, escape by emerging into space and return toward the centrum diverges again etc... (see fig.23).

For the same reasons, we can expect that, from a torus T^3 (3 frequencies), the

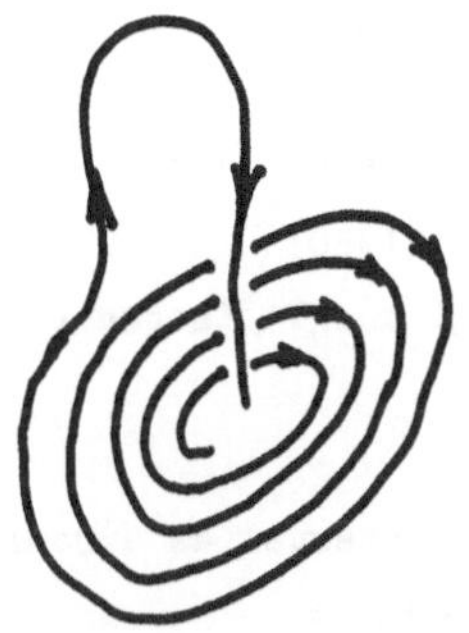

Fig.23. Scheme of the Rossler attractor.

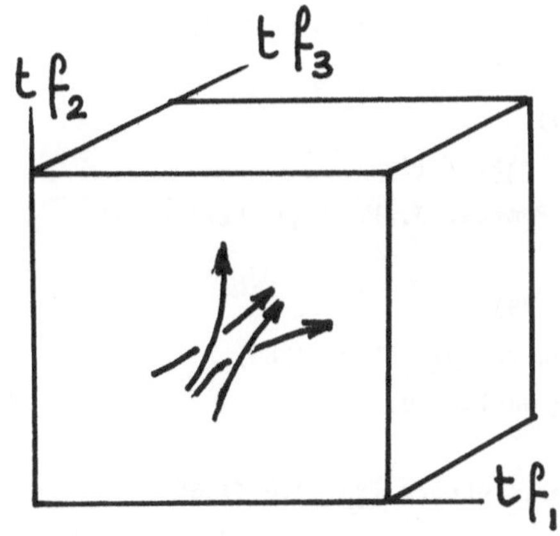

Fig.24. Hyperbolicity in a T^3 phase space.

attracting region being in 3 dimensions, an instability may lead to trajectories which can be attracted along a direction but diverge along the perpendicular other one (hyperbolicity) (see fig.24).

Then, on the contrary to what happens on a torus $T_2(f_1,f_2)$ whose instability leads to synchronisation (limit cycle), an instability on a torus $T_3(f_1,f_2,f_3)$ may lead to a strange attractor.

It is through topological consideration of this kind that one can understand the highly non intuitive idea that a deterministic system with 3-independent frequencies (3 degrees of freedom) may lead to turbulent behaviour.

We do not want to leave the reader on the (false) idea that deterministic chaos in hydrodynamic instabilities always occurs through strange attractors and according to the Ruelle-Takens mechanism (or -at least- through biperiodism). Añother kind of route -through intermittencies- has been proposed by Pomeau-Manneville [16] which has been experimentally checked up to a quantitative level by the Saclay group [17] . In this later case, there is not a strange attractor, though the chaos is deterministic (phase space at 3 dimensions [7]). A third kind of route has been proposed by Feigenbaum [18] and corresponds to a cascade of sub-harmonics bifurcation. The more precise experimental evidence of this kind of route is due to the ENS Paris group [19] and CISE (Milano) group [20].

REFERENCES

[1] L.D. Landau and E.M. Lifshitz "Fluid Mechanics", Pergamon Press, London 1959.

[2] D. Ruelle and F. Takens, Comm. Math. Phys. 20, 167 (1971).

[3] P. Bergé and M. Dubois "Systems far from Equilibrium" Sitges 1980, Pringer-Verlag ed. by L. Garrido, p.381.

[4] P. Bergé "chaos and order in nature" Schloss Elmau 1981, Springer-Verlag ed. by H. Haken p.14.

[5] P. Bergé and M. Dubois, J. Physique Lettres 40, L505 (1979).

[6] M. Dubois and P. Bergé, Phys. Letters A76, 53 (1980)

[7] M. Dubois, this volume and M. Dubois, P. Bergé and V. Croquette, Comptes Rendus Acad. Sci Paris, Sept. 1981.

[8] E. Lorenz, J. Atmos. Sci 20, 130 (1963).

[9] P. Collet, J.P. Eckmann "Iterated maps on the interval as dynamical systems" P. Ph. ed. A. Jaffe, D. Ruelle, Birkhaüser 1980.

[10] M.G. Velarde in "Non Linear Phenomena at Phase Transitions and Instabilities" NATO Advanced Study Institute, Norway (1981).

[11] M. Dubois in "Symmetries and Broken Symmetries in Condensed Matter Physics" ed. by N. Boccara (IDSET Paris) 1981.

[12] M. Dubois and P. Bergé, J. Physique 42, 167 (1981).

[13] P. Bergé, M. Dubois and V. Croquette "Convective Transport and Instability Phenomena" Euromech 138, Karlsrule 1981 ed. by H. Oertel and J. Zierep.

[14] M. Henon, Commun. Math. Phys. 50, 69-77 (1976).

[15] O.E. Rossler, Phys. Letters A57, 397 (1976).

[16] P. Manneville and Y. Pomeau, Phys. Letters 75A, 1 (1979).

[17] P. Bergé, M. Dubois, P. Manneville and Y. Pomeau, J. Physique Lettres 41, L341 (1980).

[18] M.J. Feigenbaum, Phys. Letters 74A, 375 (1979).

[19] J. Maurer and A. Libchaber, J. Physique Lettres 41, L515 (1980).

[20] M. Giglio, S. Musaggi and U. Perini, to be published.

[21] R. Shaw, Z. Naturforsch 36a, 80 (1981).

[22] A. Arneodo, P. Coullet and C. Tresser, Phys. Letters 79A, 259 (1980).

EXPERIMENTAL ASPECTS OF THE TRANSITION TO TURBULENCE IN RAYLEIGH-BÉNARD CONVECTION

M. Dubois
Service de Physique du Solide et de Résonance Magnétique
CEN-Saclay - 91191 Gif-sur-Yvette Cedex
France

Abstract

In a small box (aspect ratio $\simeq 2$), a high Prandtl number fluid submitted to Rayleigh-Bénard convection undergoes a cascade of different spatial order, i.e. presents different stable structures when the Rayleigh number is increased. Inside each structure, we observe an evolution of the temporal order, from generally a stationary to a turbulent state, through a small number of bifurcations. Some experimental examples illustrate the properties of these temporal states, in particular Poincaré sections have been drawn in the phase space and have provided new dynamical information.

During the last few years, there has been a considerable evolution in the concepts of turbulence, in particular those related to the appearance of turbulence in systems with a small number of degrees of freedom. The experimental results we describe in this chapter, are all devoted to such systems and we will show how the Rayleigh-Bénard convection in finite geometry furnishes good examples of the approach of such turbulent states. The degrees of freedom, which are directly related to the independent variables of the considered system, may be of very different nature. In the case of the convection in a small box, we are dealing essentially with temporal degrees of freedom ; we hope that spatial modes will be stabilized by boundary effects and then that the phase turbulence will be inhibited [1]. The experiments confirm in great part this expectation, and, further, the study of convection in a small box is shown to be a good candidate for examining periodic time dependent phenomena and turbulent ones when the Rayleigh number R_a is increased.

A. EXPERIMENTAL CONDITIONS

The major part of the experimental results which are given below have been obtained in a small cell with dimensions $Lx = 2d$, $Ly = 1{,}2d$ where $d = 2$ cm is the depth of the fluid layer. (see fig.1) A temperature difference ΔT is applied between the two horizontal plates. The used fluid is silicone oil, with viscosity $\eta = 0.1$ stoke at 25°C and thermal diffusivity $D_T = 7.65\ 10^{-4}\ \mathrm{cm}^2\mathrm{s}^{-1}$; so that the Prandtl number Pr is 130.

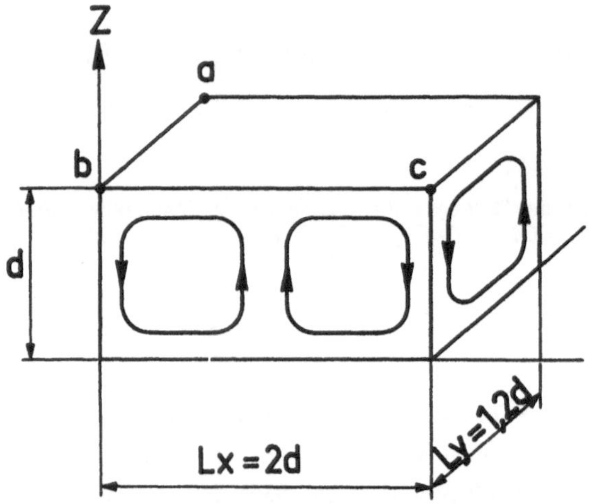

Fig.1. Scheme of the small box, used for the reported experiments and in which the convective structure -2x rolls, 1y roll- has been drawn.

The survey of the convection state is performed using different technics, using measurements of one or the other of the convective variables i.e. the local velocity or the temperature perturbations [2]. Local velocity measurements were performed by laser Doppler anemometry. Differential interferometry pictures, realized in a simple way [3], give temperature isogradients $\nabla_z T$ or $\nabla_x T = f(x,z)$ or $f(y,z)$ (see for example fig.2). The great interest of such pictures is that : 1) they give instantaneously a global view of the spatial arrangement in the whole cell (number of rolls, ascending or descending motions, etc.) and 2) they enable the temperature perturbations giving rise to the time-dependent properties to be visualized. We also make use of the deflection of a light beam, due to the temperature gradients. So that, a large parallel beam,

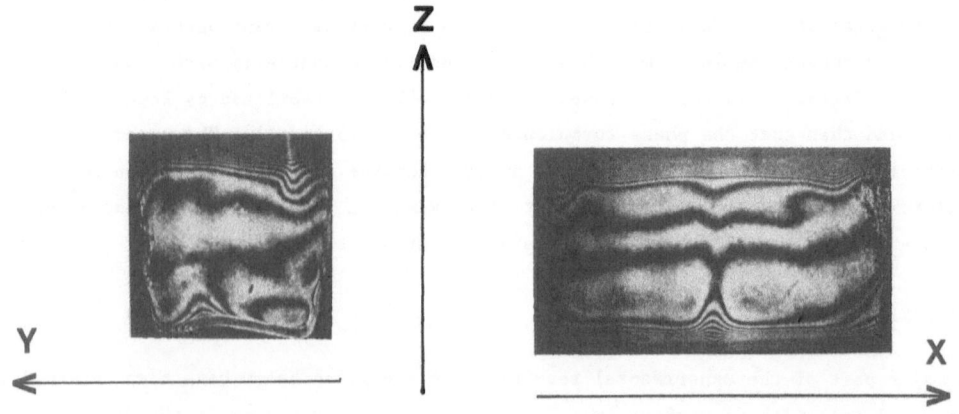

Fig.2. Differential interferometric images $\nabla_z T = f(xz)$ and $f(y,z)$, viewed at $R_a/R_{a_c} \simeq 250$ with the structure 2x rolls, 1y roll. We can recognize clearly the thermal boundary layers.

after having crossed the convective fluid layer, gives a "shadowgraphic" image [2], which reflects the spatial and dynamical properties of the fluid ; the measurement of the intensity in a point of this image, easily obtained using a photodiode, then gives detailed informations concerning the actual time-dependent properties.

B. *SPATIAL ORDER*

One of the properties of convection is its spatial order. So, near the threshold R_{a_c} of the appearance of the convective motions, we may observe parallel straight rolls in large boxes under certain geometrical conditions [1]. A wavelength Λ may be defined as the distance between two adjacent upwards or downwards motion. At R_{a_c}, $\Lambda_c \simeq 2d$. When R_a is increased, the variation of Λ depends strongly on the geometry of the cell confining the fluid, but in large boxes, the natural tendancy is always towards an increase of the wavelength [4]. In a small box, as mentioned above, the lateral boundaries act to limit the possible spatial structures ; nevertheless, when R_a is varied, many structures are available. At the threshold, we observe two rolls of axis parallel to the short side Ly of the cell (2x rolls), then at $R_a/R_{a_c} \simeq 20$, the motion becomes three-dimensional : two new rolls appear perpendicularly to the first rolls (2y rolls). This well defined structure is stable and stationary until $R_a/R_{a_c} \simeq 200$. (fig.3).

Beyond this value, the situation is more complicated and different spatial structures may be observed at the same R_a number, depending on the thermal history of the convection and also on some unidentifiable perturbations, which often prevent the formation of the desired structure. Nevertheless, from the different situations we have observed, we can deduce a general tendancy, a scheme of which is shown on fig.3 where have been drawn for comparison the curve of marginal stability and the Busse balloon [5] , calculated for the case of infinite geometry.

First, each structure is generally stable in a relatively great range of R_a numbers. Secondly, the higher the value of the R_a number, the higher is the actual wavenumber (if we can speak of wavenumber with only a very small number of rolls), i.e. there is a tendancy for the structure to contain a greater number of rolls. Then, the structure I is composed of two x rolls and one y roll, the structure II of two x rolls and two y rolls and then the structures III and IV have respectively three and four x rolls with two y rolls. A remarkable fact is that the transition from one structure to another is always performed, in particular at high R_a numbers, through a strong spatial chaos, where the spatial organisation is destroyed and the structure continuously changes. But always, at fixed R_a number, this transitory chaotic state leads to a stable organised structure after a time which may last from a few hours to a few days.

In fact, on the scheme of the figure 3, the evolution of the convective structure is highly simplified. Structures, which seem to be roughly identical, present specific small differences (sign of the velocity, amplitude of the spatial harmonic modes, perturbations of the velocity field at the boundaries) which have a great influence on their stability and on their dynamical behaviour. So at the same R_a number, a great number of structures -and also of time dependent properties- may be observed depending on the spatial arrangement, "locked" from the spatial chaotic phase.

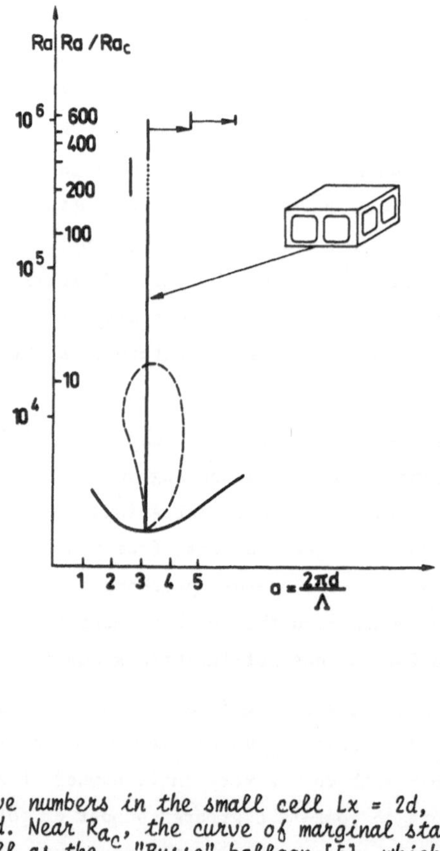

Fig.3. Observed wave numbers in the small cell $Lx = 2d$, $Ly = 1.2d$, when R_a is varied. Near R_{a_c}, the curve of marginal stability has been drawn as well as the "Busse" balloon [5], which bounds the stability of bidimensional structures in large boxes.

C. TIME-DEPENDENT PROPERTIES

The observation of the time dependent properties of the different structures reveals that each one has its own dynamical evolution from a stationary state (or periodic one) to a low turbulent one or to the spatial chaos, which destroys the structure. So, as mentioned previously, at the same Rayleigh number, we may observe a stationary convection, or different periodic regimes or turbulence depending on the actual spatial structure.

Nevertheless, for a fixed and given structure, the evolution with R_a is well determined and reproducible. Our experiment with fluid of high Pr number, has shown essentially two types of evolution (fig.4) : 1) a monoperiodic regime, turbulent intermittencies, 2) a monoperiodic regime, biperiodic regime then turbulence preceded or not by a locking state between the two initial frequencies. We have to note that the turbulent states, following temporal bifurcations in a fixed spatial order, are not stable over a wide range of Ra numbers and are always followed by a new temporal order in a new structure. We will now describe with two examples the specific properties of these two types of evolution.

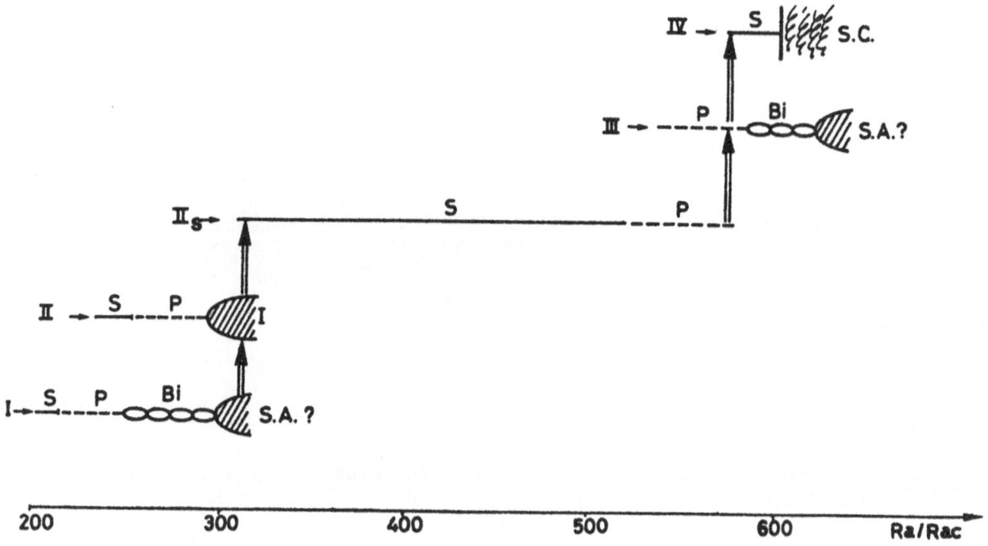

Fig.4. Cascade of successive spatial and temporal order in the small box (L_x = 2d, L_y = 1.2d) with R_a/R_{a_c}.

1) *Turbulent intermittencies*

When the structure can be schemed by 2Y rolls and 2X rolls, one being greater than the other ($a_x \simeq 2.6$), (structure II on the fig. 4) the convection is stationnary until $R_a/R_{a_c} \simeq 250$, i.e. local velocity and temperature perturbations are constant with time at fixed R_a number. Then, when R_a is increased, the velocity becomes monoperiodic at a frequency f_o. At $R_a/R_{a_c} \simeq 295$, some interruptions appear in the regular oscillations of the velocity, in the form of perturbations of the velocity amplitude.

see fig.5 and [6] . These perturbations occur at random but the mean time lapse between two events varies as $(R_a - Ra_I)^{-0.5}$ near the threshold Ra_I ($295\,Ra_c$) of this intermittent behaviour. So the turbulence appears continuously following the increase of the number of irregular events [7].

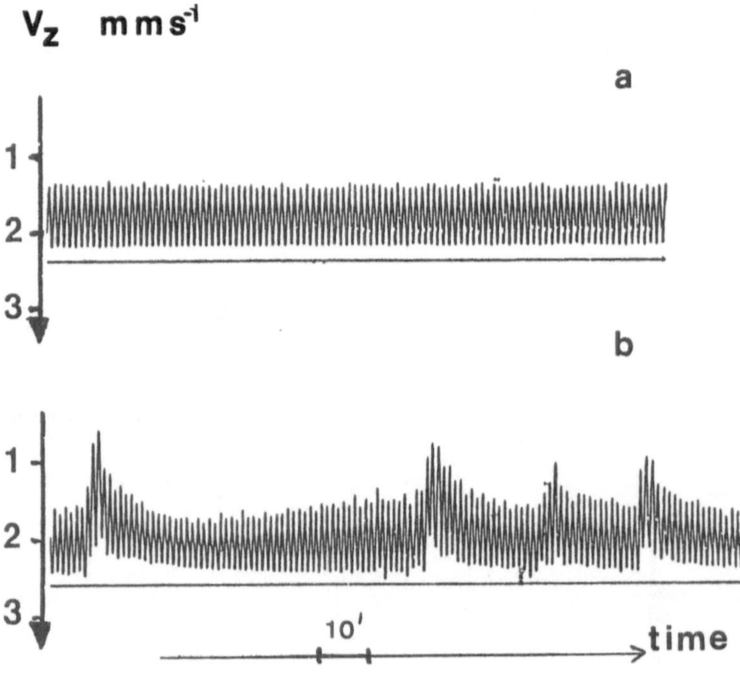

Fig.5. Time dependence of the vertical velocity component V_z measured near the center of the cell : a) Ra/Ra_c = 270 ; b) Ra/Ra_c = 300.

The differential interferometric images clearly demonstrated the physical mechanisms of the periodic behaviours. In the actual structure, the periodic variation of the velocity at the frequency f_o is due to the periodic formation of a droplet near a lateral boundary of the cell, then this droplet is advected and a new droplet is formed and so on. When the intermittencies are present, they are related to the formation of an abnormal droplet.

The dimensions of the phase space in this case of turbulence is probably 3 : below the threshold of intermittencies, we are in the presence of a limit cycle ; beyond the threshold, a new unstable direction is necessary to account for the experimental results. Otherwise these results are in good agreements with calculations performed by Y. Pomeau and P. Manneville [6].

2) *Temporal turbulence through biperiodic regime*

Many structures exhibit the familiar sequence : monoperiodic regime - biperiodic regime - low turbulent state. One of these evolutions corresponding to the strcuture I has been described in detail in [8]. Here we report the observed behaviour of the convection with the structure III, schemed by 3x rolls and 2y rolls. In this case, the time variation of the velocity is monoperiodic (frequency f_1) until $R_a/R_{a_c} \simeq 570$, when R_a is increased, we then observe a biperiodic regime (frequencies f_1 and f_2), followed by a frequency locking ($f_1/f_2 = 6$) at $R_a/R_{a_c} \simeq 610$. A further increase of R_a leads to a low turbulent state, with a threshold $R_{a_T} \simeq 625\ R_{a_c}$. These different time properties have been studied by the Fourier spectra of the velocity variations which can be seen on fig.6. It is obvious that the two oscillators, at the frequencies f_1 and f_2, are coupled for the spectra are composed of many combinations $nf_1 + mf_2$.

The first oscillator, responsible for the frequency f_1, in the monoperiodic regime, is given by a local instability in the cold boundary layer, near the lateral boundary in descending motions. This instability gives rise to the formation of a cold plume which carries away the convective flux with it, enhancing strongly the velocity amplitude on its stream line. The second oscillator could be given by a structural instability between two harmonic spatial modes. The coupling is due to the interactions velocity amplitude-temperature perturbations.

When the turbulence appears in the spectrum (broad band noise near zero frequency), a memory of the preceding order remains, as shown by the presence of some peaks which are not too broadened. But we cannot say anything about the mechanism which gives rise to its erratic contribution. It can be asked if there is a new third frequency, in agreement with the Ruelle and Takens model or if there are complex intermittencies. More precisely, what is the new destabilizing degree of freedom ?

A present, we have not definitely answered these questions. But it seems highly probable that a great deal of information could be obtained from direct measurements in the phase space.

D. PHASE SPACE DIAGRAMS

An ideal way to construct the phase diagrams would be to measure separately the independent variables of the motion, which however requires prior the knowledge of these variables (or at least how many they are) and then to look at the trajectories of the representative point in the phase space. In a biperiodic regime, with two coupled oscillators, we may expect that a good representation of the motion would be given at least by three variables, the representative point describing a torus T^2 (see P. Bergé this volume). It is not easy to draw trajectories in a 3d phase space and if we look only at two variables, we obtain complex curves, which are the projections on a plane of all the trajectories on the torus T^2 which represents the observed dynamical state. We can see one example on fig.7. On this curve recorded on a xy plotter the abscisse is a measure of a local temperature gradient, T (by looking at the intensity in a point x_1, z_1 of the shadow-graphic image as previously

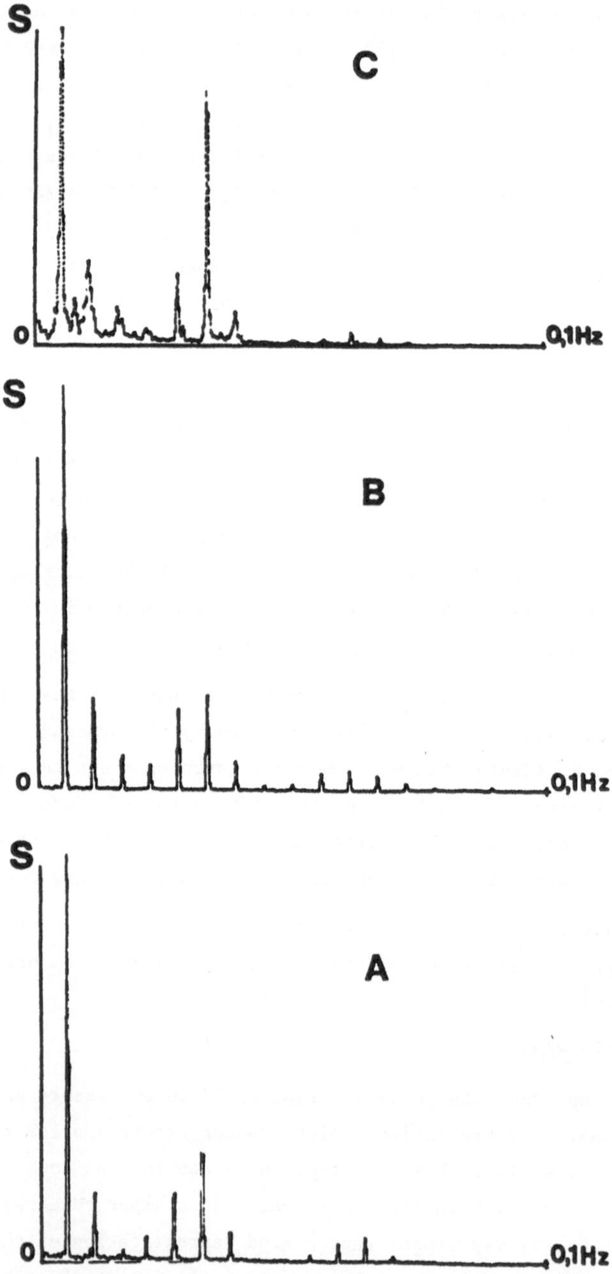

Fig.6. Fourier spectra of the time variations of the velocity component v_z measured in a point of the structure III. A) R_a/R_{a_c} = 592 quasiperiodic regime. B) R_a/R_{a_c} = 610 frequency locked regime C) R_a/R_{ac} = 636 turbulent regime.

described) and the ordinate is the time derivative $\dot{T}$. The diagram is constructed as the time is flowing.

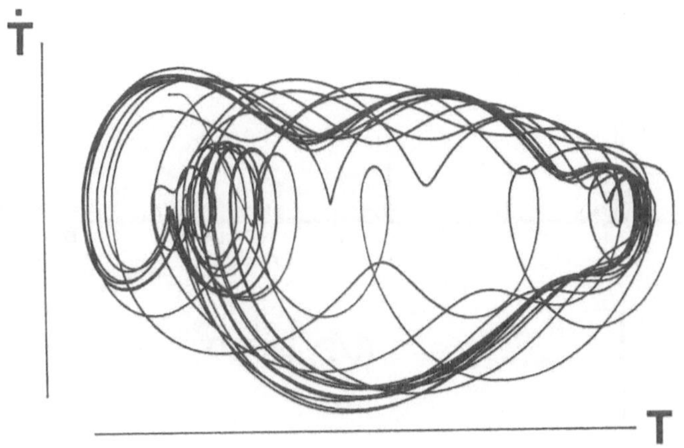

Fig.7. 13 successive phase loops $\dot{T} = f(T)$, recorded in a quasi periodic regime at R_a/Ra_c = 569. f_1 = 26.2 10^{-3} Hz ; f_2 = 9 10^{-3} Hz.

To reduce the dimensions of the phase space and obtain diagrams which can be more easily interpreted, we have constructed Poincaré sections of the preceding curves. To do this, points of the curve were taken only at discret times, t_o, $t_o+\tau_1$, $t_o+2\tau_1$, ...$t_o+n\tau_1$ where τ_1 is the period of one of the actual oscillator. Our aim is to obtain in this manner a section in the phase space by a plane perpendicular to the axis which represents the dynamical variable of the corresponding oscillator. (the phase φ_1 of the first oscillator for example).

Approach of a locking state [9]

In the evolution of two coupled oscillators, many examples may be found of a locking state observed in the course of the evolution of the two oscillators with the R_a number. In a given structure, which presents some similarity with the structure II_s, such a biperiodic regime was observed. Let us look at the convective state by the Fourier spectra (fig.8) of the temperature perturbations, measured through a local intensity in the shadowgraphic image (T). At R_a/R_{a_c} = 569, the spectrum is composed of the two frequencies f_1(26,2 mHz) and f_2(9 mHz) and some combinations $nf_1 + mf_2$. We can remark that f_1 is near $3f_2$; the lower frequency (0.8 mHz) corresponds to the difference $\delta f = 3f_2 - f_1$ and represents the distance to the locking state. When the R_a number is increased, this lower frequency begins smaller and smaller, particularly at R_a/R_{a_c} = 590, δf is inferior to the frequency resolution of the spectrum. At R_a/R_{a_c} = 593, the two frequencies are locked and all the peaks of the Fourier

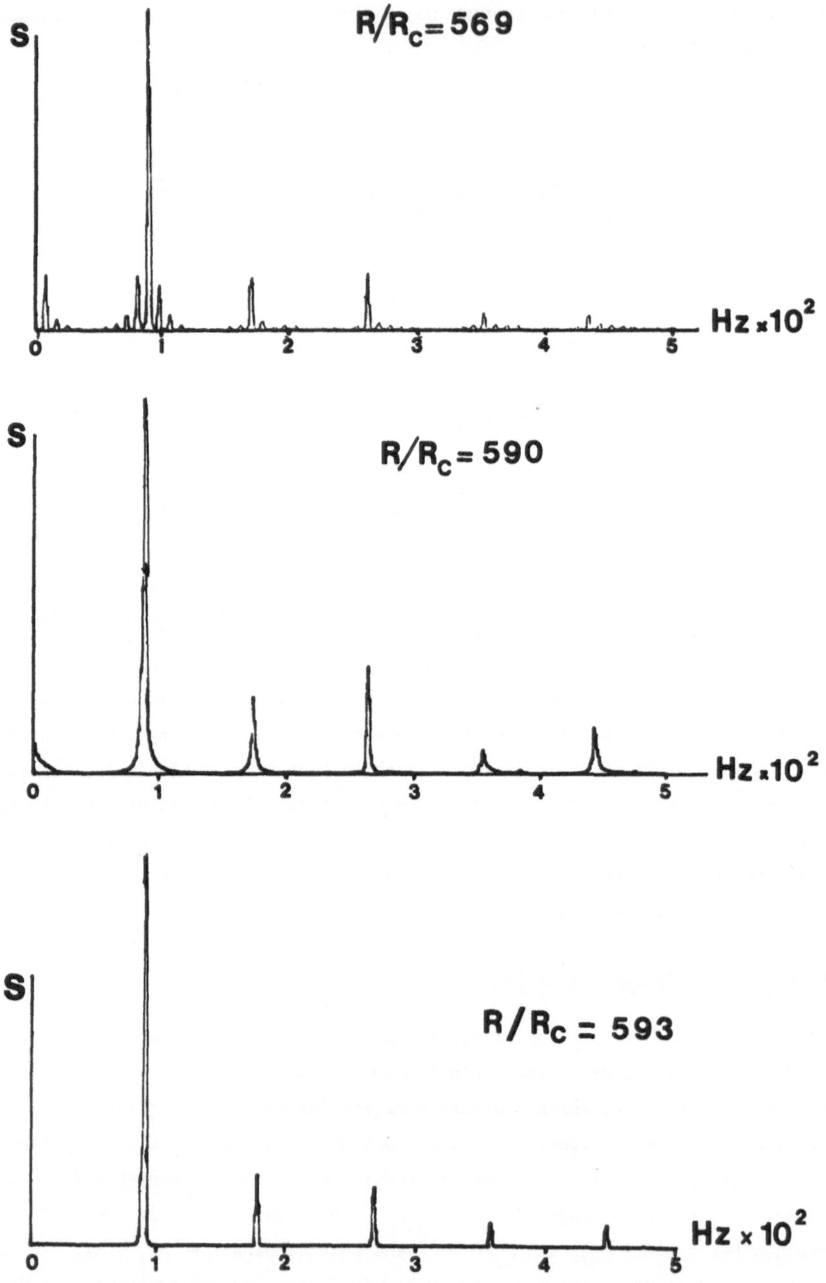

Fig.8. Fourier spectra of the time variations of the temperature perturbations T. 1) Ra/Ra_c = 569, f_1 = 26.2 10^{-3} Hz ; f_2 = 9 10^{-3} Hz.
2) Ra/Ra_c = 590; f_1 = 26.7 10^{-3} Hz, f_2 = 8.9 10^{-3} Hz.
3) Ra/Ra_c = 593; f_1 = 26.8 10^{-3} Hz, f_2 = 8.9 10^{-3} Hz.

spectrum are harmonics of f_2. The evolution of the two frequencies f_1 and f_2, when R_a is varied, is shown on fig.9 and is quite classical.

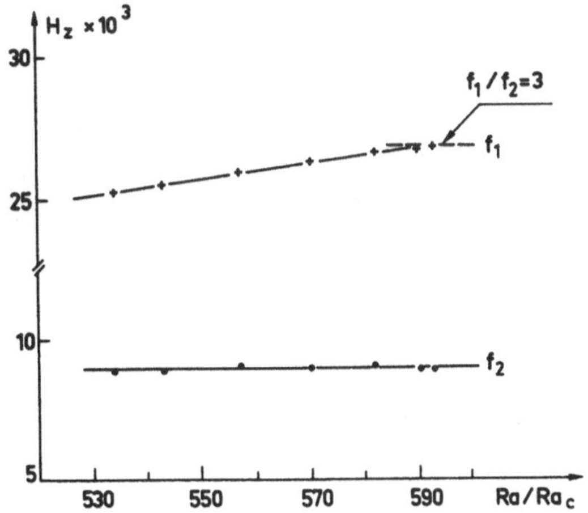

Fig.9. Variations of the frequencies f_1 and f_2 with R_a/R_{a_c} in the structure studied by Poincaré sections.

Let us look now in the phase space, through Poincaré sections. The first point is to obtain a good temporal reference at the frequency f_1, choosen for the "stroboscopy" of the measurements. As mentioned above, we were generally able to point out the thermoconvective oscillators by the study of interferometric images. Here, the frequency f_1 is due to the periodic formation of a droplet from the cold boundary layer ; the frequency f_2 is probably due, as in a previously described case, to a structural instability between harmonic spatial modes. The first oscillator is well localized and it is easy to do velocity measurements, which practically see this oscillator alone. After filtering, we obtain then a pure sinusoïdal signal at the frequency f_1, which will serve as reference for the discrete time measurements of the Poincaré sections. The variables of these sections (T and $\dot{T}$ defined previously) contain all the informations concerning the convective dynamics. T and $\dot{T}$ were fed respectively to the x and y input of a plotter.

A first result is shown fig. 10. It represents the Poincaré section of the trajectories given in fig.6, a section which provides a well defined loop from a very complicated information. This section corresponds to the described periodic regime at $R_a/R_{a_c} \simeq 569$, (see fig.8a). Although the loop is very simple, we have to remark the higher density of the points in the regions named A, B, C. When Ra is increased, this particularity is reinforced, as we can see on fig.11. The section -drawn at $R_a/R_{a_c} \simeq 590$ and corresponding to the spectrum fig. 8b- shows three very dense

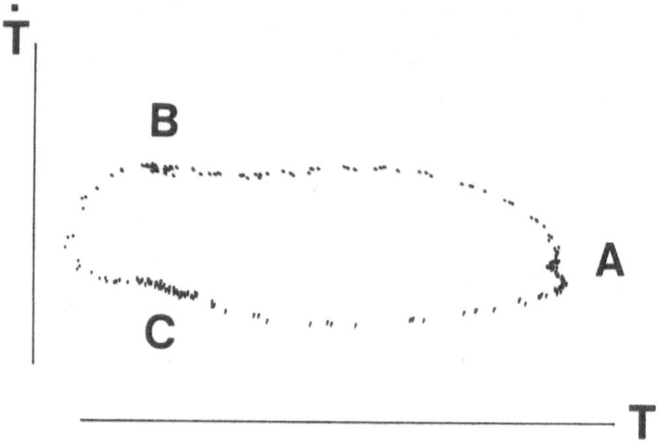

Fig.10. Poincaré section corresponding to the loops of the fig.7, sampled at the frequency f_1 (200 points) the regions A,B and C are that of higher density.

Fig.11. Poincaré section obtained with the same conditions at that of the figure 11 but at Ra/Ra_c = 590 and with 1500 points.

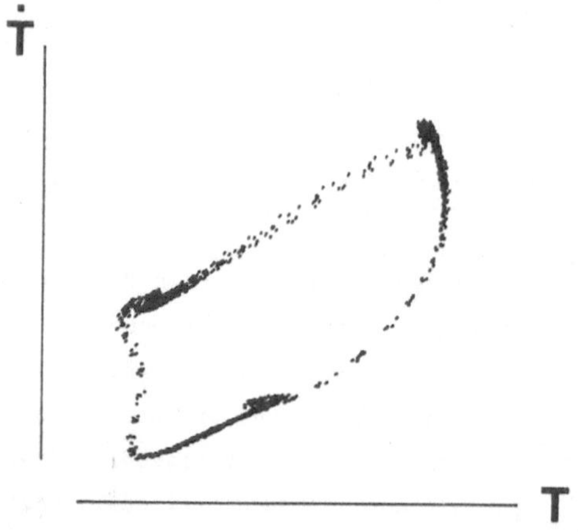

regions, joined together by scarce points. It is obvious that the presence of the three regions correspond to the approach of the locking state $f_1/f_2 = 3$. But moreover, it is fascinating to look at the construction of the loop. During a long time, the points lie in the regions A, B, C and then, rapidly, the rest of the loop is described, the representative point going from A to B, B to C and C to A, where it remains again for a long time. If we translate this behaviour in term of phase, we can say that when the points are in A, B and C there is some temporary locking : the phases of the two oscillators are close and vary very slowly. When the phase difference between the two oscillators is sufficiently high, this difference varies rapidly by $2\pi/3$ to reach a new semi-locked state as we can see on fig.12 where have been recorded the successive ordinates $\dot{T}$, measured, as in the Poincaré section, at the discrete times t, $t+\tau_1$, $t+2\tau_1 \ldots t+n\tau_1$. The "unlockings" appear periodically at the frequency $\delta f = 3f_2 - f_1$.

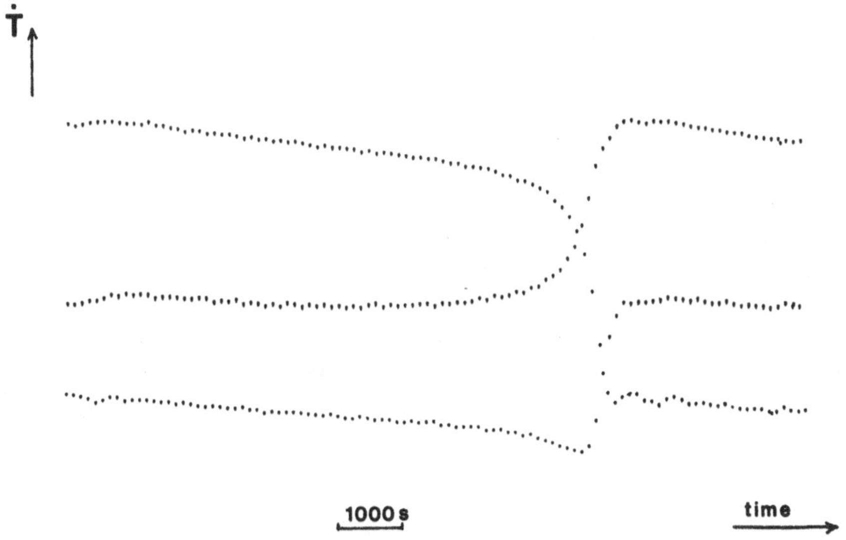

Fig.12. Time successive ordinates Ṫ of the Poincaré section of the fig.11. $Ra/Ra_c = 590$.

At $Ra/Ra_c = 593$, $\delta f = 0$, the locking state is achieved and the Poincaré sections involve only three points.

DISCUSSION

It is clear that direct measurements in the phase space have brought specific informations that could never have been obtained using Fourier spectra ; so the state of dynamical locking that we have pointed out have never been mentioned. Moreover, as previously mentioned, the study of a dynamical state requires the knowledge of the number of the independent variables and the Poincaré sections is of a great help for this. In the example that we have given on the approach of a locking state, it seems highly probable, the loop being unic and without width outside the experimental noise, that the representative phase space has three dimensions. It is not always the case ; indeed in different biperiodic regimes -obtained with different structures-, we observed complicated Poincaré sections, although the Fourier spectra of the velocity or the temperature perturbation do not show any noise. These sections could be explained by a phase space in four dimensions.

It is probably the case also in the Poincaré section we draw in the temporal disorder, which follows the locking state observed in the structure III* .(discussed in the paragraph C2). This diagram is shown fig.13. It is somewhat complex, and its interpretation requires complementary information but we can recognise some topological properties, specific of certain strange attractors.[10][11].

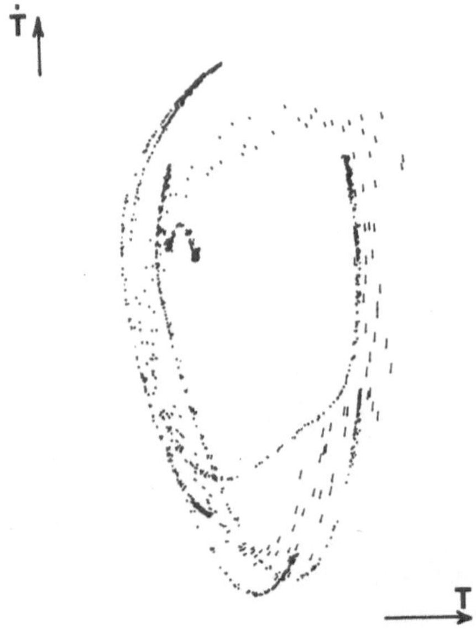

Fig.13. Poincaré section $\dot{T}=f(T)$ sampled at the frequency f_1 at $R_a/R_{a_c}=636$ in the low turbulent state of the structure III, the corresponding spectrum velocity fluctuations is that of the fig.6c.

From the examples we have shown from convection with high Prandtl number fluid in confined geometry, we have seen the approach of turbulent states involving a small number of dynamical variables, three in the case of intermittencies, three or four in the case of biperiodic regimes. It is important to note at this stage that another route to turbulence has been observed by other groups in convection with low Prandtl number fluids : the route through subharmonic bifurcations. This one is well described in [12][13][14]and benefits of many theoretical models [11][15][16] where the phase space dimension is generally three.

We have described the approach of turbulence by the evolution of convective states, but we should not forget that many other systems - chemical reactions [17] electrohydrodynamical instabilities [18] forced oscillator [19] and so on, exhibit similar behaviour.

----

*Unfortunately, the structure which was present in the study-previouly described of the approach of a locking state disappeared soon after the locking state, when R_a was increased.

BIBLIOGRAPHY

[1] P. Bergé, in "Chaos and Order in Nature" Schloss Elmau 1981 ed. by H. Haken (Springer-Verlag) p.14.

[2] P. Bergé, M. Dubois "Scattering techniques applied to supramolecular and non-equilibrium systems" ed. by S.H. Chen, B. Chu - Plenum Press 1981.

[3] P. Bergé, M. Dubois, J. Phys. Lettres 40, L.505 , (1979).

[4] E.L. Koschmieder, S.G. Pallas, Int. J. Heat Mass Transfer 17, 991 (1974).

[5] F.H. Busse and J.A. Whitehead, J. Fluid Mech. 47, 305 (1971).

[6] P. Bergé, M. Dubois, P. Manneville, Y. Pomeau, J. Phys. Lettres 41, L341, (1980).

[7] M. Dubois in "Symmetries and Broken Symmetries in condensed matter Physics" edited by N. Boccara (IDSET Paris) 1981.

[8] M. Dubois and P. Bergé, J. de Phys. 42, 167 (1981).

[9] M. Dubois, P. Bergé, V. Croquette, Comptes Rendus Acad. Sci. to be published.

[10] M.I. Rabinovich, Sov. Phys. Usp. 21 (5), 443 (1978).

[11] B.A. Huberman and J.P. Crutchfield, Phys. Rev. Letters 43, 1743 (1979).

[12] J. Maurer and A. Libchaber, J. de Phys. Lettres 40, L419 (1979).

[13] J.P. Gollub and S.V. Benson, J. Fluid Mech. 100, 4119 (1980).

[14] M. Giglio, S. Musazzi and U. Perini, to be published.

[15] J.B. Mc Laughin, J. of Stat. Physics 24, 375 (1981).

[16] M.J. Feigenbaum, Phys. Letters 74A, 375 (1979).

[17] J.C. Roux, A. Rossi, S. Bachelart and C. Vidal, Physics Letters 77A, 391 (1980).

[18] B. Malraison, P. Atten, Comptes Rendus Acad. Sci. 292, 267 (1981).

[19] V. Croquette, C. Poitou, Comptes Rendus Acad. Sci. 292, 1353 (1981).

HEAT FLUX IN CONVECTIVE INSTABILITIES

M. Zamora
Departamento de Termología
Universidad de Sevilla
Sevilla (Spain)

I. INTRODUCTION

The study of thermodynamics systems in situations very far from equilibrium is at present one of the more important branches of research in macroscopic physics. A system designed for remaining in a steady state in a situation next to equilibrium, i. e. bearing the fluxes induced by the generalised forces to which it is submitted, can be considered to lose equilibrium when the quantity of these forces increases step-by-step. In such circumstances, the expected behaviour of the system must be modified by the interaction between the applied forces and those forces due to external actions (such as field forces). The result is a set of very important phenomena that are called instabilities. These instabilities are shown in essential changes of the internal structure of the system and by important modifications in the fluxes that pass through it.

Some questions can already be established if we continue the logical sequence of the facts:

a) What value of the generalised force is necessary to initiate the instability?.

b) When the generalised force increases, does the instability begin slowly or abruptly?.

c) What role do the internal forces of the system play in these phenomena?.

During this seminar, we shall answer some of these questions, particularly for a very interesting and instructive type of instability.

The examples to be found of this type of phenomena we are describing vary greatly, from the famous chemical instability of Zabotinsky to the no less surprising electrical instability of the ballast resistor, and the convective instabilities of fluids, examination of which is a preliminary step for the study of one of the oldest and most important problems in physics: the study of turbulence. In this seminar, we shall be occupied specifically with this last problem, i. e. the instabilities provoked by a temperature gradient in the presence of a gravitational field. In this case, both forces have the same tensorial character and, by means of an adequate arrangement, they can interact with each other when they are applied to the mass unit in opposite directions. As we shall see, the internal forces play an outstanding role. We will centre our attention on the resulting flux from this interaction of forces; that is, the heat flux and its changes, always taking as a reference point the steady state next to equilibrium, extrapolated to situations very far from it, which are those that most interest us.

As this seminar is an initiatory one, we will be confined for the moment to sketching the current state of the problem and we will indicate the future possibilities. We intend to do all this with the following outline:

- The theoretical basis of heat flux in thermal convection, which includes a rigurous treatment of the problem, as well as an intuitive approach which will allow us to explain the physical significance of all the parameters that appear.
- The most outstanding experimental results to date, with a little discussion of their significance with relation to known theories.

II CONVECTIVE INSTABILITIES

Let us consider a layer of fluid, with expansion coefficient α positive, placed horizontally on a solid surface, with high thermal conductivity, and contained by adiabatic lateral walls. If this layes of fluid is heated from below, two clearly different phenomena are induced:

a) a vertical heat flux due to the temperature gradient,

b) a density gradient, also vertical, due to the character of the expansion coefficient and to the distribution of established temperatures.

The second effect provokes decantation forces in the fluid, due to the gravitational field to which it is submitted. Consequently, we observe a tendency to internal motion trying to regenerate a density distribution consistent with the gravitational field. Against these forces, and consequently against the internal motion that they produce, we find the opposing internal dissipation forces of the fluid, that must be overcome for the internal motion to take place.

We see, therefore, two separate situations characterised by the following facts:

a) Near-equilibrium situation (temperature gradient applied near zero): The internal motion in fluid is not produced, since the internal dissipation forces are higher than those derived from gravity. There is only vertical heat flux due to pure conduction.

b) Far from equilibrium situation (high temperature gradient) : The internal motion is carried out since the gravitational forces are higher than dissipational forces. The heat flux is strengthened by the movement of the mass, and what Rumford called thermal convection appears.

Briefly, the second situation is a convective instability that is called by different names, depending on the nature of the top surface which holds the fluid. The passage between both situations is a transition to instability, which is customarily characterised by a critical value of the temperature increment applied, corresponding to the equality of gravitational and dissipative forces.

When the top surface of the liquid is free, in contact with air, the instability is called Bénard's problem, since he was the first to study it experimentally. The internal motion in the instability is the upward or downwrd current located in the geometrical centres of regular hexagons, around whose perimeters the corresponding currents necessary for conservation of the mass are formed. When the top surface is a rigid boundary the instability is called Bénard-Rayleigh's problem, since Lord Rayleigh, in trying to explain the experimental results of the Bénard problem, actually interpreted this case. The internal dissipation force is the viscous force and the pattern that generates the internal motion consists of uniformly spaced upward and downward currents that generate horizontal rolls composed of very different forms. The distance between two upward or downward currents is known as the wavelength of the structure.

The peculiar variables of the system, apart from the gravity acceleration, g, and the applied difference of temperature, ΔT, are as we have said, the lateral geometry and the physical properties of the fluid. The former is characterised by the aspect ratio, L/h, that is the quotient between characteristic lateral length dimension, L, and the height of the fluid layer, h. The interveningproperties of the fluid are the mean density, ρ_o, the thermal diffusivity, κ, and the viscosity coefficient, η, with special mention of the kinematic viscosity, ν (= η/ρ_o).

Nevertheless, the physical phenomenon can be more easily described with nondimensional combinations of the above mentioned variables or "numbers". These numbers are essentially two, the Rayleigh number:

$$Ra = \frac{\alpha\, g\, h^3\, \Delta T}{\kappa\, \nu}$$

and the Prandtl number

$$P = \frac{\nu}{\kappa}$$

We must add a third number related to heat flux, the Nusselt number:

$$Nu = \frac{q_{cond} + q_{conv}}{q_{cond}}$$

where q_{cond} and q_{conv} are the heat flux, i. e. heat transported per unit of surface and time, due only to conduction and convection effects, respectively.

III. HEAT FLUX IN THE BENARD-RAYLEIGH PROBLEM

A. Exact development. When the equations of hydrodynamics are applied to fluid submitted to Bénard-Rayleigh problem conditions, we frequently accept a very good aproximation developed by Oberbeck-Bussinesq. In this approximation it is assumed that forces provoking the convection, derived from the gravitational field, are, in general, relatively small. Therefore, we suppose that all the physical properties of the fluid are independent of temperature, except the density, whose variation with it is only taken into account in terms of the gravitational force. Thus the complicated system of equations is simplified.

Consider the equation of energy conservation:

$$\rho\, c_p \left(\frac{\partial T}{\partial t} + \vec{v}\, \nabla T\right) = \mathrm{div}\,(k\, \nabla T) + \sigma'_{ik} \frac{\partial v_i}{\partial x_k} \qquad (1)$$

where σ'_{ik} represents the viscous tensor that arises from the classical partition of the strength tensor

$$\sigma_{ik} = -\,\delta_{ik}\, p + \sigma'_{ik}$$

where p is the hydrostatic pressure and δ_{ik} the unity tensor. The last term of the right-hand side in (1) represents the energy dissipated by internal friction in the fluid and, since the velocities are small, it is negligible in the first approximation. Combining the resulting equation and the continuity equation, $\nabla\vec{v} = 0$, and keeping in mind the constance of the physical properties of the material, we obtain:

$$\rho\, c_p \frac{\partial T}{\partial t} + \nabla\,(\rho\, c_p\, \vec{v}\, T - k\, \nabla T) = 0$$

which is a classical balance equation with zero energy production. In steady state the local variation of energy with time is zero, and we have:

$$\nabla\,(\rho\, c_p\, \vec{v}\, T - k\, \nabla T) = 0$$

This equation assures us that in the Bénard-Rayleigh problem there is no energetic accumulation in any point of the fluid, or, what is the same, that the heat flux is independent of the surface height if the lateral walls are adiabatic:

$$\iint (\rho\ c_p\ \vec{v}\ T - k\ \nabla T)\ dx\ dy = \frac{1}{h} \iiint (\rho\ c_p\ \vec{v}\ T - k\ \nabla T)\ dx\ dy\ dz$$

In other words, the fluid constitues a "tube" of current. Therefore we can write for the heat flux in all the system:

$$q = \frac{1}{h} \iiint (\rho\ c_p\ \vec{\lambda}.\vec{v}\ T - k\ \vec{\lambda}.\nabla T)\ dx\ dy\ dz \tag{2}$$

where $\vec{\lambda}$ is the upward unity vector.

To simplify the calculation we can express the temperature distribution, $T(x,y,z)$, in terms of the uniform gradient corresponding to the static situation, pure conduction, plus a certain perturbation $\theta(x,y,z)$. Consider that T_2 and T_1 are the temperatures in the lower and upper boundaries respectively, thus

$$T(x,y,z) = T_2 + \frac{T_1 - T_2}{h} z + \theta(x,y,z) = T_o(z) + \theta(x,y,z) \tag{3}$$

It is of interest to consider that the θ perturbation in our system is zero in each and every boundary point of the fluid: in the upper and lower surfaces obviously because it is in contact with solid thermostated materials, and the lateral boundaries because they transport heat by pure conduction only, then:

$$\{\theta(x,y,z) = 0\}_{\text{boundaries}} \tag{4}$$

The substitution of (3) in (2) gives four integrals, two of which (those that contain $T_o(z)$ and $\nabla\theta$) are anulled by the conservation of the mass and by condition (4), respectively. Rearranging the other two integrals we obtain:

$$q = \frac{1}{h} \iiint \rho\, c_p\ \vec{\lambda}.\vec{v}\ \theta\ dx\ dy\ dz + k\ S\ \frac{T_2 - T_1}{h} \tag{5}$$

where S is the total section of fluid. It can be observed that in the case qhere $\vec{v} = 0$, the classical conduction equation results, thefore the first term of the right-hand side must represent the convective heat flux.

Equation (5) can be nondimensionalised using κ/h and $(T_2 - T_1)/Ra$

as scale factors, i. e. substituting the old variables for:

$$\vec{v}^* = \frac{\vec{v}}{\kappa/h} \quad \text{and} \quad \theta^* = \frac{\theta}{(T_2 - T_1)/Ra}$$

We thus obtain

$$\frac{q}{S} \frac{Ra\ h}{\rho\ c_p\ \kappa\ (T_2 - T_1)} = \frac{1}{V} \iiint \vec{\lambda}.\vec{v}^*\ \theta^*\ dx\ dy\ dz + Ra$$

where V is the total volume of fluid.

If H is the mean heat transport per unit surface and time nondimensionalised, and if the integral is represented by an average value, we have:

$$H = Ra + \langle\vec{\lambda}.\vec{v}^*\ \theta^*\rangle$$

From this equation we obtain the following results: The total heat flux can be expressed by two components, a pure conduction component and a convection component. The Rayleigh number can be interpreted as the energy flow density by pure nondimensionalised conduction. The convection heat flux can be expressed as an average value, in all volume, of the product of the velocity distribution by temperature perturbances. Therefore, Koschmieder {1} establishes that "the experimental investigation of the convective motions of a fluid on a plane plate heated from below has the following aspects:

a) The observation of the onset of convection and of the pattern of the motion (the planform).

b) The measurement of the wavelength of the motion.

c) The measurement of the heat flux through the convecting layer.

d) The measurement of the temperature distribution in the layer, and

e) The measurement of the velocity distribution of the motion."

In conclusion we can show the relation between the Nusselt and Rayleigh numbers as:

$$Nu = \frac{H}{Ra} = 1 + \frac{\langle\vec{\lambda}.\vec{v}^*\ \theta^*\rangle}{Ra}$$

B. Qualitative development. In order to get the physical significance of the cuantities which appear in the problem and to attain some qualitative relations between them, the Bénard-Rayleigh problem can be studied using the dynamics and energetic requierements necessary for the convection to begin.

Supposing that T_2 and T_1 ($T_2 > T_1$) are the temperatures of the lower and upper boundaries, respectively, and we define the mean temperature of the layer, T, as:

$$T = \frac{T_2 + T_1}{2} \tag{6}$$

and assuming a linear dependence of the density on temperature:

$$\rho = \rho_o \left(1 + \alpha \left(T - T_o\right)\right) \tag{7}$$

The total force per unit of volume that moves a fluid particle upward is:

$$F_p = \Delta\rho\ g = c_o\ \rho_o\ \alpha\ g\ (T_2 - T_1) \tag{8}$$

where c_o is a numerical constant depending on the boundary conditions. Using the height, h, of the layer as the typical scale length and v as the characteristic velocity of the internal motion, the dissipating force that opposes the motion due to the density gradient can be expressed as:

$$F_d = c_1\ \rho_o\ \nu\ v/h^2 \tag{9}$$

where c_1 is another numerical constant depending on the geometrical conditions.

Now we can determine, with Busse {2}, the energetic requirements for the onset of convection. The energy lost per unit of time and volume by viscous dissipation can be obtained from (9), and we have

$$E_d = c_1\ \rho_o\ \nu\ v^2/h^2 \tag{10}$$

and the potential energy can be deduced from (8) in the same way, if we know that a fluid particle which begins its ascent from a lower average temperature is capable of retaining its original density, the ratio

$$v\ h/\kappa,$$

where κ/h can be regarded as the typical velocity with which distorted isotherms on an h scale return to their equilibrium position. For small values of v we have:

$$E_p = c_o \rho_o \alpha g (T_2 - T_1) v^2 d/\kappa \tag{11}$$

The convection requires $E_p/E_d \geq 1$, and we can write;

$$Ra \equiv \frac{\alpha g (T_2 - T_1) h^3}{\nu \kappa} \geq \frac{c_1}{c_o} \equiv Ra \tag{12}$$

where Ra_c represents the critical Rayleigh number and is the tipical parameter of the onset of convection, independent of the individual proper ties of the fluid under study.

Returning to the dynamic condition, the static situation in the layer can be described by the conditions:

$$F_d/F_o \geq 1$$

or

$$\frac{\nu v}{\alpha g (T_2 - T_1) h^3} \geq \frac{c_o}{c_1} = \frac{1}{Ra_c}$$

then

$$\frac{v}{Ra} \frac{h}{\kappa} \geq \frac{1}{Ra_c} \tag{13}$$

and we can follow that in the onset of convection the fluid can move with a local velocity,

$$v_o = \frac{\kappa}{h} \tag{14}$$

which means that this velocity corresponds to the relaxation of the local fluctuations at the limit of the fluid steady state.

Now consider the convective state of the fluid layer. In this case, all the fluid particles have two independent velocities, the former, v_o, is the local velocity of fluctuation, and the latter, υ, is the actual velocity of convection. The actual motion of convection is a steady motion with velocity υ, and therefore the maximum potencial energy applied per unit volume and time to support the density gradient in an element of fluid is:

$$E'_p = c_o \rho_o \alpha g (T_2 - T_1) \kappa/h$$

and, at the same time, the energy density per unit of volume and time dissipated by viscous effect in two independent motions must be:

$$E'_d = c_1 \rho_o \nu (\upsilon^2 + (\kappa/h)^2)/h^2$$

both energy densities should be equal, so that by solving the resulting equation we find

$$\upsilon = \frac{\kappa}{h} \left(\frac{Ra}{Ra_c} - 1\right)^{1/2} \qquad (15)$$

Now consider the Nusselt number in terms of the Rayleigh number. The total heat transported by the unit of volume from bottom to top, is:

$$Q = \rho_o c_p (T_2 - T_1)$$

where c_p is the specific heat of the fluid. The actual heat flux will be the product of this value multiplied by the climbing velocity, plus the lateral losses during the climbing. As we know, the periodical character of the convective pattern ($\Lambda = 2h$) makes the lateral conduction of heat important, there fore the effective velocity of mass transport of the heat will be larger than the actual convective velocity.

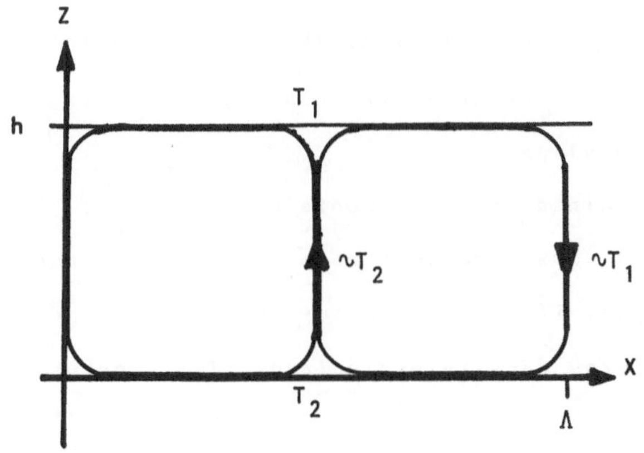

Fig. 1. Scheme of the periodical character of the convective pattern.

The effect of pure conduction can be interpreted as mass conduction due to fluctuations, which, as we have already seen, have a characteristic velocity given by κ/h. In fact, for pure conduction:

$$q_{cond} = \rho_o c_p (T_2 - T_1) \frac{\kappa}{h} = k \frac{T_2 - T_1}{h} \qquad (16)$$

where k is the heat conductivity, this is the classical conduction law.

So then, the increase factor of mass transport velocity with respect to convective velocity can be estimated as the quotient $\upsilon h/\kappa$, with is the factor that we have used already in (10) and now it assumes its full significance. Therefore the heat flux due to convection is:

$$q_{conv} = \rho_o c_p (T_2 - T_1) \upsilon \upsilon h/\kappa \tag{17}$$

From (16) and (17) we arrive at the important relation, valid for Rayleigh number values close to the critical one:

$$Nu = 1 + \frac{h^2}{\kappa^2} \upsilon^2 \tag{18}$$

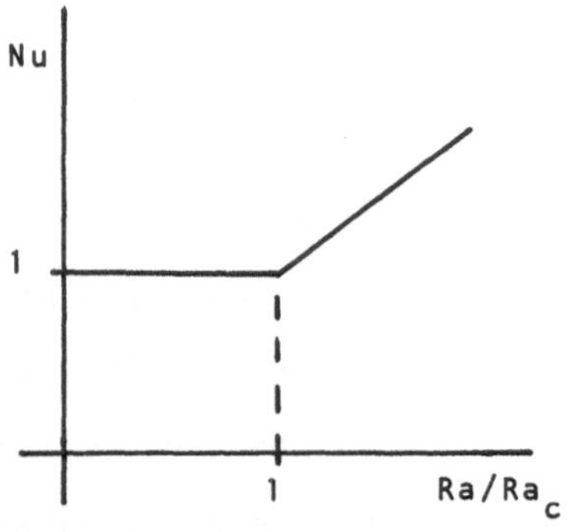

Fig. 2. The onset of convection.

that allow us, by the use of (15) to obtain the welcome result

$$Nu = 1 + (\frac{Ra}{Ra_c} - 1) \tag{19}$$

This approximate result confirms that the onset of instability begins with a discontinuity of the heat flux at the critical point, which we can assume, at a first aproximation, to have a constante slope.

The result that we have obtained does not contain the Prandtl number because in the theoretical considerations have not included nonlinear considerations. For example, if in working out the former energetic balances we include the energy density per unit of time, corresponding to the rotational motion

$$E_r' = c_2 \rho_o \upsilon \upsilon^2/h$$

the equation

$$\frac{Ra}{Ra_c} = 1 + (\frac{h}{\kappa} \upsilon)^2 + \frac{c_2}{c_1} \frac{1}{P} (\frac{h}{\kappa} \upsilon)^3$$

can be obtained for the relation between υ and Ra, where P appears as a relevant parameter.

IV. EXPERIMENTAL RESULTS

The heat flux measurements have allowed us to determine the Ra_c values, using, for this purpose, the sudden increase in the heat transfer through the liquid at the onset of convection. Schmidt and Milverton {3} obtained the value 1770 ± 140 and Silverton {4} 1700 ± 51, both of which sufficiently coincide with the theoretical value of the critical Rayleigh number, 1701. Later on, Schmidt and Saunders {5} observed a second transition at a Rayleigh number of about 45000 and Malkus {6}, by measuring up to $Ra = 10^{10}$, obtained four more transitions at Ra = $= 1.7\ 10^3$; $11.0\ 10^3$; $55.0\ 10^3$; $17.0\ 10^4$; $42.5\ 10^4$; $86.0\ 10^4$ and $17.0\ 10^5$. These transitions, which have been justified as steps to turbulence, are not so clear as the first one.

A. Nusselt number dependences. Apart from the Nusselt number dependence on the Rayleigh number, there are other variables that influence the heat flux. Thus the convective pattern of the motion can, in principle, affect energetic transport in the system. With respect to this subject, Koshmieder stated in 1974 that "a measurement of the heat flux through a steady convective liquid with a defined pattern has not yet been published." Recently, Zamora, Cordoba and Moreno {7} have studied this influence, utilising a rectangular geometry with a convective pattern which was very accurately determined by Bergé {8} and Dubois {9} using Doppler annemometry. The convective pattern is made up of rolls parallel to the shorter lateral boundary, and by means of various heating rates several wavelengths can be obtained (Λ = 2.0 h; 2.2 h; 2.5 h). The first conclusions show that such changes in Λ do not appreciably affect the heat flux. Further measurements {10} show that the heat flux differences between the wavelengths Λ = 2.0 h and Λ = 2.5 h are detectable in the interval Ra/Ra_c = 1 to 26, aproximately, and this difference vanishes from the last value on.

Another significant influence on the heat flux is the Prandtl number, set forth by Rossby {11} and other authors. Unfortunately, this dependence has not been systematically studied from a experimental view point, although it is accepted that at the onset of convection, $Ra \simeq Ra_c$, the theoretical results of Schluter, Lortz and Busse {12}, are confirmed, and that in the case of the parallel rolls, takes the form:

$$\frac{\bar{H}}{Ra - Ra_c} = (0.69942 - 0.00472\ P^{-1} + 0.00832\ P^{-2})^{-1}$$

where $\bar{H} = q_{conv} = (Nu - 1)\ q_{cond}$.

One last dependence of the Nusselt number, set forth by Ahlers, corresponds to the aspect ratio $\Gamma = D/2h$ (D is the diameter, since he used a circular geometry). For high values of Γ the experimental results ifconcur appreciably with the theoretical results if the fluid layer is laterally infinite, substantially decreasing the initial slope when Γ diminishes. Unfortunately there is no theoretical structure to justify this fact, although the results coincide with the forecasts of the numerical models.

<u>B. The onset of convection.</u> We have already noted that the heat flux measurements can show the onset of convection. Making measurements of heat flux, Ahlers {13} has demonstrated that the Bénard-Rayleigh problem can be considered as a continuous phase transition, wich confirms Landau's theory. He worked with low temperatures, using normal liquid and gaseous helium and a cylindrical geometry. What follows is, in general terms, a summary of his reasoning.

If we take a system whose upper temperature remains constant throughout all the experiences, from (6) we can characterise such a system by its mean temperature T. If we consider our system as being made up of two phases, one symetrical or static for $T < T_c$ and the other non-symetrical or convective for $T > T_c$, we can represent it by means of a order parameter, η, to be zero in the former and not in the latter. This parameter can obviously be the characteristic velocity. If we accept the existence of a potencial function in the system, $\Phi(T, p, \eta)$, we can suppose, with Landau, that it can be developed in η, and at a constant pressure we can write:

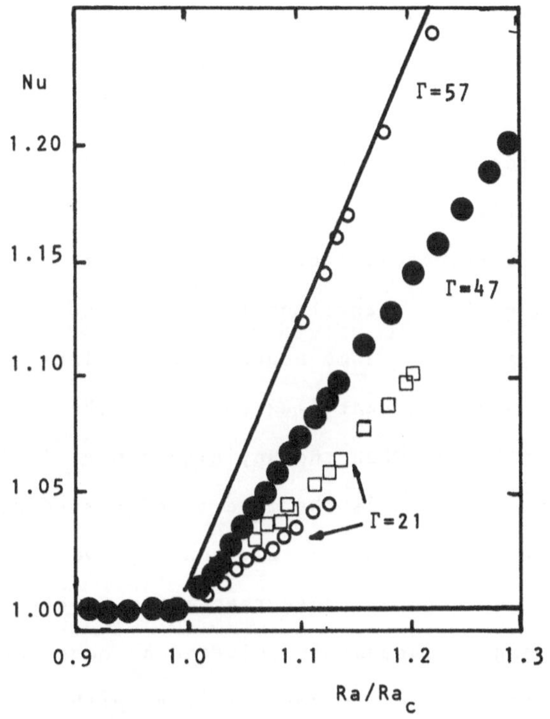

Fig. 3. Experimental results of Ahlers {13}.

$$\Phi(T,\eta) = \Phi_o(T) + A(T)\ \eta^2 + C(T)\ \eta^4 + \dots \qquad (20)$$

where C(T) > 0 and A(T) < 0 in the non-symetrical phase. Taking the classical approximations

$$C = \text{constant} \quad \text{and} \quad A(T) = -a\ (T - T_c) = -b\ \frac{Ra - Ra_c}{Ra_c} \qquad (21)$$

where a and b are positive constants, now we can find the steady state of the system by simply minimizing Φ with regard to η,

$$\left.\frac{\partial\Phi}{\partial\eta}\right)_T = 0 \qquad (22)$$

and we arrive at

$$\eta^2 = -\frac{A}{2C} = \upsilon^2 = \frac{b}{2C}\ \frac{Ra - Ra_c}{Ra_c} \qquad (23)$$

that is a similar dependence to the one expressed in (15).The equation (18) gives us

$$Nu - 1 = \frac{d^2}{\kappa^2}\ \frac{b}{2C}\ \frac{Ra - Ra_c}{Ra_c} \qquad (24)$$

a linear relation definitively confirmed by Ahlers (Fig. 3). With this paper Ahlers introduces phase transition techniques in the study of convective instabilities.

C. The transition to turbulence. When the temperature difference applied to the convective system is raised, a new set of phenomena appear: tridimensional structures, oscillatory processes, etc., that empty into the chaotic turbulence. This transition to turbulence seems to present itself in several ways for the same boundary conditions. As it has been noted, the first measurements heat flux suggested the existence of up to seven successive transitions when the Rayleigh number is increased. However, recent experiences, such as the more precise findings of Koshmieder {14} and Ahlers {13} in circular geometry show continuous curves, although the Rayleigh number values are not so high as those reached by Malkus. The experimental curves obtained can be adequately adjusted by means of a potencial form, or by a power series. The former coincides with the form of the asyntotic solutions obtained through numerical calculations. In general, the first dependence is ussually accepted for Bénard problem, with an exponent of 1/3, while the results obtained by Ahlers {15} for Bénard-Rayleigh problem are:

$$Nu - 1 = 1.034\ \varepsilon + 0.981\ \varepsilon^3 - 0.866\ \varepsilon^5$$

where $\varepsilon = (Ra - Ra_c)/Ra_c$. However, very recent experiences show that there is a second transition, known as bimodal, which entails the placement of two structures of rolls one over another in perpendicular directions. Zamora et al {7} in rectangular geometry show a clear influence of this transition on the heat flux (Fig. 4) and find it in $Ra_{II} = 21200$. They agree with the theoretical result of Busse {16} $Ra_{II} = 22600$. Also Ballesteros {17} finds that this transition can have effects on the form of the thermogram, which determines the heat flux, thus in some first experiences with the measurement of heat flux in a little box ($\Gamma = 2$) he has found a "shoulder" in the thermograms that he has justified, in principle, as a result of this transition.

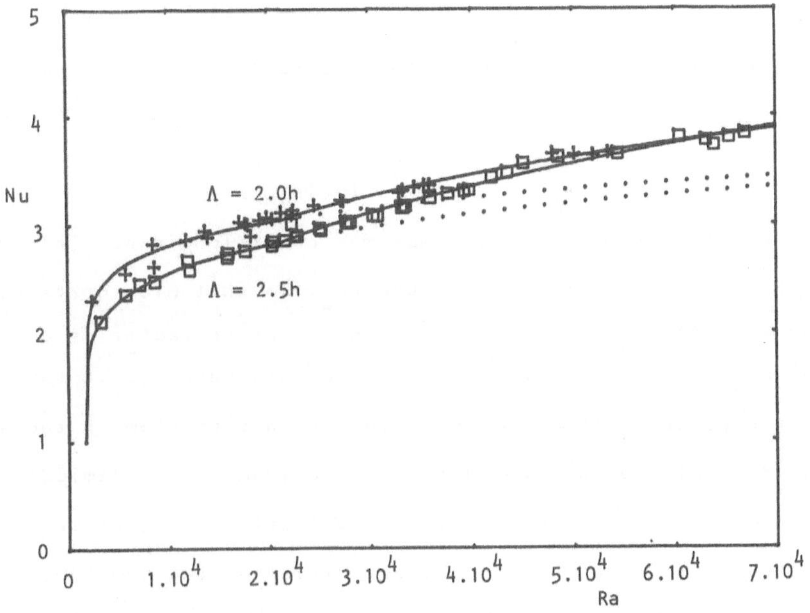

Fig. 4. Experimental results of Zamora et al. {10}.

Nevertheless, sufficiently accurate measurements of heat flux for very high Rayleigu numbers and well known physical conditions have not as yet been carried out.

V. CONCLUSIONS

I have tried to give a superficial expanation of the current situa tion of research on heat flux in convective systems. The importance of this flux has been exposed systematically ever since Rumford considered it a new transport mechanism, until Malkus {18} tried with his "maximum heat flux hypothesis" to discern the planform of the convective structure which arises in each case using the heat flux as a potential function. Malkus's ideas can be represented schematically, from the calculations already made, as follow. In (20) we have accepted the existence of a potential function in the system with respect to the order paramenter. Substituing the approximations (21) and the result (18), we obtain:

$$\Phi(T,\eta) = \Phi_o(T) - b\,\frac{\kappa^2}{h^2}\,(Nu - 1) + C\,\frac{\kappa^4}{h^4}\,(Nu - 1)^2 + \ldots$$

where Nu is considered a function of η, the order parameter that characterises the type of structure formed, for the same boundary conditions. In the linear approximation

$$\Phi(T,\eta) = \Phi_o(T) - b \frac{\kappa^2}{h^2} (Nu(\eta) - 1)$$

the minimum condition of Φ becomes a maximum condition in Nu. This determines that the emergent type of structure, η_o, is that wich corresponds to a maximum heat flux. As can be seen, the linear character is quite a rough approximation, especially at high Rayleigh's numbers. It thus seems that Malkus's hypothesis can be interpreted in this problem in the same sense as the Berthelot principle in the thermodynamics of chemical reactions, as an important part of a real extremal principle that must characterise the structure and depend on the definition of the complementary parameter that it configurates. This parameter must be sought in the convective system's own energy, which must characterise the thermic structure of the system in its evolutive process. A study of this type would permit the problem to neatly approach the Thermodynamics of Irreversible Processes, enlarging the purely mechanical description now submitted.

VI. REFERENCES

{1} E.L. Koschmieder, Adv. Chem. Phys. 26, 177 (1973).
{2} F.H. Busse, Rep. Prog. Phys. 41, 1929 (1979).
{3} R.J. Schmidt and S.W. Milverton, Proc. Roy. Soc. A152, 585 (1935).
{4} P.L. Silveston, Forsch. Ing. Wes., 24, 29, 59 (1958).
{5} R.J.Schmidt and O.A. Saunders, Proc. Roy. Soc. A165, 216 (1938).
{6} W.V.R. Malkus, Proc. Roy. Soc. A225, 185 (1954).
{7} M. Zamora, A. Córdoba and J. Moreno, in "Symmetries and Broken symmetries in Condensed Matter Physics" ed. N. Boccara, IDSET, Paris 1981.
{8} P. Bergé, in "Fluctuations, Instabilities and Phase Transitions",ed. T. Riste, Plenum, New York, 1975.
{9} M. Dubois and P. Bergé, in Synergetic, Vol. III, "Far from Equilibrium", Springer-Verlag, 1979.
{10} M. Zamora, A. Córdoba and J. Moreno, to be published.
{11} H.T. Rossby, J. Fluid Mech., 36, 309 (1969).
{12} A. Schluter, D. Lortz and F.H. Busse, J. Fluid Mech., 23, 129 (1965).
{13} G. Ahlers, Phys. Lett., 62A, 329 (1977).

{14} E.L. Koshmieder, J. Fluid Mech. 30, 9 (1967).

{15} G. Ahlers, in "Fluctuations, Instabilities and Phase Transitions", ed. T. Riste, Plenum, New York, 1975.

{16} F.H. Busse, J. Math. Phys. 46, 140 (1967).

{17} J. Ballesteros, Ph. Thesis, University of Sevilla, 1981.

{18} W.V.R. Malkus, Proc. Roy. Soc., A225, 199 (1954).

UNSTABLE FLOWS
OF CONCENTRATED SUSPENSIONS

D. Quemada
Laboratoire de Biorhéologie
et d'Hydrodynamique Physico-chimique
Université Paris 7 - 75005 Paris / FRANCE

1 . INTRODUCTION

Complex systems as concentrated suspensions, emulsions, polymer solutions and melts, exhibit macroscopic behaviours which are generally considered as resulting from some average of material properties at the microscopic scale. However many cases (or else all cases ?) exist where large scale properties differ from simple superposition of the small scale ones : at the macroscopic level, one often observes *collective* properties which can be entirely new ones and in which microscopic details are completely over-shadowed. So that these collective properties can appear as *universal* ones (as, for instance, many bulk characteristics of disperse systems).

Nevertheless, as the system evolves, structural modifications may occur, that changes the scale of heterogeneity. Starting from a macroscopically homogeneous system (although microscopically heterogeneous), a transition to large scale heterogeneous states (as floculation in a suspension, for instance) may result from interactions between small scale elements. Such states should be unstable, noticely if external constraints maintain the system far from equilibrium (as under flow conditions) and it can be expected that more homogeneous subsystems can appear, separated by frontier regions where inhomogeneties are concentrated. It is believed that such subsystems behave as *dissipative structures,* the formation of which should reduce the entropy production in the whole system.

A great number of (classical) hydrodynamic instabilities - as, for instance the Taylor or Rayleigh Benard ones - have been observed in simple fluids when some characteristic (dimensionless) numbers N (as Ta , Ra ...) exceed critical values N_c. Curves for neutral stability have been calculated from linear theory and many extensions to include non linear effects are still in progress now.

Within the validity of description of a suspension as a *continuous media** it could be expected that many of such hydrodynamic instabilities would also be observed with (at least) qualitative similarities in physical processes involved in the instability onset, provided that suitable changes in critical values N_c of the dimensionless numbers were performed. However, as it will be recall later, the suspension behaviour is dominated by the *particle volume fraction* ϕ *, and* (excepted in the limit of extreme dilution) by the flow conditions, i.e. by the *deformation rate* (shear rate $\dot{\gamma}$ or elongational rate $\dot{e}$, in simple cases).

It is only in the limit of very diluted disperse systems ($\phi \ll 1$) that we can expect the classical results on hydrodynamic instabilities to be valid. Indeed, adding a small amountof particles to a simple fluid under unstable conditions should slightly modify the characteristics of the instability, noticely N_c . In this limit $\phi \to 0$, particle-particle interactions are completely negligible and then, one has to take only into account particle-fluid interactions, that leads to linear ϕ-variation of *transport coefficients*, as viscosity**, thermal conductivity...

As soon as ϕ reaches values above few per cent, *particle - particle* interactions result in non linear ϕ - dependence of transport coefficients. Similar extension of classical results on stability, then becomes impossible because of the importance of *internal structure* in the rheological behaviour. Structural changes can occur if one modifies ϕ and/or $\dot{\gamma}$ (or $\dot{e}$). For instance as $\dot{\gamma}$ is increased shear-thinning behaviour results from breaking down of the structure which exists at low shear rate by viscous forces exerced by the fluid on particles. Moreover, in confined flows, for example through a circular tube (and especially if $\xi = a/R$ becomes significantly different from 0), effects of *particle - wall* interactions arise which lead to concentration inhomogeneties, mainly localized close to the wall. The validity of continuous description of such a complex behaviour can be recovered if one introduces a two-phase structure, for example a particle rich core surrounded by the so-called particle depleted *marginal layer*. In such a

---

* the smaller the ratio ξ = (particle size a)/(characteristic lengh of the flow L), the better the approximation.

** for instance the well-known Einstein expression for the shear viscosity, $\eta = \eta_F(1 + k\phi)$ where η_F is the shear viscosity of the suspending fluid and k , the (shape dependent) intrinsic viscosity (k = 2.5 for spheres).

dissipative structure, each phase is assumed homogeneous and having properties which can be represented by *effective transport coefficients*. Nevertheless these coefficients now depend on the structure of the system and therefore can be only defined under well-defined conditions. As an example, the structural (shear) viscosity (a concept which was first introduced by OTSWALD (1926)) requires the flow conditions to be precised.

However, from the following coupling circle

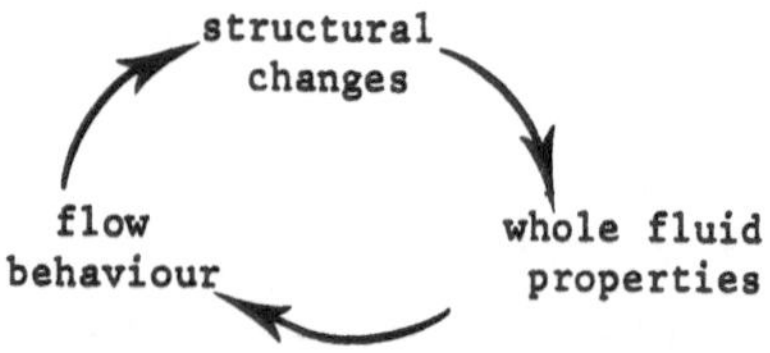

it can be predicted that such flow induced structural changes may initiate flow instabilities and that control by non-linearities could promote the existence of steady states far from equilibrium.

A preliminary study of main features of these *structural instabilities*, leading to expect some flow instabilities in concentrated suspensions, is the aim of these lectures.

In order to clarify the structure-behaviour relation in concentrated disperse media, a brief recall in basic rheology of these media will be given in the next section.

2 . RHEOLOGY OF CONCENTRATED DISPERSE SYSTEMS

2.1. Basic definitions (see Appendix)

As for simple fluids, rheological properties of concentrated disperse media can be characterized by material functions entering into the *constitutive relation* of the system (i.e. the relation between stress and rate of deformation).

The usual measurement is the response to uniform plane shear. It gives the *shear viscosity* as the ratio of the viscous shear stress σ to the shear rate $\dot{\gamma}$, $\eta = \sigma/\dot{\gamma}$. More difficult are the measurements of normal stresses and elongational viscosity noticely in the case of concentrated suspensions.

Unsteady properties, as viscoelasticity are currently measured but their interpretation is not so easy since it requires the use of non linear viscoelastic models.

For the sake of simplicity, we will limit to the study of shear viscosity in the following sub-sections.

2.2. Steady shear viscosity of suspensions

Attempt to predict flow behaviour of concentrated suspensions require to introduce "structural parameters" into rheological equations in order to interelate the rheological properties of the material and its molecular or particulate structure. Progress in choosing such parameters can be gained using a dimensional approach similar to that given by KRIEGER (1963) for colloid suspensions. As an example, in the simplest case of neutrally buoyant particles suspended into a newtonian fluid in laminar steady flow, the relative viscosity is given by

$$\eta_r = \frac{\eta}{\eta_F} = \eta_r(\phi, \dot{\gamma}_r) \qquad (2.1)$$

where $\phi = (4\pi/3)na^3$ = volume fraction of spheres (2.2)

(a = radius ; n = number/unit volume)

$\gamma_r = \tau\dot{\gamma}$ = reduced shear rate (2.3)

η_F = suspending fluid viscosity.

In (2.3) τ is a characteritic (internal) time of the system. In the case of colloïdal particles τ can be taken as the Brownian diffusion time

$$\tau \sim a^2 D_{tr}^{-1} \sim D_{rot}^{-1} \qquad (2.4)$$

where $D_{tr} = KT/6\pi\,\eta_F\,a$ and $D_{rot} = KT/8\pi\,\eta_F\,a^3$ are the translational and rotational Brownian coefficients for diffusion, respectively. Therefore

$$\tau = b\,\eta_F\,a^3/KT \quad . \qquad b = Cte \qquad (2.5)$$

In the case of more complex systems, each additional variable will appear as a new dimensionless variable in the viscosity equation, as shown in the following examples. For a suspension of ellipsoidal particles, the axial ratio $a_{\|}/a_{\perp}$ of semi axis of rigid spheroid. Purely elastic particles (with a shear modulus G) suspended in a fluid of viscosity η_F have a relaxation time $\tau_m = \eta_F/G$ that will be used for large deformable particles to define $\dot{\gamma}_r = \eta_F \dot{\gamma}/G$, as the ratio of shear force to the elastic one*. An emulsion involves two additional parameters, the interfacial tension Γ and the internal viscosity of droplets η_i that leads to the relaxation time $\tau_D = \eta_F a/\Gamma$ and to the new dimensionless variables $\tau_D \dot{\gamma}$ and η_i/η_F . Furthermore, accounting for particle interactions involves the ratio W/KT of some characteristic energy W to the thermal one : for instance, short range Van der Waals interaction (energy $\sim A/6(\frac{r}{a})$, where A is the Hamaker constant) leads to take A/KT as new variable ; long range forces as electrostatic repulsion are governed by the surface potential (or surface charge density) and by screening effects from the ionic double layer, the thickness κ_D^{-1} of which (the so-called Debye lengh) enters in the dimensionless group $\kappa_D a$. Comparing experiments performed at different temperatures, ionic strengths... allow estimation of the effects that these dimensionless parameters would have on the rheological behaviour.

2.2.1. Viscosity of extremely diluted and dilute suspensions.

In a very dilute system, the suspension viscosity η has been observed higher than the suspending fluid viscosity η_F . The relative viscosity η_r , as a function of the *volume* concentration ϕ , is given by :

$$\eta_r = \frac{\eta}{\eta_F} = 1 + k_1\phi \qquad (2.6)$$

where k_1 is a particle shape dependent factor, thus which depends on particle deformability (through the deformation level reached in the experiment). For hard spheres, EINSTEIN (1906), found theoretically $k_1 = k_E = 2.5$.

As eq. (2.6) contains only *particle-fluid interactions*, discarding any hydrodynamic action of one particle upon another, this equation only holds at extremely low concentrations. Above $\phi \simeq 0.01$ deviation from linearity is observed and attributed to particle-particle interactions. The simplest interaction involves two single particles and gives the ϕ^2 term in a power series for relative viscosity :

$$\eta_r = 1 + k_1 \phi + k_2 \phi^2 + \ldots \qquad (2.7)$$

For suspensions of rigid spheres, number of theories calculated the ϕ^2

---

* τ_m is the Maxwell relaxation time of a viscoelastic liquid (which shows how viscosity and elasticity mix to give visco-elastic properties) (See Appendix).

coefficient , k_2 , by different methods, but leading to very different values. The most rigorous calculation of shear viscosity recently performed (BATCHELOR and GREEN, 1972) gave $k_2 = 5.2 \pm 0.3$.

Increasing concentration requires more terms in the series (2.7). No theoretical calculation of coefficients k_i exists and their determination could be only carried out by data fitting (see later).

2.2.2. Viscosity of concentrated suspensions.

a) - ϕ-dependent viscosity. For highly concentrated suspensions, phenomenological approaches have been developped. They lead to non-linear viscosity equations, $\eta = \eta(\phi)$ which must represent the strong increase in viscosity up to infinity, as ϕ tends towards its maximum (packing) value ϕ_M (Fig. 2.1)

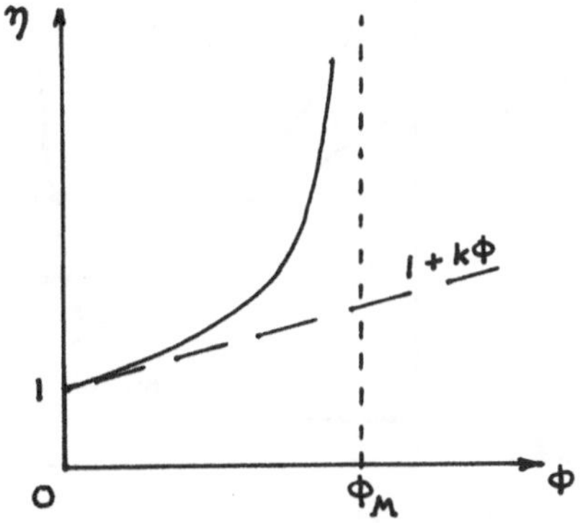

Fig.2.1 Relative viscosity η_r versus volume fraction ϕ of particles, in concentrated suspensions. Infinite viscosity is found at packing concentration ϕ_M .

Nevertheless, most of classical relations (see a revue in QUEMADA, 1982) does not satisfactorily represent the behaviour at very high concentration, close to ϕ_M . For instance the well-known Mooney equation (MOONEY, 1951) is

$$\eta_r = \exp \frac{k_1 \phi}{1 - \lambda\phi} \tag{2.8}$$

where λ is a crowding factor, the reciprocal of which should be equal to ϕ_M . Better agreement with experimental data is obtained using the Krieger-Dougherty equation (KRIEGER and DOUGHERTY, 1967)

$$\eta_r = (1 - \lambda\phi)^{-k_1/\lambda} \tag{2.9}$$

b) - Non-newtonian behaviour. Concentrated disperse systems generally exhibit non-newtonian properties. Especially, the shear viscosity becomes shear rate (or, alternatively, shear stress) dependent, $\eta = \eta(\dot\gamma)$

or $\eta = \eta(\sigma)$. Number of attempts have produced a very great variety of viscosity equations, most of them remaining free of any explicit concentration dependence. Indeed, as it has been stressed above, one has to consider both dependences, $\eta = \eta(\phi,\dot{\gamma})$ because of missing one of them would limit very much the validity of the model. Bearing this in mind, it is nevertheless interesting to recall some main features of the shear dependence of viscosity, in relation with the structure of the system.

Fig. 2.2 displays different types of shear rate dependence both for the stress σ and the viscosity η .(Recall that, according to the dimensionless analysis, the variable is a reduced shear rate, $\dot{\gamma}_r = \tau\dot{\gamma}$, involving some characteristic time , τ).

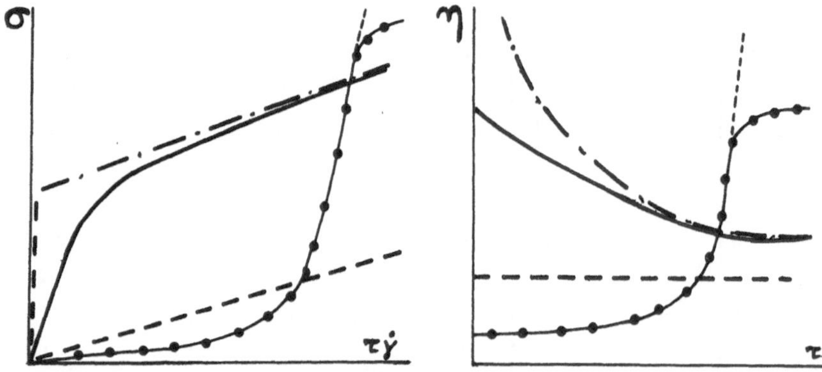

Fig.2.2 Different types of non-newtonian variations $\sigma(\tau\dot{\gamma})$ and $\eta(\tau\dot{\gamma})$
---- Newtonian | ——— pseudo-plastic (shear thinning)
-●-●-● Shear-thickening | —·— plastic (Bingham for ex.)

Number of phenomenological or empirical relations $\eta(\dot{\gamma})$ have been proposed, the simplest been the well-known power law $\eta = \kappa\,\dot{\gamma}^m$ which can represent only a part of the involved range of $\dot{\gamma}$. Such a relation seems limited in interest since it appears very difficult to relate fitted values of the two parameters κ and m to the structure. On the contrary, relations which enable one to display both low and high shear behaviour require using at least three structural parameters. Such relations can be written in the general form

$$\eta = \eta(\dot{\gamma}) = \eta_\infty + \frac{\eta_0 - \eta_\infty}{F(\dot{\gamma})} \qquad (2.10)$$

where η_0 and η_∞ are the limiting viscosities as $\dot{\gamma} \to 0$ and $\dot{\gamma} \to \infty$, respectively. $F(\dot{\gamma})$ is a dimensionless function of the shear rate such as $F(0) = 1$ and $F(\infty) = \infty$. Classical examples are

$F = 1 + A\,\dot{\gamma}$ (empirical, with A = Cte , WILLIAMSON, 1929)
$F = 1 + A\,\dot{\gamma}^m$ (semi-empirical, with $m = \frac{2}{3}$, $\frac{4}{5}$, CROSS,1965)
$F = \tau\dot{\gamma}/\sinh^{-1}(\tau\dot{\gamma})$ (theoretical, (2 components), with τ = relaxation time, REY-EYRING, 1955).

KRIEGER and DOUGHERTY (1959) obtained a relation $\eta(\sigma)$, similar to (2.10)

$$\eta = \eta_1 + \frac{\eta_2 - \eta_1}{1 + A\,\sigma}$$

in which low and high shear viscosities η_1 and η_2 are assumed to verify eq (2.9) : then their model $\eta = \eta(\phi,\sigma)$, depends on five parameters.

Although all these relations were proposed for the (general) case of pseudo-plastic behaviour (if $\eta_0 > \eta_\infty$), it is worth noting that the form (2.10) contains the cases of newtonian (if $\eta_0 = \eta_\infty$) and dilatant (if $\eta_0 < \eta_\infty$) fluids, but the plastic case. The latter is represented (in the simplest situation) by the (linear) BINGHAM equation

$$\sigma = \sigma_0 + \eta_B\,\dot{\gamma} \qquad (2.11)$$

where σ_0 is the yield stress, under which the material behaves as a solid, and η_B is the Bingham viscosity, or, more generally a non-linear yield-pseudo-plastic equation $\sigma = \sigma_0 + \kappa\,\dot{\gamma}^m$. For pigment oil suspensions, Casson's theory (CASSON, 1959) gave the expression

$$\sigma^{1/2} = \sigma_0^{1/2} + (\eta_C\,\dot{\gamma})^{1/2} \qquad (2.11a)$$

which was found in fair agreement with various data of concentrated suspensions.

The relation between the model parameters and the structure is not quite clear in most cases, however at least qualitative arguments may be presented. In general, the structure at rest is believed to result from more or less strong association of "particles", forming a structure whose compacity depends on the history of the material. Increasing the shear stress breaks down this structure, either immediately (then the behaviour is pseudo-plastic) or after reaching some threshold (the so-called yield stress in plastic behaviour). Different processes are involved in the (more or less reversible) destruction of the structure as $\dot{\gamma}$ increases : disaggregation, and then orientation and (eventually) deformation of different structural units. However, some disperse systems (noticely highly concentrated suspensions of rigid particles having a small sedimentation volume) exhibit a shear induced building up of a structure, that leads to a shear-thickening behaviour, which can turn to a dilatant one if the formed structure corresponds to a very loose packing of particles. As the suspending fluid is drawn into the widening interstices, this may promote a transition "wet → dry" if there is just enough liquid present in the system. Such a picture provides a very plausible explanation for both shear-thickening and dilatancy and their interdependence. On the other hand, shear thickening may also result from a shear induced flocculation process, due to increase in both the frequency of particle collisions and the force involved in them.

In the following, a structural model based on a viscosity relation $\eta = \eta(\phi,\dot{\gamma})$ will be studied in some details.

c) - A structural model. As coupling between structure and behaviour is achieved through flow conditions, special attention will be given to the following viscosity equation

$$\eta_r = (1 - \frac{1}{2} k\phi)^{-2} \qquad (2.12)$$

which involves an *effective intrinsic viscosity* k , as a structural parameter.

The ϕ-dependence which appears in (2.12) was found theoretically by applying a minimum energy principle to a two-phase flow model for newtonian fluids (QUEMADA, 1977). Fairly good fitting of this equation, using various k values, was obtained from several data including blood and Red Cell suspensions.

In very dilute systems, $\phi \to 0$, eq.(2.12) tends to the Einstein equation (2.6) , k then appearing as the true intrinsic viscosity. As the volume fraction of particles ϕ is raised, the volume fraction value which enters in (2.12) is the volume fraction of effective particles, as for instance particle flocs, incliding the fluid trapped on the inside. Alternatively to take into account such structural changes in the suspension, one may keep ϕ unchanged in (2.12) as the true concentration, then taking k as an effective intrinsic viscosity which will involve all changes in the effective volume of particles. As the latter depends on shear rate $\dot{\gamma}$, k appears both ϕ- and $\dot{\gamma}$-dependent.

It is possible to consider the effective intrinsic viscosity k (deduced by (2.12) from a viscometric measurement at given ϕ and $\dot{\gamma}$) as characterizing the structure the suspension has at ϕ and $\dot{\gamma}$. Indeed k is related to the packing concentration ϕ_p which would reached if particles were packed keeping unchanged their actual state. Since the viscosity would tend to infinity if $\phi \to \phi_p$ (2.12) gives

$$\phi_p = \frac{2}{k(\phi,\dot{\gamma})} \qquad (2.13)$$

In fact, because of structural changes induced during rising of the concentration, such a packing is not the real packing that one could observed but an *actual packing concentration* which can be related to the actual structure in the following way.

For a given experiment (given sample at fixed concentration ϕ_1 and shear rate $\dot{\gamma}_1$) , this packing value ϕ_p is the one which would be reached if the particles (single ones, multiplets or flocs of multiplets) were packed (by fictive increase of their number per unit volume) without changing the state they possess under the experimental conditions, that is *keeping unchanged the characteristics of particles* (such as the degrees of aggregation, deformation and orientation) *that the particles have for the actual volume fraction* ϕ_1 *and under the actual shear rate* $\dot{\gamma}$. Therefore ϕ_p , hence k by (2.13), are believed to be representative of the actual structure of the system, at con-concentration ϕ_1 and shear rate $\dot{\gamma}_1$, $\phi_p = f(\phi_1,\dot{\gamma}_1)$.

The possibility to extend the validity of (2.12) to the non-newtonian behaviour of concentrated suspensions was found (QUEMADA, 1978a) in considering that the evolution of k as a structural variable is governed by a rate equation for building up and breaking down of the structure, i.e. for thixotropic behaviour. It is assumed that such two opposite "processes" can be roughly described by characteristic (relaxation) times τ_A and τ_D respectively. For instance, if one assumes that the system contains N "particles" of two kinds, aggregates and single (dispersed) particles, a rate equation which describes the dynamic equilibrium between these two populations, can be written

$$\frac{dn_A}{dt} = \tau_A^{-1}(N - n_A) - \tau_D^{-1}\, n_A \qquad (2.14)$$

where n_A and $n_D = N - n_A$ are the number of aggregates and single particles, respectively. Using the intrinsic viscosities k_A of aggregates and k_D of single particles suspension, k in (2.12) is given by $k\phi = k_A\phi_A + k_D\phi_D$, using the corresponding volume fractions $\phi_i = n_i v_p/V$. At equilibrium, $n_A = N(1 + \theta)^{-1}$, $\theta = \tau_A/\tau_D$. Writing $k_A \equiv k_0 = k(0)$ = intrinsic viscosity at zero shear rate and $k_D \equiv k_\infty = k(\infty)$ = intrinsic viscosity at infinite shear rate leads to the *shear dependent intrinsic viscosity*,

$$k = k(\dot{\gamma}) = k_\infty + \frac{k_0 - k_\infty}{1 + (\tau\dot{\gamma})^p} \tag{2.15}$$

where τ = critical time which characterizes the evolution of the structure (as for instance a time for particle aggregation or, more generally, a mean relaxation time).

The reduced shear rate which appears in (2.15), $\dot{\gamma}_r = (\tau\dot{\gamma})$ is directly related to the ratio τ_D/τ_A . For concentrated systems, an empirical relation is taken as

$$\tau_D/\tau_A = (\tau\dot{\gamma})^p \tag{2.16}$$

In (2.16) p-values close to 0.5 have been found for blood and suspensions of flexible particles PAM gel, red cells ...) and p = 1 for colloidal spheres (QUEMADA, 1978b).

- <u>Pseudo-plastic and plastic behaviour</u>

Inserting (2.15) into (2.12) leads to a non-newtonian viscosity equation for pseudo-plastic systems (when $k_0 > k_\infty$). However, for p = 1/2, *plastic* behaviour is recovered in the two limits $\tau\dot{\gamma} \gg 1$ and $\phi \to \phi_M$. (see the Casson equation (2.11a)).

<u>A. If $\tau\dot{\gamma} \gg 1$</u> (nevertheless not to high in order to avoid any turbulence effects), eq (2.12) and (2.15), to the first order in $(\tau\dot{\gamma})^{1/2}$, can be written in the form (2.11a) with

$$\eta_C \to \eta_\infty = \eta_p \left(1 - \frac{1}{2} k_\infty \phi\right)^{-2} \tag{2.17}$$

$$\sigma_0 \to \sigma_{py} = \frac{\eta_\infty}{\tau} \left(1 - \sqrt{\frac{\eta_\infty}{\eta_0}}\right) \tag{2.18}$$

(2.18) appears as a pseudo-yield stress , σ_{py} , defined by the intersection of the σ axis and the straightline of slope η_∞ which represents the high shear behaviour (see Fig. 2.3, Curve I).

<u>B. If $\phi \to \phi_M$</u> (assuming there exists some(real) packing concentration ϕ_M at rest). Then , the zero shear viscosity $\eta_0 \to \infty$.

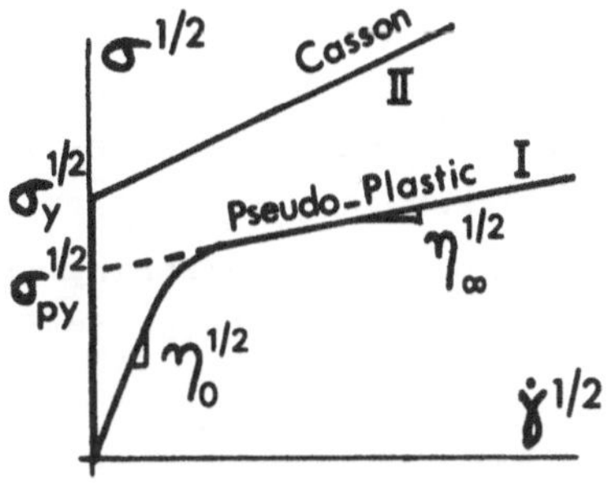

Figure 2.3 Casson Plots
- Curve I
- Curve II ($\phi \to \phi_M$)

Therefore (2.11a) is recovered without any approximation with

$$\eta_C \to \eta_\infty = \eta_p \left(1 - \frac{k_\infty}{k_0}\right)^{-2} \tag{2.19}$$

$$\sigma_0 \to \sigma_Y = \frac{\eta_\infty}{\tau} \tag{2.20}$$

Thus (2.12) and (2.15) exhibit a true yield stress σ_Y simply related to the high shear viscosity η_∞ and the characteristic time τ of the system (see Fig. 2.3, Curve II).

- Shear thickening behaviour

Finally, it is worthy of noting that in some systems, contrary to the usual situation of shear thinning behaviour ($k_0 > k_\infty$), it may occur that $k_0 < k_\infty$. Then, the system exhibits a *shear thickening behaviour* : some examples will be given later.

3. SHEAR VISCOSITY AT VERY HIGH CONCENTRATION

3.1. Dilatant and discontinuous viscosity behaviour.

HOFFMAN (1972) observed in concentrated suspensions of monodisperse polymeric resins, a dilatant viscosity behaviour which is tranformed into a discontinuous viscosity behaviour when the volume fraction of particles is raised above about $\phi = 0.50$. Fitting a parallel plate viscometer, consisting of an upper polished glass and a lower mirrored platen, this author was able to carry out simultaneously
(i) measurements of shear stress and white light diffraction , both at various solid fraction levels and shear rates, and
(ii) photographs of the packed spheres close to the glass plate.
Careful analysis of diffraction patterns exhibit an abrupt change when

the discontinuity in viscosity occurs.

In order to understand such a flow behaviour, he hypothesized that, under shear forces and mutual interactions, particles form surfaces of 2-dimensional, hexagonally packed spheres, which pass one over another in the direction of flow (see Fig. 3.1). Increasing the relative velocity of these layers increases the shear stress inside the

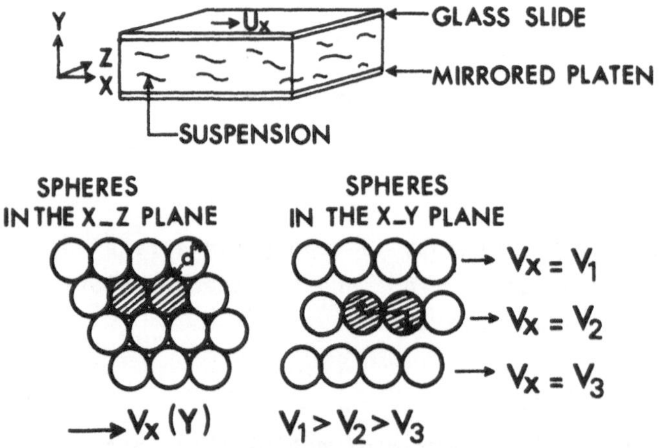

Figure 3.1 Arrangements of packed spheres in plane shear flow. (from HOFFMAN, 1974)

interstitial fluid. At some shear rate value , $\dot{\gamma}_C$, a critical value of shear stress σ_C is finally reached, at which the ordered surface breaks down into parts which roll up into eddies, leading to the onset of disordered flow. Far above the critical shear, ordered flow occurs once again. It is assumed that, after a further increase in shear rate, the above-mentionned eddies are broken into smaller units, which ultimately reduce to individual spheres rotating in ordered surfaces of hexagonal packing.

Viscosity curves $\eta(\dot{\gamma})$ at various volume fractions are shown on Fig. 3.2. At $\phi \leqslant 0.525$ the discontinous behaviour is replaced by a dilatant (or shear-thickening) one. Shear thinning preceeds and follows the critical behaviour.

Fig 3.3 displays variations of the discontinuity shear rate $\dot{\gamma}_{CR}$ against the volume fraction of spheres, with the two limits
$\dot{\gamma}_{CR} \to 0$ as $\phi \to \phi_H = 0.605$ (2d-hexagonal packing) and
$\dot{\gamma}_{CR} \to \infty$ as $\phi \to \phi_C = 0.52$ (cubic packing)

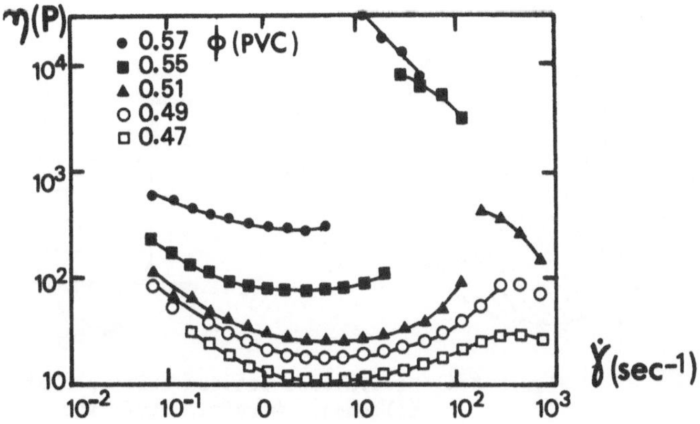

Figure 3.2 Dilatant and Discontinuous Viscosity Behaviour in suspensions of monodispersed PVC spheres (1μm) at various volume fractions. (From HOFFMAN, 1972).

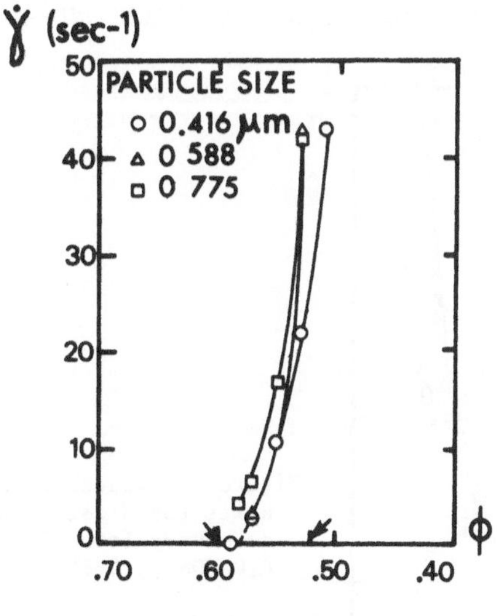

Figure 3.3 Critical shear rate $\dot{\gamma}_{CR}$ for discontinuity in viscosity, as a function of volume fraction ϕ of particles. (From HOFFMAN, 1972).

Similar findings have been obtained by CHENG and RICHMOND (1978) on solid-liquid mixtures at very high concentration, which exhibit "granulo-viscous" behaviour, the main features of which are recalled in the following.

(i) The *stick-slip* phenomenon, illustrated in a rotational viscometer at steady rotation, by a slow rise followed by a rapid fall of the torque, with regular periodicity. Although acceptable theory of stick-slip does'nt exists yet, it is quite obvious that any mechanism capable to explain this phenomenon must account for the presence of a *marginal (or wall) layer*. Indeed the presence of the latter is the more plausible explanation of slip at the wall, as the appearance at large scale (Fig. 3.4b) of an axial shear flow between the wall and the frontier of the core, moving with the velocity v_β

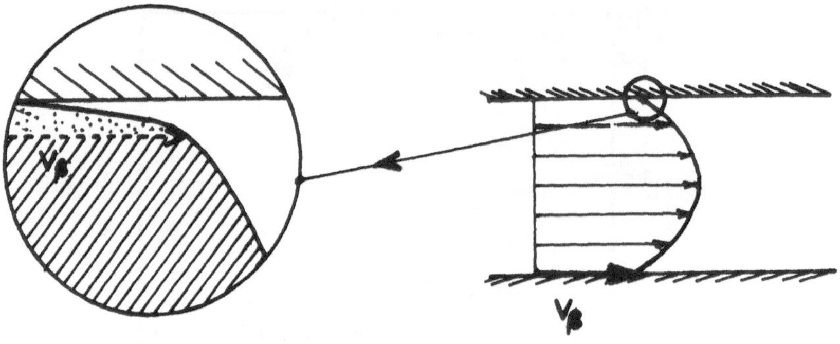

Figure 3.4 Equivalence between existence of
(a) shear marginal layer and
(b) wall-slip with slip velocity v_β .

(Fig. 3.4a).

(ii) The dependence of flow curve on the *current packing concentration* is illustrated by the torque curves (see Fig. 3.5). Although reproductibility of measurements were not complete, CHENG and RICHMOND showed

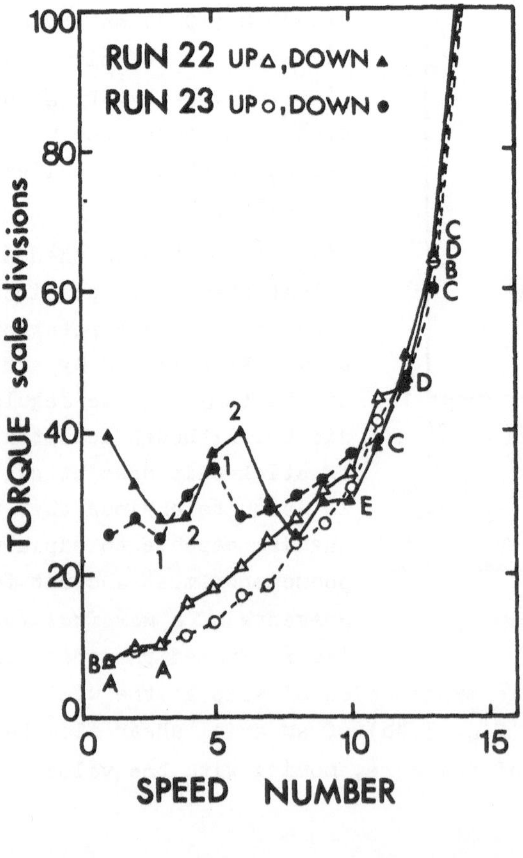

Figure 3.5 Viscometric results for a red lead oxide slurry (from CHENG and RICHMOND, 1978).

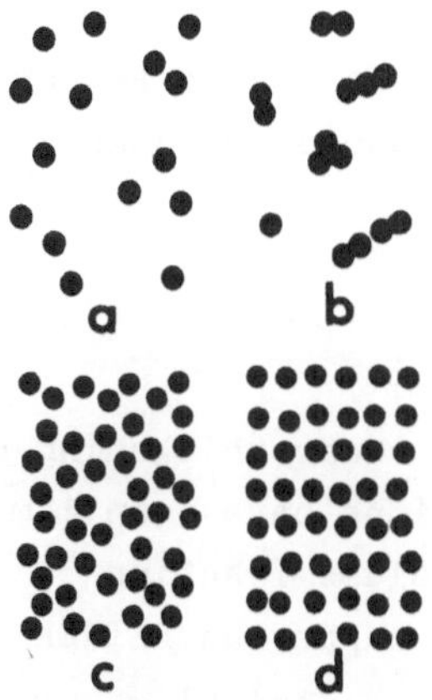

Figure 3.6 Particle arrangements
(a) dilute, random
(b) dilute, ordered
(c) concentrated, random
(d) concentrated, ordered
(from CHAFFEY, 1977)

that curves of peak torque measurements during the stick-slip phenomenon are quite distinct according to the shear rate (∿ speed number) is raised or is lowered. WELTMANN (1960) found very similar behaviour in rheopectic pigment suspensions.

Such a behaviour can be explained in terms of shear induced changes in *particle arrangement* which depends on volume fraction and on shear rate, as CHAFFEY (1977) pointed out (Fig. 3.6). Deviations from random (a,c) to ordered arrangements, can be obtained by clustering (b) or alignment into layers (d). Concentrated systems (c) and (d) exhibit very high resistance to flow. Even in the later system(it is obvious for the former), the arrangement can be perturbed by a small disturbance, the higher the concentration, the smaller the disturbance.

It seems plausible that increasing the shear rate will lead to form *open structures*, as loose flocs which immobilize part of the suspending fluid. As these flocs, with the entrapped fluid, behave as single entities, a strong increase in the effective volume fraction will give a strong increase in viscosity, i.e. a *shear-thickening* behaviour (or a discontinuous one at higher volume fraction). Nevertheless, if the cohesive forces in such open flocs are not too weak , they can survive lowering the shear rate, and then, one observes a jump to a different flow curve corresponding to a change in packing density, as shown in Fig. 3.5. As a consequence, such a surviving may result in plastic behaviour. In the case of coarse materials, with broader distribution in particle size and shape, the ajustment of packing concentration likely takes place at every step of speed change (CHENG and RICHMOND, 1978), what may be considered in agreement with the actual packing concentration, ϕ_p , introduced in section 2. It must be stressed that the exact response in shear stress depends on instrumental conditions, noticely the rate at which the shear rate is applied and the precise state of the initial particle packing. Generally,after some irreversible change in packing at the initial stages of the measurements, reproducible results may be obtained if the speed cycle is repeated. Under such conditions, one may observed a cyclic response, i.e. a *structural hysteresis* which is exhibited by number of highly concentrated disperse systems.

3.2. Rheological models.

Theoretical investigations into the shear-thickening-dilatant behaviour of very highly concentrated suspensions have been limited to phenomenological equations viscosity-versus volume fraction ϕ and viscosity-versus shear rate $\dot{\gamma}$. Many studies supported REINER's diagrams depicting a flowing dilatant suspension, with parallel layers of spheres sliding over one another after occuring of some volumetric expansion (Fig. 3.7, (α)→(β)), which is evidenced by surface drying. Nevertheless, as the latter was observed at shear stresses well below those at which the flow curve began to exhibit rheological dilatancy, some authors claimed that drying and rheological dilatancy (as shear-thickening) are not clearly related. This will be certainely true for systems of particles with strong interaction, for which rheological dilatancy may result from a more or less strong and abrupt increase in the number of bonds between particles, as the shear rate increases (i.e. shear induced particle aggregation). This leads to a maximum in viscosity beyond which shear-thinning is sometimes recovered. For example, GILLESPIE (1966) gave a model of this type, based on an extension of the impulse theory of viscosity of GOODEVE (1939) who assumed that hydrodynamic (newtonian) effects, σ_N and particle interaction effects, G , are simply additive, leading to the constitutive relation

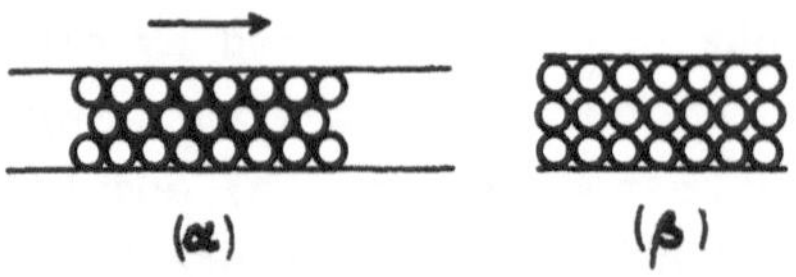

Figure 3.7 Volumetric dilation of a close packing of spheres under shear.

$$\sigma = \sigma_N + G \quad , \quad \sigma_N = \eta_\infty \dot{\gamma} \tag{3.1}$$

from which a "dilatant" viscosity was derived such as

$$\eta = \eta_\infty \left[1 + \left(\frac{\eta_0}{\eta_\infty} - 1\right) \frac{1 + \lambda \beta \tau \dot{\gamma}}{(1 + \beta\tau\dot{\gamma})^2} \right] \tag{3.2}$$

where $\eta_0 = \eta(\dot{\gamma}=0)$, $\eta_\infty = \eta(\dot{\gamma}=\infty)$

β = rate constant for shear induced link-rupture at unit shear rate,

τ = an average relaxation time,

$\lambda = N_\infty/N_0$, N_∞ and N_0 being the number of links per cubic centimeter at infinite and zero shear rate, respectively.

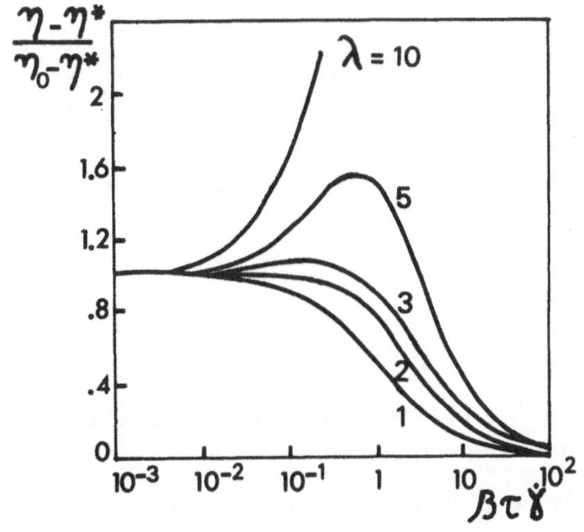

Figure 3.8 Shear thinning and shear thickening behaviour of the Goodeve-Gillespie's model (from GILLESPIE, 1966).

Fig. 3.8 shows continuous change from a pure shear thinning behaviour to a shear thickening or a dilatant one. Shear-thinning is obtained when $\lambda < 2$. But at higher values, $\lambda > 2$ relatively moderate shear thickening leads to a maximum in viscosity followed by a η-decrease.

However, the particle concentration dependence of viscosity does not appear in this model, though this dependence plays a crucial role in shear thickening and dilatancy as Hoffman's measurements have shown (see Fig. 3.2). In order to take these concentration effects into account one can use, tentatively, the relation (2.12)-(2.13)

$$\eta_r = (1 - \frac{\phi}{\phi_p})^{-2} \tag{3.3}$$

where ϕ_p is the "actual packing concentration", the shear dependence of which leads to non-newtonian behaviour of the system. From (2.15) and (2.13), one can take ϕ_p as a function of $\theta = \tau\dot{\gamma}$, in the following form

$$\phi_p = \phi_\infty \frac{1 + \theta}{\frac{\phi_\infty}{\phi_0} + \theta} \tag{3.4}$$

where $\phi_0 = 2/k_0$ and $\phi_\infty = 2/k_\infty$ are the (actual) zero shear and infinite shear packing concentrations, respectively. As it has been seen in section 2, a shear-thickening system is obtained when $k_0 < k_\infty$, i.e. if

$$\phi_\infty < \phi_0 \qquad (3.5)$$

In terms of actual packing concentration, eq.(3.5) means that increasing the shear rate lowers the compacity of the packing which would observed at low shear rate, what is in complete agreement with the dilatancy concept.

As the viscosity (3.3) remains only finite if $\phi < \phi_p$, a semi-quantitative interpretation of results of Hoffman (Fig. 3.2) can be given, especially the discontinuous behaviour in viscosity. From (3.4), the condition $\phi < \phi_p$ becomes

$$\theta\left(\frac{\phi}{\phi_\infty} - 1\right) < 1 - \frac{\phi}{\phi_0} \qquad (3.6)$$

(3.5) & (3.6) lead to consider two domains in volume fraction

(i) if $\phi < \phi_\infty$, (3.6) is always fulfilled and the suspension is a shear thickening system, with continuous increase in viscosity

(ii) if $\phi_\infty < \phi < \phi_0$, (3.6) defines a critical value of shear rate, $\dot\gamma_{CR}$, at which the suspension becomes a solid ($\eta \to \infty$)

$$\dot\gamma < \dot\gamma_{CR} = \frac{1}{\tau}\left[\frac{1 - \frac{\phi}{\phi_0}}{\frac{\phi}{\phi_\infty} - 1}\right] \qquad (3.7)$$

Fig. 3.9 illustrates these two domains, for two typical concentrations $\phi_1 < \phi_\infty$ and $\phi_\infty < \phi_2 < \phi_0$.

Remark. In the case of shear thinning behaviour, $\phi_0 < \phi_\infty$, there exists a critical shear rate, but a minimum, defined by $\theta(1 - \phi/\phi_\infty) > \phi/\phi_0 - 1$, when $\phi_0 < \phi < \phi_\infty$, which is more difficult to reach since it requires to increase the volume fraction above the packing value at rest, ϕ_0 , (that is to add particles to the flowing suspension, that is quite unusual).

It is believed that, in terms of *apparent viscosity*, the over-simplified description given above corresponds to the main features observed by Hoffman. Nevertheless, these observations (see Fig.3.2) exhibit shear thinning at both low and high shear rates. The former can be considered as resulting from the progressive disruption of transient open structures formed at low shear rate (which would correspond to a looser packing $\phi_0' < \phi_0$), reaching to a more regular arrangement of particles

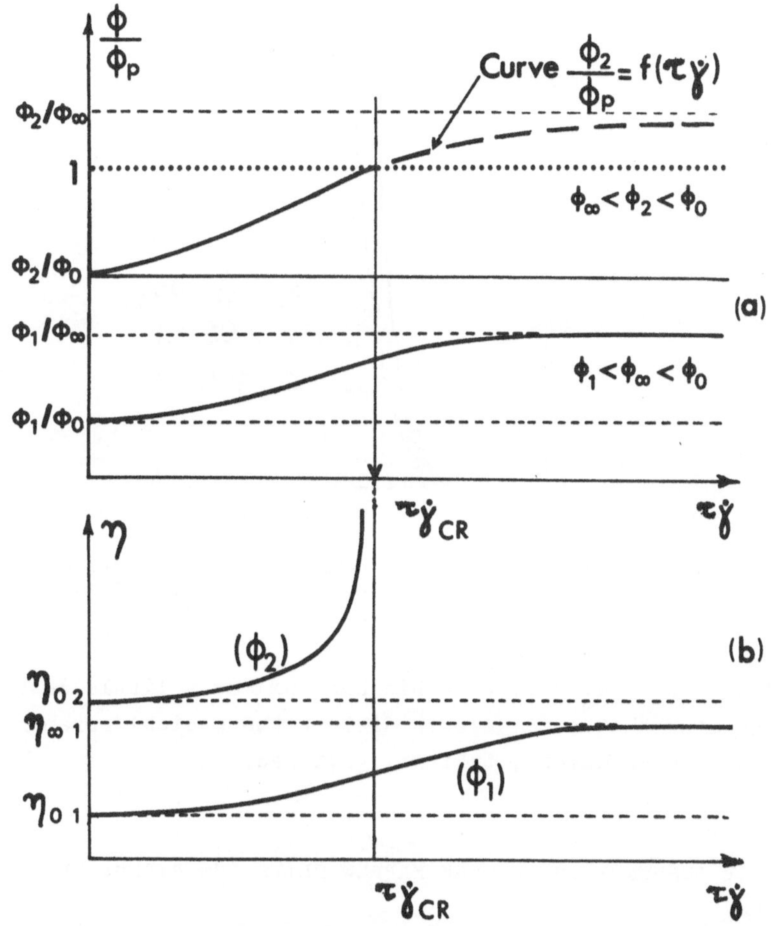

Fig. 3.9 **Model for dilatant (ϕ_1) and discontinuous (ϕ_2) viscosity behaviour.**
(a) actual packing concentration and critical shear rate
(b) viscosity curves

(as the 2-D hexagonal layers described by Hoffman). The latter involves a dramatic structural change, likely the development of marginal layers, since at $\dot{\gamma} \gtrsim \dot{\gamma}_{CR}$, the scale of the dispersed entities becomes comparable with the characteristic length of the apparatus, followed, when $\dot{\gamma}$ increases, by progressive reduction in size of these structural units (i.e. Hoffman's eddies), resulting ultimately in a more or less regular array of single particles, as Hoffman suggested. Therefore, precise comparaison between the expectations of our model and the observations of Hoffman are impossible. Nevertheless, the ϕ-dependence of $\dot{\gamma}_{CR}$, as shown on Fig 3.3, is in fair agreement with eq (3.7), where ϕ_0 and ϕ_∞ are chosen such as $\phi_0 \equiv \phi_H = 0.605$ (2D-hexagonal

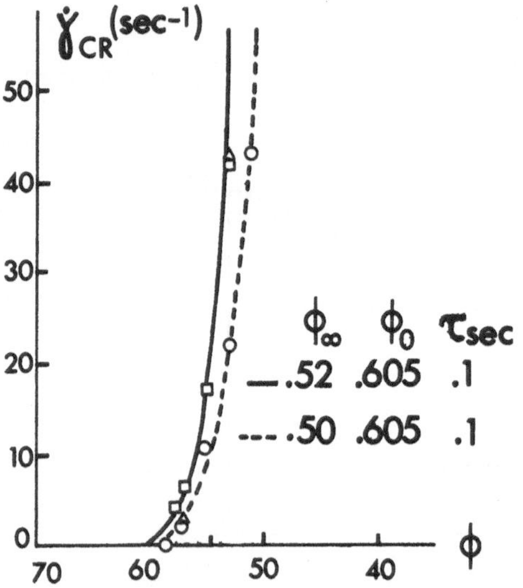

Figure 3.10 Critical shear rate $\dot{\gamma}_{CR}$ against ϕ according to eq (3.7). Comparison with Hoffman's data (see Fig. 3.3).

packing) and $\phi_\infty \equiv \phi_C = 0.52$ (cubic packing) or a little less ($\phi_\infty = 0.50$, assuming that some defects disturb the cubical arrangment). Fig. 3.10 shows such agreement, taking $\tau = .1$ sec.

4 . FLOW OF SUSPENSIONS THROUGH NARROW SLITS AND PIPES. FLOW INSTABILITIES.

After recalling some characteristic properties of flowing suspensions, special emphasis will be given on structural hysteresis. In special circonstances, the latter might lead to flow instabilities, the (speculative) prediction of which will be therefore presented.

4.1. Concentration redistribution in shear flow of suspended particles.

BRANDT and BUGLIARELLO (1966) studied experimentally the concentration redistribution and the stability in the shear flow of monolayers of suspended particles. The suspension consists of a single layer of rigid spherical particles flowing through a long rectangular channel, the section of which being a narrow slit (minor dimension = 1.6 d , d = particle diameter), with moderate Reynolds numbers $100 \leqslant Re \leqslant 400$,

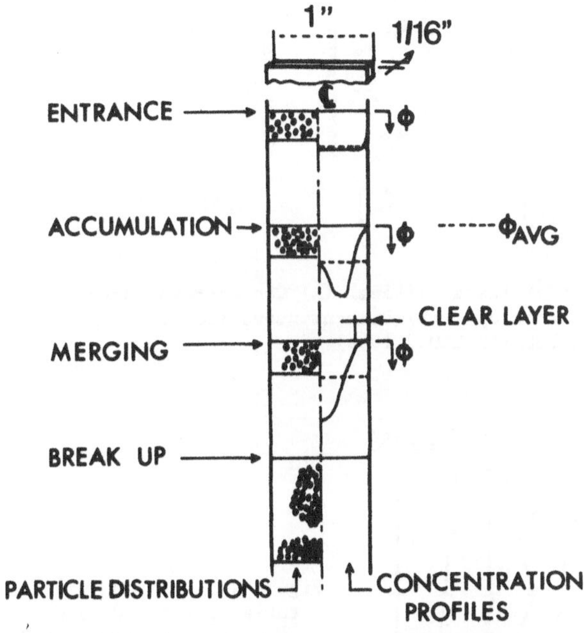

Figure 4.1 Broad outline of the sequence of the events from the channel entrance up to the region where the pattern becomes instable (From BRANDT and BUGLIARELLO, 1966).

and particle volume fraction between 1.7 and 5% .

Fig. 4.1 shows consecutive regions which have been found by detailed analysis of stroboscopic photographs of particle motions. Measurement of average particle distribution in the direction of the broad dimension of the channel were carried out after subdivision of the channel width into a fixed number of bands and counting the total number of particles within each band as a function of the down-flow distance Y .

a) A uniform particle disbution was found in the initial region.

b) Further down the channel, a second region is characterized by the formation of a particle-free wall layer by particle motion towards the center of the channel. Non-uniform distribution in the core was observed, with concentration peaks at the outer margins of the core.

c) A third region exhibits a wider particle-free layer surrounding a nearly uniform concentrated core, after merging of the marginal peaks.

d) A last region characterized by the onset of instabilities in the core configuration, followed by break up of the core, with large groups of particles separated by axial gaps of suspending fluid .

The development of these instabilities is shown schematically on Fig. 4.2. Asymetric expansion of the core is followed by the oscillation about the centerline of the channel. Then the sinuous core breaks up in groups of particles which often span the entire channel width and which are separated in the longitudinal direction by regions containing only suspending fluid.

This development is accelerated increasing the flow velocity

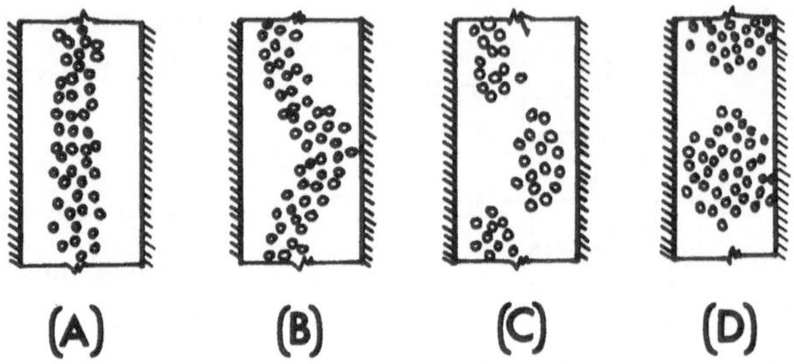

Figure 4.2 Development of instabilities. (A) Concentrated core ; (B) Wavy core ; (C) Break up ; (D) Discontinuous flow . (From BRANDT and BUGLIARELLO , 1966).

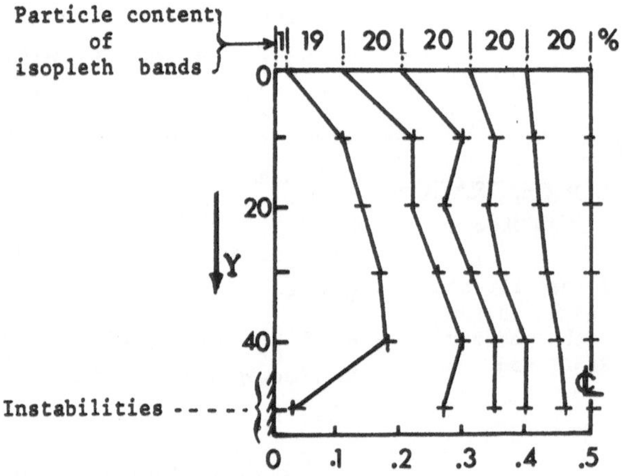

Figure 4.3 Isopleth trajectories . $\phi = 0.05$; $Q = 24.9\ cm^3/sec$. (From BRANDT and BUGLIARELLO, 1966).

or decreasing the concentration, as shown in Fig. 4.3, which is a usefull synthetic description of particle migrations. The half-channel profile of initial concentration is subdivided into six bands from the center line to the wall : four 20 %-internal bands, one 19 %-band and a 1 %-band, close to the wall, which has been arbitrarily choosen as a measure of the particle-free, or particle depleted wall-layer. Further down the channel, one can see the progressive increase in the 1 %-layer thickness, up to a location where the onset of instabilities occurs simultaneously with a rapid expansion of the core. Two domains in flow rates can be distinguished. At the lower ones, the onset of instability exhibits a kind of yield value of core concentration, while at the higher flow rates,once such a limiting value has been reached, the onset becomes flow rate dependent.

As the above-described instability of the wall layer was observed at low concentration ($\phi \sim 5\%$), the extension to (very high) concentrations needed for dilatancy would be questionable. Nevertheless these observations point to the presence of hydrodynamic forces(as a wall-particle interaction) leading to the formation of a marginal layer which, increasing the flow rate (hence the wall shear rate), becomes unstable, i.e. undergoes a more or less abrupt reduction in thickness, in consequence of the core expansion under particle-particle interactions. It is believed that, the same processes are at least qualitatively involved in the order-disorder transition of the wall layer, observed by Hoffman. In particular, increasing the concentration reduces the wall layer thickness and flattens the velocity profile (i.e. increases the wall shear rate).

4.2. Laminar flow regimes for suspensions of large rigid spheres.

Flows of suspensions of spheres through circular pipes were studied experimentally by SACKS and TICKNER (1966). In the case of small values of tube-to-sphere diameter ratio ($1.1 < \xi < 5$), they observed different regimes according to particle concentration ϕ and tube Reynolds number Re. Omitting from the present discussion the case of the larger particles, the low ϕ regimes are

(i) the classical Poiseuille flow at low Reynolds number Re, and

(ii) as Re is increased, the "streak flow" (with tubular pinch).

Unsteady flow with both radial and axial particles motions at low Re and intermediate concentrations ($0.01 < \phi < 0.30$) showed a flat velocity profile associated with a relatively large axial core of particles surrounded by few particles rolling along the wall ("core-tumbling flow"). Radial motions of the core as a whole lead to appearance of a snaking axial motion with some wall particles as ball bearings for the core. Increasing Re left the core almost unchanged, but no particles remained close to the wall ("core-skimming flow"), leading to a reduction in apparent viscosity.

Increasing the concentration above .45 produced a plug-flow with a core formed by packed particles, at low Re or higher Re . Under certain conditions, relatively steady plugs were observed, separated by more or less regular intervals of unsteady tumbling flow.

Although extension of continuous description to such small ξ-values appears difficult, these obervations again underline the importance of the wall layer characteristics in analysing flows of suspensions, noticely its dependence with concentration and flow rate.

Number of experimental studies of flows of suspensions of spheres were carried out, especially to elucidate two main features of blood circulation in narrow vessels : the formation of a marginal layer and the concomitant reduction of the tube averaged concentration ϕ_t in comparison with the feed concentration ϕ_a (the so-called Fahraeus effect in blood flows).

SESHADRI and SUTERA (1970) also demonstrated the importance of the tube-to-sphere diameter ratio ξ for the interpretation of flow-pressure drop data of coarse, concentrated suspensions in tube flow. They found that ξ-values larger than about 50 must be reached before the suspension can be treated as an homogeneous medium characterized by a ϕ-dependent viscosity, or a ϕ-dependent apparent fluidity $F_a = (\eta_{ra})^{-1}$ such as $F_a = F_a^\infty(\phi)$, where F_a^∞ is the limiting value of F_a as $\xi \to \infty$. At lower ξ-values, wall effects led to radially inhomogeneous concentrations, that is both wall layer formation and concentration reduction. These authors measured the concentration reduction as a function of ξ for various ϕ_a-values (Fig. 4.4a). Simultaneous measurements of F_a were compared to the F_a-values which would be predicted by inserting ϕ_t (instead of ϕ_a) into the $F_a^\infty(\phi)$ equation. Fig. 4.4b illustrates this comparison for $\phi_a = .40$, showing that for ξ-values greater than about 20 , F_a can be approximated by F_a^∞-values. In this region, as ξ decreases, the fluidity increases by wall layer lubrication up to a point at which it begins to decreases. At lower ξ values F_a exhibits a marked

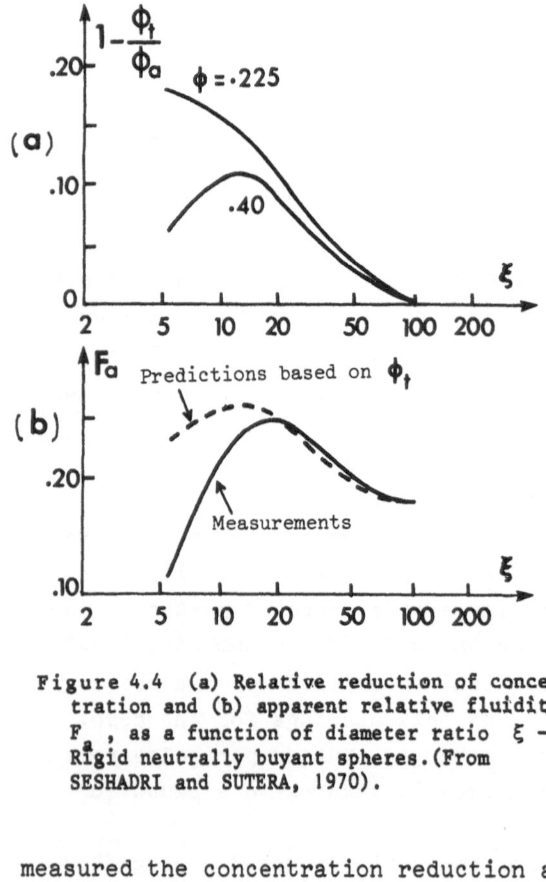

Figure 4.4 (a) Relative reduction of concentration and (b) apparent relative fluidity F_a , as a function of diameter ratio ξ - Rigid neutrally buyant spheres. (From SESHADRI and SUTERA, 1970).

lowering (more rapid than that of $F_a^\infty(\phi_t)$) as ξ is lowered. This can be interpreted as the effect of gradual development of plug flow which progressively fills the tube, increasing the wall-particle interactions, up to a point at which "particle at the fringe of the plug are seen to be dragged, rolling or slipping along the wall".

4.3. Multi-branched flow curve.

A very important feature of the above-discussed connexion between flow and structure is the possibility for the system to reach some conditions at which the flow curve is multi-branched. For example, as seen on Fig. 3.5 yet, different shear stress values were obtained increasing or decreasing the shear rate. As it has been stressed in the previous sections, the more plausible explanation is based on flow induced structural changes.

The existence of a multi-branched flow curve requires as shown below, a sufficiently large and rapid variation of viscosity. Two classes of systems exhibit such variations.

a) Systems governed by intense interactions between particles.

In such systems, a shear-dependence of structural units leads to a pseudo-plastic behaviour. For instance, in colloidal systems, Brownian collisions promote the formation of flocs which are broken down by shear forces, the higher the shear rate, the smaller the floc size.

The relevant parameter is the ratio of high shear viscosity to the lowshear one $\mu^2 = \eta_\infty/\eta_0$, the value of which must be lower than some critical value μ_c^2 in order to obtain a multivalued shear rate at given shear stress. Fig. 4.5 displays such a result in the case of the model eq.(3.3, 3.4) previously discussed. Therefore, a S-shaped flow curve (Fig. 4.5b) will be obtained if particle aggregation at low shear rate is strong enough, a condition which is not very restrictive. It is worthy of note that shear-thickening ($\mu>1$) plays a contrary role. Similar result is observed using the Gillespie's model (3.2) : getting a S-shaped curve requires both large values of η_0/η_∞ and small values of the shear thickening parameter λ.

Returning to the expression (2.15) for the structural parameter k, it must be stressed on the fact that more complicated behaviour could be taken into account using more general shear dependences of the rate constants τ_A^{-1} and τ_D^{-1}. In this direction, JOLY (1958) proposed

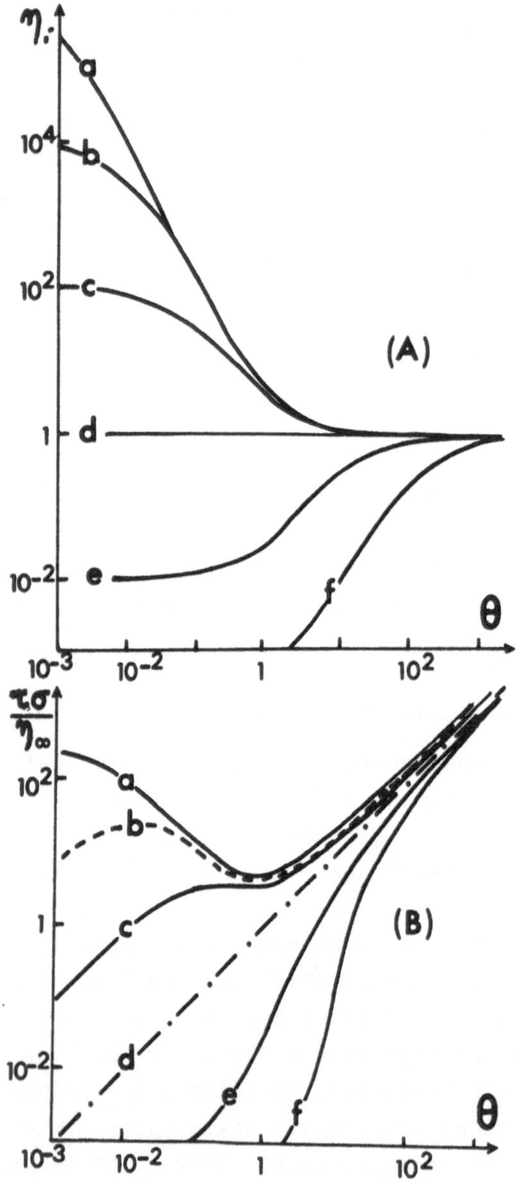

Figure 4.5 Model Viscosity (A) and Flow curves (B) according to eq. (3.3, 3.4) .
$\sigma = \frac{\eta_\infty}{\tau}\,\theta\left(\frac{1+\theta}{\mu+\theta}\right)^2$, $\theta = \tau\dot{\gamma}$, $\mu = \left(\frac{\eta_\infty}{\eta_0}\right)^{1/2}$
$\mu_c = 1/9$. Curves for different μ-values : $\mu = 10^{-3}$(a) ; 10^{-2}(b) ; 10^{-1}(c) ; 1(d) ; 10(e) ; 10^2(f).

a kinetic model for the flow dependence of aggregate size.

b) Systems with weak particle interactions.

In this case, the viscosity ratio η_0/η_∞ cannot reach a large enough value, then a monotonic increasing flow curve is observed, leading to a single-valued shear rate. However as shown on Fig. 3.2, dilatant effects, at very high volume fraction, provide us with a way of reaching very large viscosity values, at $\dot{\gamma}$ close to $\dot{\gamma}_{CR}$. Beyond this critical value, shear thinning behaviour is recovered, as resulting from some re-formation of marginal layers and/or rearrangement within the core. (It must be stressed on that the presence of such lubricant layers will avoid the *apparent* viscosity to reach an infinite value at $\dot{\gamma} = \dot{\gamma}_{CR}$). Fig. 4.6 schematically displays the occurence of multi-valued rate associated with dilatant viscosity.

As CHAFFEY (1977) discussed, continuous changes in the viscosity curves must be observed increasing the volume fraction ϕ that tends to transform a pure shear-thinning curve into one showing gradual or abrupt shear-thickening (see also Fig. 3.2). But due to great differences between the systems under

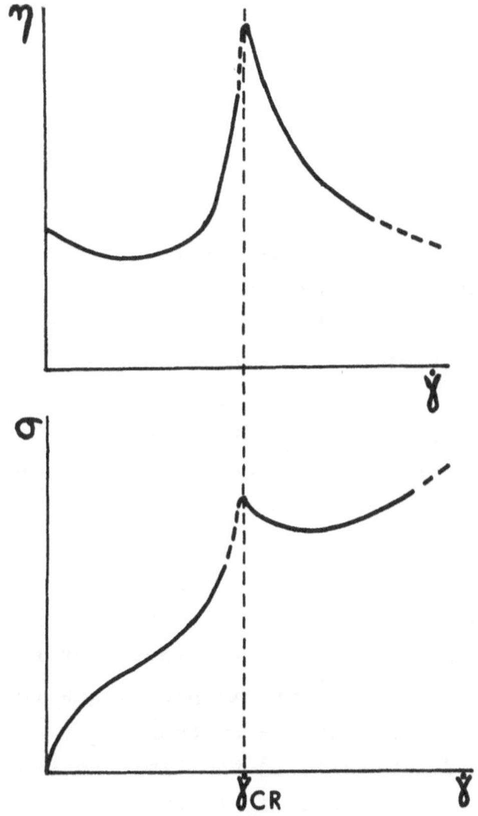

Figure 4.6 Shear rate variations of viscosity and corresponding flow curve. Dilatant Systems.

study, on the one hand, and the methods of measurement on the other hand, only parts of these curves can be seen through the whole $\dot{\gamma}$-range.

Many authors reported anomalous behaviour of shear thickening systems. For instance, GILLESPIE (1966) observed that beyond a critical shear rate the suspensions climbed out of the cup in a more complex manner than in the well-known Weissenberg effect, due to normal stresses. Moreover, some fraction of the sample was pumped out the gap. Others, as STRIVENS (1976), found that hysteresis effects almost always existed with, raising $\dot{\gamma}$, a steep increase of σ at first, then more slow, followed by an abrupt drop to a relatively low value, and then falling during the remainder of the cycle. Drying effects have also been observed (see, for ex. METZNER and WHITLOCK, 1958, who found visual and audible fracture of the material). Therefore, most of investigations using Couette viscometers were carried out avoiding these instrumental difficulties, i.e. by not exceeding the critical shear rate. Generally, this limitation prevents any hysteresis effect to appear in flow curves.

Nevertheless, as shown before, some studies were devoted to analysis of abnormal behaviour of disperse systems. It has been mentionned yet that drying effects have been found in granulo-viscous materials, by CHENG and RICHMOND (1978) who reported observations of visual fracture inside the sample, effects which are explained as consequence of the development of a *slip surface*. Using either a smooth or a rough bob, these authors proposed to locate the slip surface close to the wall (where the shear rate is highest) in the former case, but in the bulk of the sample in the later case, since solid particles are "positively captured" on the rough bob. As these authors stressed on, such an interpretation calls for the great importance played by normal stresses which could reduce the "thickness" of the slip surface.

4.4. Structural hysteresis.

Whatever the processes involved into the structural changes may be, the existence of different behaviour as $\dot{\gamma}$ is increased or decreased will lead to *hysteresis phenomena*, with two different branches in the flow curves, as it has been yet observed in certain polymer systems.

Under *dynamic* conditions, the system does not follow the equilibrium flow curve, because of thixotropic (and viscoelastic) effects which lead the system to exhibit some delay in response to any modification of flow conditions. Moreover, these structural changes are governed by relaxation times (as τ_A and τ_D in (2.14)) which generally appear as structure-dependent, i.e. shear-dependent times. As a consequence this may result in very complex behaviour, due to the fact that rising the shear rate from zero up to a value $\dot{\gamma}_1$ for example, leads to a structural state (S_1) which not only depends on the successive $\dot{\gamma}$ levels met before the $\dot{\gamma}_1$-value is attained, but also on the time the system spent at these different levels. For instance, if all relaxation times are very large in all the $\dot{\gamma}$-range, but a narrow domain close to a certain value $\dot{\gamma}_A$, some structural change would be observed only, and only if, sufficient time was spent to cross this narrow domain, whatever long was the total duration of experiment. If on the contrary, the devation near $\dot{\gamma}_A$ is short enough then the new shear structure will be maintained at high shear rate, as it was "frozen". Conversely, if after reaching high shear rates, one returns slowly to a lower value $\dot{\gamma}_0$, a (new) state, (S_0) can be achieved and maintained at low shear rate because of the shear forces are then not strong enough to break down this new $\dot{\gamma}_0$-state. This may explain the existence of two separate branches in the flow curve, that is the *structural hysteresis* phenomenon.

Since most of the anomalous effects discussed above play a role - although more or less unknown - on shear viscosity, it can be expected that multi-branched flow curve would be a valid model for structure hsyteresis and, as a consequence, for flow instability as it will be conjectured in the next sub-section.

4.5. Flow instability.

The above-described rheological properties of highly concentrated disperse systems leads to a prediction (in some extent speculative) on the onset of flow instability in such systems.

First at all, it is worth noting that qualitative features of the torque-angular speed curves in rotationnal viscometers are also exhibited by the pressure drop-flow rate curves in capillary tubes. The essential difference results from the shear stress inhomogeneity (since $\dot{\gamma}$ must vary from zero on the tube axis up to its (higher) value on the wall, while in Couette flow, the shear stress is constant throughout the gap. In fact the presence of wall layers, which reduce $\dot{\gamma}$ near the center line and, on the contrary, increase $\dot{\gamma}$ near the walls lowers these differences. Therefore, flow curves, as pressure drop P versus flow rate Q, in circular tube of radius R and length L, can be expected similar to the corresponding flow curves shear stress vs. shear rate in rotational viscometers, through the relations giving the apparent shear stress σ_a and shear rate $\dot{\gamma}_a$ in the tube

$$P = \frac{2L}{R}\sigma_a \qquad Q = \frac{\pi R^3}{4}\dot{\gamma}_a$$

A S-shaped curve, as shown on Fig 4.7, will be drawn carrying out flow rate controlled experiments, as it may be achieved using a constant speed plunger which pushes out the suspension from a reservoir through the capillary. This allows us to reach any point of the curve. Conversely, a pressure drop controlled experiment (i.e. using given values of the pressure head) exhibits an unstable behaviour, after the pressure is raised up to $P = P_1$. Just beyond this value, a jump from branch (I) to branch (II), i.e. from A_1 to A_3 occurs. However, such a jump corresponds to a large increase in flow rate which lowers the pressure head at the tube entrance as resulting from up-stream hydraulic resistance. Then the

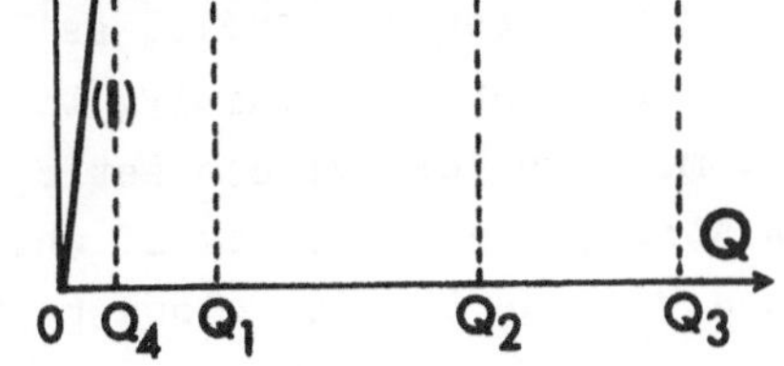

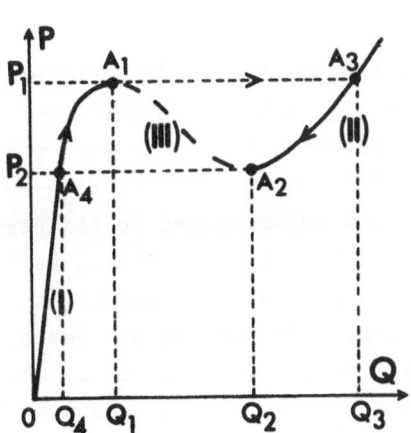

Figure 4.7 Double-branched flow curve, showing a relaxation cycle, $A_1A_3A_2A_4$.

system moves along the portion II of the curve, from A_3 to A_2. At this point, unstable conditions are once again achieved, and further decrease of pressure, below $P = P_2$, leads to a jump $A_2 \rightarrow A_4$, where the flow rate Q_4 is too small and allows us to increase,then to recover the applied pressure P. Hence, the pressure increases up to P_1 (from A_4 to A_1), leading the system to begin again to follow the cycle* $A_1A_3A_2A_4$. Therefore, an *oscillatory flow* is expected, with pressure and flow rate amplitudes $\Delta P = P_1 - P_2$ and $\Delta Q = (Q_3 - Q_4)$, respectively. These oscillations belong to the general class of *relaxation oscillations*.

Such a phenomenological model might explain number of "abnormal" features observed in flows of highly concentrated suspensions, and in some extent of polymer solutions and melts. For instance, VINOGRADOV and MALKIN (1966) compared flowing polymers and flowing plastic disperse systems, and found in both systems comparable time dependences (Fig. 4.8).

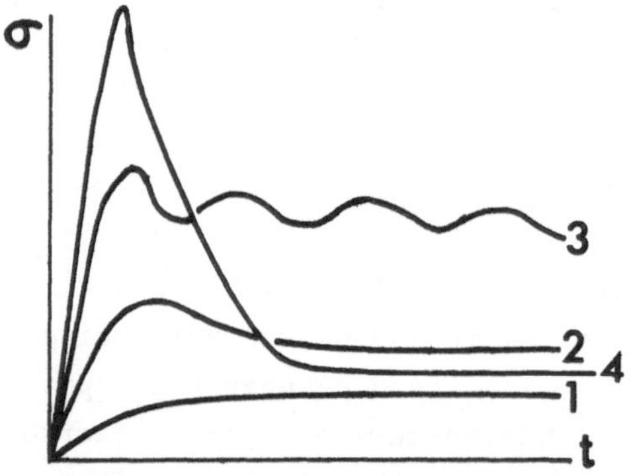

Figure 4.8 Time curves of torque at different constant angular speed. (1) at low speed - (2) at moderate speed - (3) periodic oscillations of torque - (4) sharp peak followed by strong lowering of torque due to progressive formation of lubricant wall layers. (From VINOGRADOV and MALKIN, 1966).

The low speed time curve of the torque (Fig. 4.8,1) corresponds to the linear portion of branch I in Fig. 4.7, while the overshoot in time curve (2) at moderate speed involves the non-linear portion of the same branch I. As evidence, initial parts of both time curves (1) and (2) could include elastic effects superimposed to thixotropic ones. Time curve (3) is believed related to a (small) cycle, as $A_1A_3A_2A_4$ in Fig. 4.7, leading to relaxation oscillations. Lastly, the time curve (4) begins as a sharp response, more or less viscoelastic, followed by an irreversible jump** to a new state, the viscosity of which is strongly lowered by the presence of less viscous layers, close to the

*Notice that the portion III (A_1A_2) of the curve is an unstable branch in pressure controlled experiments.

**however reversible on very long times, as resulting from a very slow "healing" of the homogeneous structure.

walls. These authors claimed that for such disperse systems, abrupt oscillations of the stress depend unambiguously on the thixotropic destruction and restoration of the structure, without wall effects in the case of plastic disperse systems, but in polymer systems. What seems not true, since wall effects also exist in suspensions. Especially, small lubricant effects are presumably present in the part A_3A_2 of the oscillatory cycle which belongs to the branch II.

Another feature which can be explained qualitatively by the present model is the above-mensioned *stick-slip* phenomenon. Indeed, in very highly concentrated disperse systems, the volume fraction ϕ is very close to its packing value ϕ_M, then the low shear viscosity is very high (i.e. idem for the slope of OA_4 in Fig. 4.9). Therefore, for the oscillatory cycle, the corresponding flow rate range (Q_4Q_1) corresponds to very small values, that is the flow is so slow that it seems absent. This is the *stick* part of the flow curve. Conversely, beyond A_1, the suspension starts to flow, with some lowering in flow rate, from Q_3 to Q_2, according to the decrease in head pressure. This is the *slip* part of the flow curve. As evidence, the amplitude ΔQ of the flow rate oscillation and its frequency depend on the reduction in the slope of A_2A_3 compared to A_4A_1, that is the viscosity reduction from branch I to branch II, i.e. the importance of wall effects. The higher the wall effects, the larger the amplitude and the smaller the frequency.

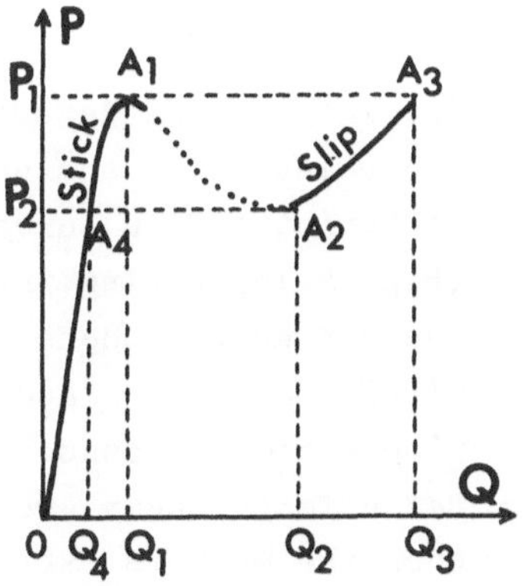

Figure 4.9 Schematic interpretation of the stick-slip phenomenon, from flow curves.

4.6. Similarities with flowing polymers.

Some analogies, at least qualitative, may be found between the present speculations on flow instabilities in concentrated suspensions and many observations on unstable flows of extruded molten polymers through capillary tubes. (see a revue by TORDELLA, 1969). Increasing the shear stresses, a critical value is reached at and above which the emerging stream of polymer exhibits irregular distortion. A number of different types of distorsions, which may involve viscoelastic effects at the entrance or/and emergence of the die, were observed.

A first type of distorsion is the gross, wavy-type shape irregularity, called *waviness,* which is exhibited at high flow rates, above a critical stress. The phenomenon responsible , may be slip, is within the capillary system. Using tracer techniques, slip was observed at the critical stress and above, but not below.

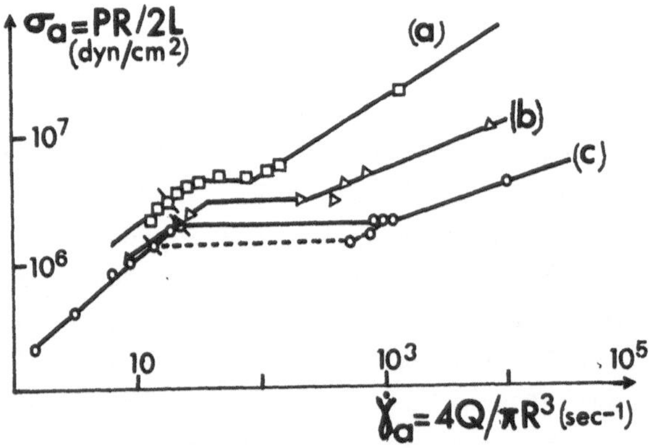

Figure 4.10 Flow data for F^4ethylene/F^6propylene copolymer.
— flow for increasing stress ;
--- flow as stress is decreased from those on the upper branch of the flow curve.
Slash indicates stress at the onset of ripple.
(From TORDELLA, 1969).
(a) L/R = 1.6 ; (b) L/R = 7.2 ; (c) L/R = 37.2

A second, less severe distorsion, called *ripple* is observed at lower flow rates, which consists of a series of peaks and valleys on the surface of the specimen. Such irregular behaviour exhibit several features.

(i) They may be described relative to the characteristic flow curves (Fig. 4.10) taking into account what happens within the tube since ripple and its associated effects occur even when the melt is extreded by piston directly from a tube, i.e. without inlet effects.

(ii) The onset of the ripple occurs at critical shear σ_A (indicated by a slash on Fig. 4.10) ; it depends on the material and finish of the tube wall.

(iii) At and above a second critical stress σ_B, about 50-150 % greater than σ_A, the shear rate tends to be double valued, hence oscillates between two limits, the upper one depending on the tube length to radius ratio L/R.

(iv) Apparent shear viscosity at high shear rate is lowered by increasing L/R , as if the higher the length, the longer the time avaible for thinning the material by shear, leading to decrease the η_∞/η_0 ratio.

These features suggest that the breakdown of uniform flow occurs within the tube , noticely by the development of a wall layer, less viscous than the bulk, which will appear as a wall slip. Such a development is a relevant, may be a determining process in the ripple-phenomenon. While ripple and waviness seem quite different, their basic cause seem to be the same, although they initiate at different site. In such a direction, WEILL (1980) showed that ripple and oscillatory flow (in his case, (high frequency) "sharskin" and (low frequency) main flow instability) both appear as relaxation oscillations, the former occurring in the die entry region contrary to the latter which occurs in the whole die itself.

5. SOME REMARKS AS A CONCLUSION

To conclude the present paper, it isworthy of note that

(i) the (conjectural) flow instability studied here in terms of *structural instability* represents a very special class of unstable flows that one might observe in very concentrated disperse systems.

(ii) Careful experimental investigations of several aspects of these provisional conclusions would be valuable, noticely using simultaneous measurements by different methods as viscometry and direct observation of structural states of the system. Special attention will be devoted to normal stresses and viscoelastic effects in future work.

(iii) Improvements in the aera of very concentrated disperse media seems very important both from the theoretical and practical points of view. On the one hand, theoretical investigations are needed for better understanding of wall layer formation - and the parameters which are involved in it - especially as a dissipative structure. On the other hand, a number of applied studies - as those on granulo-viscous materials - are connected with technological problems rising from industry. They are concerned with flows of the most possible concentrated media, for which one needs to control the conditions for flow stability, for example, pipeline transportation of many products(as heavy-oil/water emulsions, minerals, coals, animal foods,...), energy storage and exchanges...

REFERENCES

BATCHELOR G.K. and GREEN J.T.(1972) - The determination of the bulk stress in a suspension of spherical particles to order C^2. J. Fluid. Mech. 56, 401-427.

BRANDT A. and BUGLIARELLO G. (1966) - Concentration redistribution phenomena in the shear flow of monolayers of suspended particles. Trans. Soc. Rheol. 10, 229-251.

CASSON N. (1959) - A flow equation for pigment-oil suspensions of printing ink type. In "Rheology of Disperse Systems", (ed. Mill C.C.) pp 84-102, Pergamon, London.

CHAFFEY C.E. (1977) - Mechanisms and Equations for shear thinning and thickening in dispersions. Colloid & Polymer Sci. 255, 691-698.

CHENG D. CH and RICHMOND R.A. (1978) - Some observations on the rheological behaviour of dense suspensions. Rheol. Acta. 17, 446-453.

CROSS M.M. (1965) - Rheology of non-newtonian fluids : a new flow equaequation for pseudo-plastic systems. J. Colloid. Sci.20, 417-437.

EINSTEIN A. (1906) - Ann. Physik, 19, 289-306. For english translation see Einstein A., "The theory of Brownian movement", pp. 36-54. Dover N.Y. 1956.

GILLESPIE T. (1966) - Application of the hydrodynamic-structural theory of non-newtonian flow to suspensions which exhibit moderate shear thickening with particular reference to "dilatant" vinyl plastisols. J. Colloid. Interface Sc. 22, 554-562.

HOFFMAN R.L. (1972) - Discontinuous and dilatant viscosity behavior in concentrated suspensions. I. Observation of a flow instability. Trans. Soc. Rheol. 16, 155-173.

HOFFMAN R.L. (1974) - II. Theory and experimental tests. J. Colloid Interface Sci. 46, 491-506.

JOLY M. (1958) - Changements de structure provoqués par l'écoulement. Rheol. Acta 1, 180-185.

KRIEGER I.M. and DOUGHERTY T.J. (1967) - Some problems in the theory of colloids. In "Surface and Coatings Related to Paper and Wood". R. Marchessault, C. Skaar ed. Syracuse Univ. Press.

KRIEGER I.M. and DOUGHERTY T.J. (1959) - A mechanism for non-newtonian flow in suspensions of rigid spheres. Trans. Soc. Rheol.3, 137-152.

KRIEGER I.M. (1963) - A dimensional approach to colloid rheology. Trans. Soc. Rheol. 7, 101-109.

MIDDLEMAN S. (1968) - The flow of high polymers. Interscience Pub.N.Y.

MOONEY M. (1951) - The viscosity of a concentrated suspension of spherical particles. J. Colloid Sci. 6, 162-170.

METZNER A.B. and WHITLOCK M. (1958) - Flow behavior of concentrated (dilatant) suspensions. Trans. Soc. Rheol. 2, 239-254.

OSTWALD W., AUERBACH R. (1926) - Über die viscosität kolloider lösungen im struktur, laminar - und turbilenzgebiet. Kolloid Z.38, 261-280.

QUEMADA D. (1977) - Rheology of concentrated disperse system and minimum energy dissipation principle. I. Viscosity-concentration relationship. Rheol. Acta 16, 82-94.

QUEMADA D. (1978a) - II. A model for non-newtonian shear viscosity in steady flows. Rheol. Acta 17, 632-642.

QUEMADA D. (1978b) - III. General features of the proposed non-newtonian model. Comparison with experimental data. Rheol Acta 17, 643-653.

QUEMADA D. (1982) - Blood rheology and its implication in blood flow. In "Arteries and Arterial Blood Flow (Biomechanical and Physiological Aspects)". pp 3-129 - 1980 CISME Summer School (Udine, Italy) - Springer-Verlag, Berlin (In press).

REE T. and EYRING H. (1955) - Theory of non-newtonian flow. I. Solid Plastic system. J. Appl. Phys. 26, 793-804.

SACKS A.H. and TICKNER E.G. (1966) - Laminar flow regimes for rigid-spheres suspensions.In "Hemorheology", pp 277-303. A.L. Copley (ed) Pergamon Press, Oxford, 1968.

SESHADRI V. and SUTERA S.P. (1970) - Apparent viscosity of coarse-concontrated suspensions in tube flow. Trans. Soc. Rheol. 14, 351-373.

STRIVENS T.A. (1976) - The shear thickening effect in concentrated dispersion systems. J. Colloid. Interface Sci. 57, 476-487.

TORDELLA J.P. (1969) - "Unstable flow molten polymers. In "Rheology : Theory and Applications". Vol V. Eirich F.R. (ed) Acad. Press. N.Y.

VINOGRADOV G.V. and MALKIN A.Y. (1966) - Comparative description of the peculiarities of deformation of polymer and plastic disperse systems. Rheol. Acta. 5, 188-193.

WEILL A. (1980) - About the origin of sharkskin. Rheol. Acta. 19, 623-632.

WELTMANN R.N. (1960) - Rheology of pastes and paints. In "Rheology : Theory and Applications" pp 189-248. EirichF.R. (ed). Acad. Press. N.Y.

WILLIAMSON R.V. (1929) - The flow of pseudoplastic materials. Ind. Eng. Chem. 21, 1108-1111.

APPENDIX

Basic rheological concepts. (see for instance MIDDLEMAN, 1968).

The constitutive relation involves (a) the total *stress tensor* T_{ij}, which is usually split into a pressure term, $-p\delta_{ij}$, and the dynamic (viscous) term σ_{ij},

$$T_{ij} = -p\,\delta_{ij} + \sigma_{ij} \quad , \quad p = -\frac{1}{3} T_{ii}$$

and (b) the *rate of deformation tensor*, d_{ij}, which characterizes the flow

$$d_{ij} = (\nabla v)_{ij} + (\nabla v)_{ji}$$

where ∇v is the velocity gradient tensor.

For a newtonian fluid, the constitutive relation is

$$\sigma_{ij} = \eta_N\, d_{ij}$$

where η_N is a constant, called the (coefficient of) viscosity.

Usual methods of measurement are based on simple flow geometries (i.e. viscometric flows). The simplest one (the easiest measurement) consists of the parallel plates forming a narrow gap whose distance is very small compared to plate width (such a geometry approximates the more realizable ones as two coaxial cylinders or cone-plate systems). This flow is the *simple shear flow*. Another flow is the *elongational* (or *extensional*) one, sometimes called *pure shear flow*) which is of very pratical importance (especially in polymer processing, for instance in fiber spinning).

Simple shear flow

The velocity field $\mathbf{v} = (v_1(x_2), 0, 0)$ gives

$$d = \begin{pmatrix} 0 & d_{12} & 0 \\ d_{21} & 0 & 0 \\ 0 & 0 & 0 \end{pmatrix}, \qquad |d| = \dot{\gamma}(x_2)\begin{pmatrix} 0 & 1 & 0 \\ 1 & 0 & 0 \\ 0 & 0 & 0 \end{pmatrix}$$

where the magnitude of $d_{12} = d_{21}$ is called the ***shear rate***, $\dot{\gamma} = \dot{\gamma}(x_2)$.

The generalized *shear viscosity* is then defined as

$$\eta = \frac{\sigma_{12}}{\dot{\gamma}}$$

which remains $\dot{\gamma}$-dependent for number of fluids (i.e. non-newtonian fluids).

In addition to shear stresses, normal stresses exist which can be characterized by material functions ψ_{ij} (for normal stress differences) such as

$$\psi_{ij} = \frac{\sigma_{ii} - \sigma_{jj}}{\dot{\gamma}^2}$$

Simple elongational flow

The rate of deformation tensor is then

$$d = \begin{pmatrix} d_{11} & 0 & 0 \\ 0 & d_{22} & 0 \\ 0 & 0 & d_{33} \end{pmatrix} = \dot{e}\begin{pmatrix} 2 & 0 & 0 \\ 0 & -1 & 0 \\ 0 & 0 & -1 \end{pmatrix}$$

taking into account the incompressibility of the fluid ($\nabla\cdot v = 0$), where $\dot{e} = \partial v_1/\partial x_1 = \dot{e}(x_1)$ is the *strain rate*.

The *elongational viscosity* is defined by

$$\eta_e = \frac{T_{11}}{\dot{e}}$$

using the *total* stress component T_{11}. In general, for non-newtonian fluids, $\eta_e = \eta_e(\dot{e})$. Nevertheless, the relation between η_e and η

is unknown, excepted in the case of newtonian fluid for which $\eta_e = 3\eta_N$.

Viscoelastic fluids

For unsteady flows, elastic effects become important (i.e. the fluid appears as an intermediate material between pure viscous fluid and perfect elastic solid). Coupling of viscous and elastic effects can be represented by the Maxwell model, where it is assumed that

(i) the flow response ($\dot{\gamma}$) to the application of the stress (σ_{12}) can be split into a viscous response (the shear rate, $\dot{\gamma}_V$) and a elastic one (the elastic deformation, γ_E) respectively related to the stress by the Newton's law ($\sigma_{12} = \eta_F \dot{\gamma}_V$) and the Hooke's one ($\sigma_{12} = G\gamma_E$, where G is the elastic shear modulus).

(ii) the total shear rate $\dot{\gamma}$ results from addition of the corresponding shear rates

$$\dot{\gamma} = \dot{\gamma}_V + \frac{d}{dt}\gamma_E$$

$$= \frac{\sigma_{12}}{\eta_F} + \frac{1}{G}\frac{d\sigma_{12}}{dt}$$

Taking $\lambda = \eta_F/G$ as a constant, leads to the constitutive (differential) equation of a maxwellian material

$$\lambda \frac{d\sigma_{12}}{dt} + \sigma_{12} = \eta_F \dot{\gamma}$$

If a constant shear rate $\dot{\gamma}_0$ is applied to the system at time $t \geqslant 0$, the shear stress will relax according to

$$\sigma_{12} = \eta_F \dot{\gamma}_0(1 - e^{-t/\lambda})$$

where λ appears as a relaxation time.

DISSIPATIVE STRUCTURES AND OSCILLATIONS IN REACTION-DIFFUSION MODELS WITH OR WITHOUT TIME-DELAY

Manuel G. VELARDE
UNED-Física Fundamental
Apdo. Correos 50 487
Madrid (Spain)

1. GENERAL INTRODUCTION

Some time ago (1967) ,I. Prigogine coined the term *dissipative structures* to designate ordering and function specifics to evolution under nonequilibrium thermodynamic conditions in open systems. Dissipative structures (limit cycles or any other nonlinear sustained oscillations, nonuniform space or space and time distribution, multiple steady states,...or even "chaos" and turbulence) cannot appear ,mathematically and thermodynamically, in the *infinitesimal* neighborhood of thermodynamic equilibrium states. Nor they can be sustained in *isolated* systems, closed to matter and energy transfer. Prigogine's seminal ideas were indeed quite fertile as the community was ready for their arrival.The era of great development in linear mathematics was almost over and scientists in practically all realms of Science ,ranging from Physics,Chemistry,Engineering,...to Ecology,Urban Studies,Sociology and Economy,had to face without escape the *actual* nonlinear description of their problems and the *fact* that a strong interaction between any system and its surroundings was at the origin of unexpected processes and evolutionary pathways.

During the past two decades or so,we have witnessed an explosion in research, understanding and great achievements in the mathematical study of nonlinear systems as well as in the philosophical understanding of living matter and some human problems (see, for instance, Eigen,1971;Eigen and Schuster,1979;Fife,1979; Haken,1977; Margalef,1980;Nicolis and Prigogine,1977;Prigogine,1974;Prigogine and Stengers,1979).

The *notes* that follow aim at illustrating such scientific development by the way of *worked* exercises from biology,ecology,biochemistry and semiconductor physics. Other examples not covered here can be found in the material covered by other authors in this book and for cases in fluid dynamics and laser theory in another set of notes by the present author (Velarde,1981). Note that I have excluded from this text all reference to waves and pseudo/kinematic waves (see, for instance, Winfree,1980 and references quoted there). Mathematical background for the notes can be found in modern textbooks on differential equations .

2.- EXAMPLE OF LIMIT CYCLE IN BACTERIAL CULTURES

i. Introduction

The continuos culture of microorganisms has been studied experimentally and theoretically because of the importance of such systems in industrial processes as well as in laboratory studies of the growth physiology. Many cases of sustained oscillations in continuous cultures have been reported(for a review see Harrison and Topiwala,1974). One early example was the finding by Harrison and Pirt (1967) of sustained oscillations of the oxygen concentration in a continuous culture of *Klebsiella Aerogenes*, under conditions of low oxygen partial pressure in the aerating gas. A model to explain these oscillations was proposed by Degn and Harrison (1969).See also Balslev and Degn (1975). This model is based on the assumption that a forward inhibition effect of oxygen on the rate of oxygen consumption is responsible for the oscillations.

Experimental evidence for the existence of a maximal oxygen consumption rate at low oxygen concentrations has been reported (Harrison and Pirt 1967). Even though no details of the regulatory mechanism in the cell responsible for this effect have been elucidated, there is,however, a model proposed by Degn and Harrison in the form of a two reactant chemical reaction scheme.:

$$A \longrightarrow Y \qquad (1.a)$$

$$B \rightleftharpoons X \qquad (1.b)$$

$$X + Y \xrightarrow{(I)} P \qquad (1.c)$$

where X and Y denote oxygen and nutrient,respectively. A and B are "sources" or external parameters whose concentrations are considered constant all over the reactor vessel; P is the final product to which the system is considered open. Step (1.c) is assumed to follow a non-linear rate equation of the type $XY/(1+qX^2)$, where X,Y and q denote concentrations and dimensionless measure of the strength of the inhibitory law (I) in (1.c) respectively. With such a drastically simplified picture of the respiratory process in the bacterial culture, there cannot be any pretension of a quantitative description to be compared with experimental data . The purpose of the model is to qualify the " inhibition by excess of oxygen " as a potential candidate to account for the time-periodic oscillation experimentally observed.

If bracketed quantities denote concentrations, k_i (i=1,...,4) reaction rates, and the k_5 the dimensional power of the inhibitory step in (1.c), under isothermal and continuous stirring conditions, the Scheme (1) leads to the following equations (Fairén and Velarde, 1979).

$$\frac{d}{dt}\{X\} = k_2\{B\} - k_3\{X\} - k_4\{X\}\{Y\}/(1+k_5\{X\}^2) = F(X,Y) \qquad (2.a)$$

$$\frac{d}{dt}\{Y\} = k_1\{A\} - k_4\{X\}\{Y\}/(1+k_5\{X\}^2) = G(X,Y) \qquad (2.b)$$

ii. Heuristic description of the limit cycle oscillations.

A simple heuristic argument can be developped in order to visualize the observed oscillations. Let us take $F(X,Y) = 0$ in (2.a). It has three branches, say $X_1(Y)$ $X_2(Y)$ which are *attractive* , and $X_0(Y)$ which is *repulsive*. The other equation $G(X,Y) = 0$ in (2.b) has two branches; only one intersects with $F(X,Y) = 0$ at a single point (Fig.1). Take now a state (A,B) in the diagram such that B = A and proceed along the A = const.line. At the beginning, a homogeneous steady solution of (2) belongs to the attractive branch $X_1(Y)$, and the vector field converges towards this state. For higher values of B, this steady state still moves along the branch $X_1(Y)$ with decreasing values of Y until the critical point $B = B_c$ is reached. B_c is assumed (see for a proof Fairén and Velarde,1979) to be an unstable focus from which bifurcation to limit cycle oscillation is expected. Slightly above B_c the steady state moves along the repulsive branch $X_0(Y)$ and the vector field (2) diverges from the homogeneous steady state without ,however, going too far as the other two attractive branches restrict the motion, thus forcing the limit cycle operation of the system around the unstable focus. As the steady state moves along $X_0(Y)$ the limit cycle increases in amplitude according to the Hopf bifurcation theorem.

Fig.1.

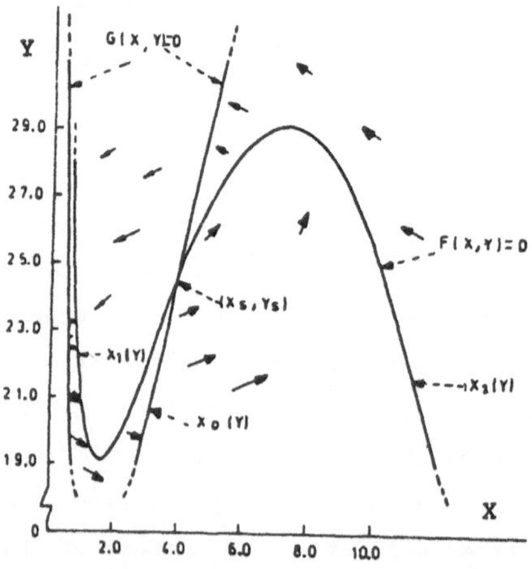

Fig.1.:Vector field generated by (2) in the case of an unstable focus.

Fig.2.:Oxygen oscillations according to the direct numerical integration of (2). Taken from Fairén and Velarde(1979) where an analytical solution can also be found, together with the complete stability analysis of the limit cycle.

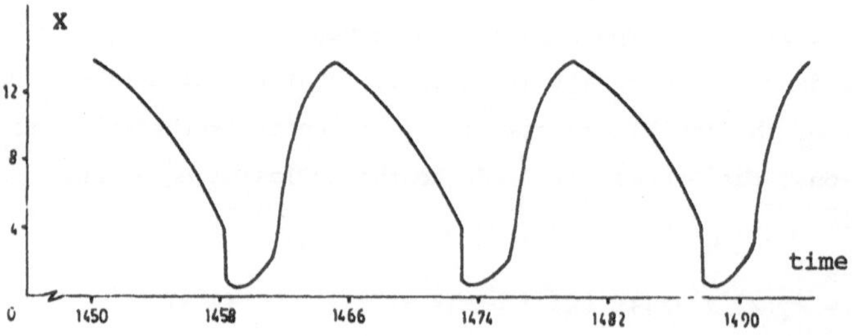

Fig.2.

Oscillations of X, say oxygen, go as follows. First X increases (see Fig.2), that is to say, it jumps from the proximity of the lower part of branch $X_1(Y)$ in Fig.1 to branch $X_2(Y)$. The quicker this jump is, the sharper will be increasing the portion of oscillation. This will strongly depend on values of parameters A,B and q. When X rounds the higher part of branch $X_2(Y)$, there happens a slow decrease in X concentration, followed by a quicker one as it jumps to branch $X_1(Y)$, where it slowly moves until it reaches the initial point. The characteristic saw-tooth shape of the oscillations is depicted in Figure 2 and qualitatively agrees with the experimental data (Harrison and Pirt,1967 ; Degn and Harrison,1969). The sharpness of the increasing and decreasing sections of the oscillations can be changed with the variation of the parameters of the model.

iii. A Stochastic Description of the Limit Cycle Behaviour

If the concentration of the reactants or the more suitable quantities here, the particle numbers $\{n\} = \{n_A, n_B, n_X, n_Y\}$ are considered as fluctuating quantities, some further understanding of the process can be obtained. Generally, the mathematical framework (Master Equation, Fokker-Planck Equations, etc) cannot be handled in practice unless some drastic approximations are made (see for instance Haken,1977; Horsthemke and Brenig,1977; Nicolis and Prigogine,1977; Oppenheim *et al.*,1977; Van Kampen,1976). For this reason we shall limit here to a Monte Carlo simulation of the scheme (1) under the following basic though not too unrealistic assumptions:

(i) The stochastic process is assumed to be Markovian.

(ii) The chemical laws in (1) are assumed invariant with time, and the particle numbers n_A and n_B are considered to be externally controllable only.

(iii) In accordance with the different reactive processes involved in (1) the following four transition probabilities are considered in a volume Ω:

$$A \xrightarrow{k_1} Y:\ W_1(n_X,n_Y;0,1) = k_1 n_A \tag{3.a}$$

$$B \xrightarrow{k_2} X:\ W_2(n_X,n_Y;1,0) = k_2 n_B \tag{3.b}$$

$$X \xrightarrow{k_3} B:\ W_3(n_X,n_Y;-1,0) = k_3 n_X \tag{3.c}$$

$$X + Y \xrightarrow{k_4,k_5} P:W_4(n_X,n_Y;-1,-1) = = (k_4/\Omega)n_X n_Y/\{1+(k_5/\Omega^2)n_X^2\} \tag{3.d}$$

which amounts to consider that at every reactive collision there is only a unit-jump per unit time.

(iv) Let $\{n\} = \{n_A, n_B, n_X, n_Y\}$ denote a state of the system and $\{r\}$ the change at every reactive process. Then if $P(\{n\},t)$ is the probability density of the state $\{n\}$ at time t, its time evolution is governed by the following Master Equation (see, for instance, Oppenheim *et al.*,1977)

$$\frac{\partial}{\partial t} P(\{n\},t) = \sum_{(r)} \Big[W(\{n-r\},\{r\})P(\{n-r\},t) - W(\{n\},\{r\})\ P(\{n\},t) \Big] \tag{4}$$

The Master Equation describes a series of discrete processes that we approximate by a continuous description.

(v) The probability density for either the steady state or the limit cycle of (1) would be given as steady solutions of the Master Equation (4). This is perhaps the least justified restriction as there is no reason to disregard the possibility of describing the limit cycle behaviour with a time-dependent solution of the Master equation. The restriction, however, turns out to be a valid working hypothese leading to reasonable description of the limit cycle behaviour in (1), as we shall describe below.

Under such restrictive assumptions the following "computer experiment" has been conducted. Assume that the system is at an initial state denoted by $\{n_0\}$. From $\{n_0\}$ there are four possible transitions, whose transition-probabilities are given by (3) plus a certain probability of remaining in $\{n_0\}$. Let us take the interval $[0,1]$ and divide it into five parts: four of which are taken proportional to the four transition probabilities (3) calculated at $\{n_0\}$, the fifth corresponding to the probability the system has to remain at $\{n_0\}$, in order to simulate the possible collision-free or elastic (non reactive) collision events. We introduce a number N such that $N > \sum_{i=1}^{4} W_i$ where W_i are given by (3). Then, a random number, R, beloging to the segment $[0,1]$ is generated. The reactive step j (j = 1,2,3,4) will take place provided the following inequality holds:

$$\sum_{i=1}^{j-1} W_i/N \leq R < \sum_{i=1}^{j} W_i/N \tag{5}$$

while the system remains at $\{n_0\}$ if

$$\sum_{i=1}^{4} W_i/N \leq R \leq 1 \tag{6}$$

Once this done we bring the system to its new position and we start again the iteration. If this manipulation is carried for a sufficiently long time we should be able to extract from the resulting random trajectory all specific properties of the stationary process.

As an illustration for a specific experiment we have taken N = 7000,Ω = 100, $k_1 = k_2 = k_3 = k_4 = 1$ and $k_5 = 0.5$. The following results have been found:

(i) For $n_A = 1100$ and $n_B = 1200$ the homogeneous steady state is a stable focus, and consequently a concentration of the density is to be expected in its neigbourhood. This is what is found in the simulation experiment and the probability density is depicted in Figure 3.a.

(ii) For $n_A = 1100$ and $n_B = 1290$ we cross the neutral stability curve and the focus passes from stable to unstable . A scattered probability density is found, as expected. There is already some indication of the eventual limit cycle behaviour which will appear later on. The result is shown in Figure 3.b.

Fig.3

(a)

(b)

(c)

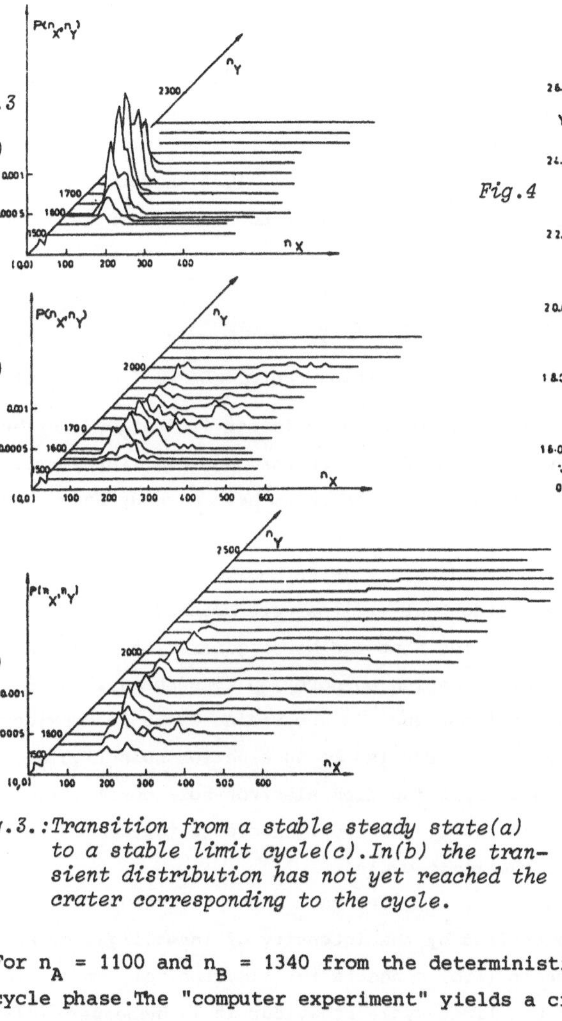

Fig.3.:Transition from a stable steady state(a) to a stable limit cycle(c).In(b) the transient distribution has not yet reached the crater corresponding to the cycle.

Fig.4

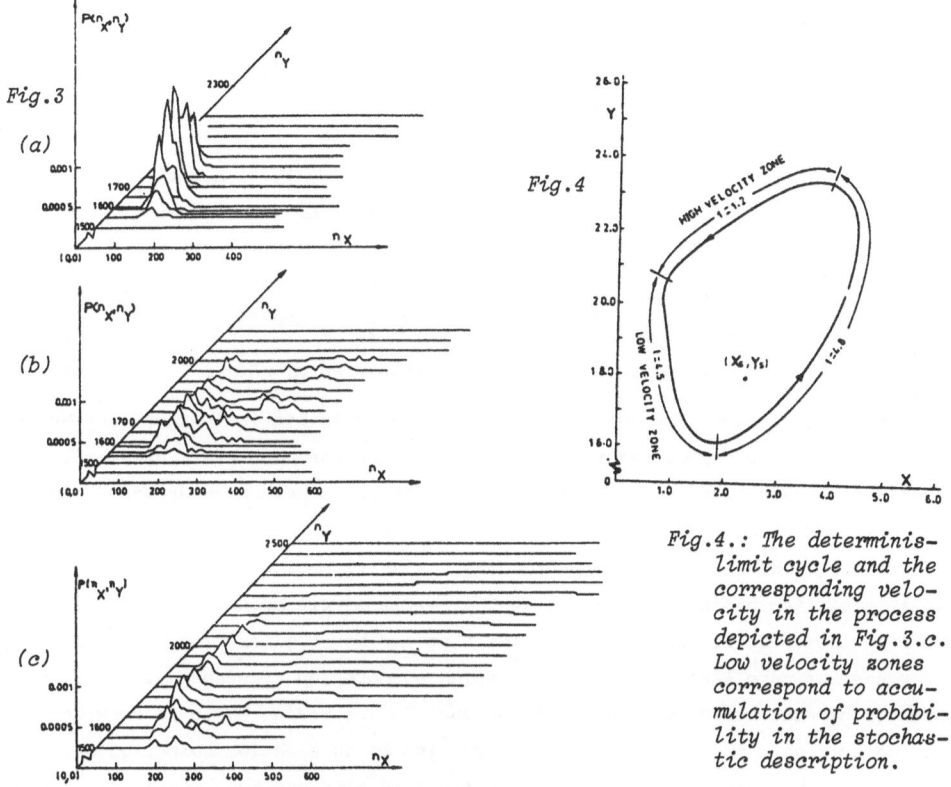

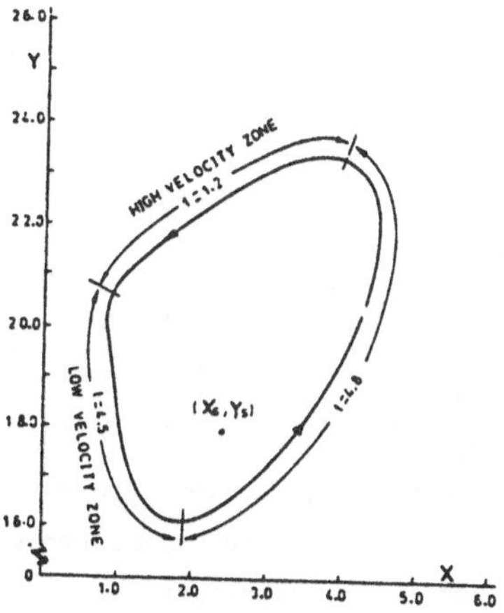

Fig.4.: The determinis- limit cycle and the corresponding velocity in the process depicted in Fig.3.c. Low velocity zones correspond to accumulation of probability in the stochastic description.

(iii) For n_A = 1100 and n_B = 1340 from the deterministic viewpoint we are in the limit cycle phase.The "computer experiment" yields a crater-like probability density(Fig.3.c) around a closed orbit which nicely corresponds to the limit cycle obtained from the deterministic equations(Fig.4).It appears that higher density of the probability corresponds to regions where the motion along the limit cycle is of lower speed. Thus Figure 3 shows the transition from the stable homogeneous steady to the stable periodic orbit (limit cycle). It is to be expected, and some more computer runs have confirmed this conjecture,that the higher the number of molecules ,N, and the volume Ω , the lower the ratio of fluctuations to Ω. Thus there is a tendency in the distribution of points defining the probability density to get closer to each other in the neighbourhood of the maxima along the crater. A better correspondence with the deterministic picture is thus obtained, in accordance with a theorem due to Kurtz (1971).

3.- EXAMPLE OF LIMIT CYCLE IN SEMICONDUCTOR PHYSICS

i. Introduction

There has been some work on the possibility of spontaneous oscillations of the electron concentration and the electron temperature due to non-linear dependence of the Auger coefficient on temperature(Degtyarenko *et al*, 1974) but this model did not contain an autocatalytic reaction.Here we consider such a process in a model with three steps (Pimpale *et al*,1981) :

photogeneration of carriers: $y \overset{G}{\rightarrow} e+h$ (1.a)

stimulated production of excitons : $e+h+X \overset{C}{\rightarrow} 2X$ (1.b)

radiative decay of excitons at a recombination center : $X \overset{k}{\rightarrow} \gamma$ (1.c)

In the case of fermions the autocatalytic process is impact ionisation. In the case of bosons it is the stimulated production of more bosons. Non-equilibrium semiconductor phase transitions based on this latter mechanism have already been considered (Landsberg and Pimpale,1976;Pimpale and Landsberg,1977).

Process (1.b) has not been observed as far as we know.Its inclusion here is however justified by the consensus view that excitons are bosons(see,e.g.Hulin *et al*, 1980). The boson distribution law implies that the boson generation due to incident energy has the form A+BX or (1+X)A, depending on notation (Landsberg,1978).Hence the process(1.b) must exist, though its importance is uncertain. While we consider an electron-hole gas, one can apply the theory also to an electron-hole liquid. In that case,however,the additional screening arising from electron-hole correlations would be expected to reduce the rate constant C. This would increase the concentration of electrons and holes required for the limit cycle.

In the above model, the process (1.b) generates excess electrons (e) and holes (h) at a rate G which can be controlled by the intensity of laser light of an appropriate frequency.The second process (1.b) connects the electron-hole concentrations with exciton concentration. For the limit cycle behaviour it is necessary (although not sufficient)that the electron-hole and exciton coupling be non-linear.The spontaneous analogue of (1.b)alone is not sufficient.

$e + h \overset{C'}{\rightarrow} X$ (1.b')

From (1.a) and (1.b) the rate of change of carrier concentration n is given by

$$dn/dt = G - Cn^2x \qquad (2.a)$$

where we have assumed the semiconductor to be nondegenerate and under high excitation conditions so that electron and hole concentrations may be taken to be equal. Reaction (1.c) is an excitonic decay process on a recombination center. This has interesting nonlinear features; the decay rate is proportional to the number of recombination centres available when there is a large number of excitons available for decay,i.e. for $x \gg 1/q$

$$dx/dt \sim -1/q$$

where $1/q$ is the concentration of recombination centers. However, when the number of excitons is much lower than the number of recombination centers, the exciton decay rate is essentially proportional to the concentration of excitons, i.e. for $x \ll 1/q$

$$dx/dt \sim -x$$

Since the Michaelis - Menten (Hinshelwood-Langmuir, Holling) S-shaped law seems a good interpolation law for the exciton decay rate, we put

$$dx/dt = Cn^2x - kx/(1+qx) \qquad (2.b)$$

where k is the rate constant appropriate to (1.c).

The exciton decay rate on recombination centres, given by the second term (2.b), is shown qualitatively in Fig.1. For small x($qx \ll 1$) one has the usual exponential decay; however for large x there is a departure from the exponential decay. Such a departure has been observed experimentally for bound excitons in qualitative agreement with Fig.1a(Klingenstein & Schmid ,1979). A generalization to an nth order reaction (Fig.1b) $dx/dt \sim -x/(1+qx)^n$, where n is a positive integer, is also possible (Ibáñez and Velarde,1977; Bonilla,Velarde,Pimpale and Landsberg, 1982)but we shall not use it here(see,however, section 2).

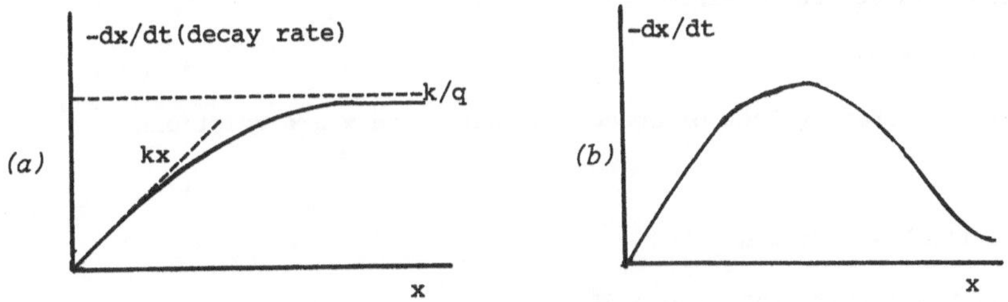

Fig. 1.: Qualitative behavior of exciton decay rate on recombination centers(of concentration $1/q$ *). (a)Michaelis-Menten law;(b) higher-order law(with inhibition by excess of product.*

Equations (2 .a,b) are the basic equations of our model. For simplicity all recombination processes, except those included explicitly in (2), are ignored.These include various electron-hole and electron (or hole)-impurity recombinations as well as excitonic processes such as the formation of excitonic molecules and electron-hole drops.The reverse processes corresponding to (1) are also ignored,since the semiconductor is assumed to be far from equilibrium.

The limit cycle would manifest itself experimentally in the form of oscillating exciton and electron-hole recombination light; and, if a small voltage is applied to the semiconductor, in the form of an oscillating current. Although such an oscillating current induced by nonlinearities in recombination has not yet been observed

Gross *et al*,(1972) have reported oscillating exciton recombination light in a CdS crystal at 4.2-77K. However, to the best of our knowledge the observations of Gross *et al*. have neither been repeated by other workers nor have they been indentified with a limit cycle behaviour. Additional experimental work on this matter would seem to be desirable.

ii. Homogeneous steady states and their local stability

If N is the concentration of the recombination centers at which excitons decay, then in the system of eqns(2) the following units will be chosen:

for $G(L^{-3}T^{-1})$:kN ;for $x(L^{-3})$:N ; for $C(L^{6}T^{-1})$:k/N^{2}; for t(T):k^{-1} ; and for $n(L^{-3})$:N (3)

In these units

$$dn/dt = G - Cn^{2}x \quad ; \quad dx/dt = Cn^{2}x - x/(1+x) \tag{4}$$

Also the reduced value of G is of order 10^{-3} for a photon absorption rate of value 10^{18} $cm^{-3}s^{-1}$, which is reasonable for CdS (Mahr and Tang,1972), an exciton life-time of order 10^{-6}s (Klingenstein and Schmid,1979) and $N \sim 10^{15}$ cm^{-3}. For meaningful solutions we have the physical constraints

$$x \geq 0,\ n \geq 0,\ G > 0, C > 0 \tag{5.a}$$

The steady states are obtained from

$$dn/dt = 0 = dx/dt \tag{6}$$

so that the steady state values of concentrations n and x are solutions of

$$G = Cn^{2}x = x/(1+x) \tag{7}$$

(7) has 2 solutions satisfying (5)

A: $x = 0,\ n = \infty$ (such that $n^{2}x = G/C$)

and

$$B:\ x = G/(1-G) \equiv x_{0},\ n = |(1-G)/C|^{1/2} \equiv n_{0} \tag{8}$$

where G must be confined to the range

$$0 < G < 1 \tag{5.b}$$

The steady state A at $x = 0$, $n = \infty$ is always unstable as can be readily seen by transforming the variable n to $y = 1/n$ and making linear approximation around the point $x = 0$, $y = 0$. To study the stability of the steady state B, put

$$x(t) = x_{0} + u(t),\ n(t) = n_{0} + v(t) \tag{9}$$

and make a linear approximation to obtain

$$d(\{u,v\})/dt = A\{u,v\} \tag{10}$$

where {u,v} is a column vector and A is a 2 x 2 matrix given by

$$A_{11} = G(1-G),\ A_{12} = 2G(C/(1-G))^{1/2}$$

$$A_{21} = -(1-G),\ A_{22} = -2G(C/(1-G))^{1/2} \qquad (11)$$

The stability character of the steady state is decided from det A and TrA. From (11) we get

$$\det A = 2G(1-G)^{3/2}C^{1/2} > 0 \qquad (12)$$

$$TrA = G(1-G)-2\{C/(1-G)\}^{1/2} \qquad (13)$$

Since det A is always positive for physically meaningful solutions (i.e. G,C. obey (5)) the Poincaré index of the steady state n_0, x_0 is +1. It appears that the steady state is stable for TrA <0 and unstable for TrA >0. Thus we can apply the Hopf bifurcation theorem and get a unique stable limit cycle for TrA > 0. From (3) we see that both TrA > 0 and TrA < 0 are possible for physical values of G and C. The bifurcation values or critical values of G and C are given by

$$TrA = 0 \qquad (14)$$

or

$$(1-G)^3 = 4C \qquad (15)$$

The region in (G,C)-parameter space where a limit cycle behaviour is possible is shown in Fig.2.

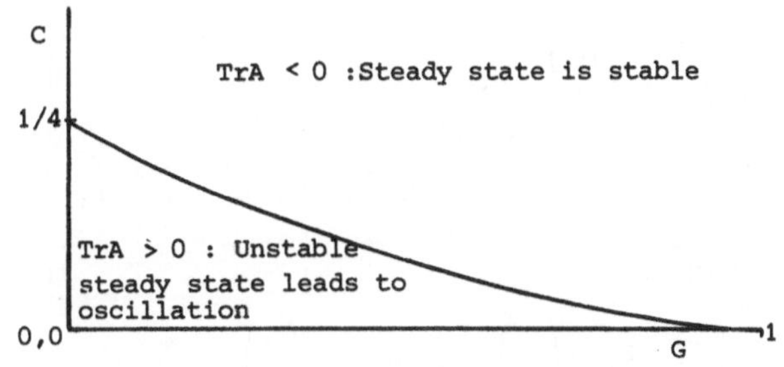

Fig.2.: Stability diagram(C,G) showing the region available for limit cycle oscillation.

iii. Construction of the limit cycle

Transfer the origin in (x,n) space to the steady state point x_0, n_0 using (9) and retain all the non linear terms to find

$$d\{u,v\}/dt = A\{u,v\} + N \qquad (16)$$

where N is the non-linear part given by a column matrix $\{N^1, N^2\}$ with

$$N^1 \equiv (1-G)^3u^2 + 2(C(1-G))^{1/2}uv + v^2CG/1-G + Cuv^2 +$$

$$+ \sum_{n=3}^{\infty} (-1)^n (1-G)^{n+1} u^n \qquad (17.a)$$

$$N^2 \equiv 2(C(1-G))^{1/2}\, uv - CGv^2/(1-G) - Cuv^2 \tag{17.b}$$

G is confined to some fixed value in the range (5.b) and the critical value C_0 of C is given by (15). As C is changed from C_0 there is bifurcation to a limit cycle (see Fig.3). We now construct this bifurcated solution by using a two-time scales method (Bonilla and Velarde,1979). Departures from the bifurcation value C_0 are given in terms of a smallness parameter ε to be determined in the analysis. Thus we set

$$C(\varepsilon) = \sum_{i=0}^{\infty} \varepsilon^i C_i \tag{18}$$

Now, the time scale is expanded in two separate, though related scales; a fast scale $\tilde{t} \equiv t$ which proceeds from the actual trajectory of the initial condition until the system reaches the limit cycle; and a slow scale, $\tau \equiv (C(\varepsilon) - C_0)t$, which transversally crosses the trajectory from the initial condition down to the limit cycle. With the notation $w \equiv \{u,v\}$, $w_i \equiv \{u_i,v_i\}$ adopt the expansion

$$w(t) \equiv w(\tilde{t},\tau,\varepsilon) = \sum_{i=0}^{\infty} \varepsilon^{i+1} w_i = \varepsilon w_0 + \varepsilon^2 w_1 + \dots \tag{19}$$

The matrix A and the non-linear part N are also expanded in powers of ε as:

$$A(C) = A_0 + \varepsilon C_1 A_1 + \varepsilon^2\{C_2 A_1 + C_1^2 A_2\} + O(\varepsilon^3) \tag{20}$$

$$N(w;\varepsilon) = \varepsilon^2 N_2(w_0) + \varepsilon^3 N_3(w_0,w_1) + O(\varepsilon^4) \tag{21}$$

where

$$A_0 \equiv A(C_0) = G(1-G)\begin{bmatrix} 1 & 1 \\ -G\,1 & -1 \end{bmatrix} \tag{22}$$

$$A_1 \equiv A'(C_0) = 2G(1-G)^{-2}\begin{bmatrix} 0 & 1 \\ 0 & -1 \end{bmatrix} \tag{23}$$

$$A_2 \equiv A''(C_0)/2 = -(1-G)^3 A_1 \tag{24}$$

$$N_2(w_0) = (1-G)^2\{u_0v_0 + Gv_0^2/2\}\begin{pmatrix} 1 \\ -1 \end{pmatrix} + (1-G)^3 u_0^2 \begin{pmatrix} 1 \\ 0 \end{pmatrix} \tag{25}$$

$$N_3(w_0,w_1) = (1-G)^2\{u_0v_1 + u_1v_0 + Gv_0v_1/2 + \frac{(1-G)}{4}u_0v_0^2 + C_1(2u_0v_0 + Gv_0^2)/(1-G)\}\begin{pmatrix} 1 \\ -1 \end{pmatrix} + \{2(1-G)^3u_0u_1 - (1-G)^4u_0^3\}\begin{pmatrix} 1 \\ 0 \end{pmatrix} \tag{26}$$

Thus the nonlinear problem (2) is transformed into a hierarchy of linear equations of which only those arising from terms in ε , ε^2, ε^3 are given:

$$Jw_0 \equiv \{\partial/\partial\tilde{t} + (-A_0)\}w_0 = 0 \tag{27.a}$$

$$Jw_1 = C(A_1 - \partial/\partial\tau)w_0 + N_2(w_0) \tag{27.b}$$

$$Jw_2 = C_1(A_1 - \partial/\partial\tau)w_1 + C_1^2 A_2 w_0 + C_2(A_1 - \partial/\partial\tau)w_0 + N_3(w_0,w_1) \tag{27.c}$$

Equation (27.a) defines J, a Fredholm operator. The solution to the homogeneous equation in the hierarchy is readily obtained as

$$w_0(\tilde{\tau},\tau) = a(\tau)\exp(iw\tilde{\tau})\{1,-1+i\omega/G(1-G)\} + c.c. = a(\tau)\phi(\tilde{\tau}) + c.c. \qquad (28)$$

where the second relation defines $\phi(\tilde{\tau})$; c.c. denotes the complex conjugate of the preceding term and

$$\omega \equiv G^{1/2}(1-G)^{3/2} \qquad (29)$$

A solution to the adjoint problem

$$J^+\psi = 0$$

is also obtained as

$$\psi(\tilde{\tau}) = \exp(i\omega\tilde{\tau})\{1,G+i\omega/(1-G)\} \qquad (30)$$

To solve the inhomogeneous eqn(27.b) we apply the Fredholm alternative to J with the scalar product defined by

$$\langle\langle f,g\rangle\rangle = \lim_{T\to\infty} \frac{1}{T}\int_0^T \langle f,g\rangle dt \qquad (31.a)$$

$$\langle f,g\rangle \equiv \sum_i f_i^* g_i \qquad (31.b)$$

where f and g are vectors with components f_i and g_i and * denotes complex conjugates. Then

$$\langle\langle\psi,C_1(A_1-\partial/\partial\tau)w_0+N_2\rangle\rangle = 0 \qquad (32)$$

It is easy to check $\langle\langle\psi,N_2\rangle\rangle = 0$, since the scalar product contains only terms with factors $\exp(\pm i\omega\tilde{\tau})$ in the integral and so the 1/T factor makes it zero.By (32) it then follows that

$$C_1 = 0 \qquad (33)$$

since

$$\langle\langle\psi,(A_1-\partial/\partial\tau)w_0\rangle\rangle \neq 0 \qquad (34)$$

The equation for w_1 is now simplified to

$$Jw_1 = N_2(w_0) \qquad (35)$$

To solve this equation subtitute w_0 from (28) into (25) and get

$$Jw_1 = B|a|^2 + B_1a^2\exp(2i\omega\tilde{\tau}) + c.c. \qquad (36)$$

where B and B_1 are column matrices independent of $\tilde{\tau}$. Thus w_1 consists of three terms one being independent of $\tilde{\tau}$ and the other two proportional to $\exp(\pm 2i\omega\tilde{\tau})$. Substituting this form of w_1 in (36) we easily obtain

$$w_1 = (1-G)\,|a(\tau)|^2\{2G,-1/2\}/G + \left[(a(\tau))^2/6G(1-G)\right]\exp(2i\omega\tilde{\tau})\{-4(1-G)^3-i\omega(1+2G),$$
$$(1-G)(3G^2-15G/2+9/2)+i\omega(8-3G-2/G)\} + c.c. \qquad (37)$$

We now turn to eqn(27.c).Using (33), the Fredholm (alternative) property of J now implies

$$\langle\langle\psi, c_2(A_1-\partial/\partial\tau)w_0 + N_3(w_0,w_1)\rangle\rangle = 0 \qquad (38)$$

Substituting w_0 and w_1 from (28) and (37), and after some algebra, (38) yields an equation for the slowly varying amplitude $a(\tau)$:

$$a'(\tau) \equiv da(\tau)/d\tau = \nu a(\tau) + \lambda\, a(\tau)\,|a(\tau)|^2/C_2 \qquad (39)$$

where

$$\nu = \langle\psi, A_1\phi\rangle/\langle\psi,\phi\rangle = -G(1-G)^{-2}+i\omega(1-G)^{-3}$$

$$\lambda = \langle\psi, N_3(\omega_0, J^{-1}\omega_0)\rangle \;/\; a(\tau)\,|a(\tau)|^2 =$$

$$= (1-G)^3/2\,\{G-11/8+i\omega\left[-G+7(1+1/G)/24-1/3G^2\right]/(1-G)\}$$

For convenience set $a(\tau)$ in polar form

$$a(\tau) \equiv \rho(\tau)\exp\{i\alpha(\tau)\} \qquad (42)$$

The stability of $a(\tau)$ can then be tested by considering $\rho(\tau)$ as $t\to\infty$. From (39) the following exactly solvable differential equations are found

$$\rho'(\tau) = \mathrm{Re}\,\nu\rho(\tau) + \mathrm{Re}\,\lambda\rho^3(\tau)/C_2 \qquad (43.a)$$

$$\rho(\tau)a'(\tau) = \mathrm{Im}\,\nu\rho(\tau) + \mathrm{Im}\,\lambda\rho^3(\tau)/C_2 \qquad (43.b)$$

The first equation yields

$$\rho(\tau) = \rho(0)\rho(\infty)\exp\left[\mathrm{Re}\,\nu\tau\right]/\{\rho^2(\infty) + \left[\exp 2\mathrm{Re}\,\nu\tau - 1\right]\rho^2(0)\}^{1/2} \qquad (44)$$

where

$$\rho(\infty) \equiv \{-\,\mathrm{Re}\,\lambda/C_2\mathrm{Re}\,\nu\}^{-1/2} \qquad (45)$$

and

$$\rho(0) = \varepsilon^{-1}\left\langle\psi(0), \begin{vmatrix} x(0)-x_0 \\ n(0)-n_0 \end{vmatrix}\right\rangle/\langle\psi,\phi\rangle \qquad (46)$$

$\rho(0)$ incorporates the influence of the initial conditions in the problem. The solution of (43.b) is

$$a(\tau) = \omega + \tau\,\mathrm{Im}\,\nu + \left[\mathrm{Im}\,\lambda/C_2\right]\int_0^{\tau}\rho^2(s)\,ds \qquad (47)$$

Consider now the asymptotic behaviour of $\rho(\tau)$ and $\alpha(\tau)$ as $t\to\infty$. Note that in the physical domain (5.b), $\mathrm{Re}\,\nu < 0$ and as $t\to+\infty$, $\tau\to+\infty$ when $C > C_0$ and $\tau\to-\infty$ when $C<C_0$. From (44) and (47) it follows that

$$\lim_{t\to\infty} \rho(\tau) = \begin{cases} 0 , & (C > C_0) \qquad (48.a) \\ \rho(\infty) , & (C < C_0) \qquad (48.b) \end{cases}$$

$$\lim_{t\to\infty} \alpha(\tau) = \begin{cases} \lim_{t\to\infty} (\mathrm{Im}\,\nu)t , & (C > C_0) \qquad (49.a) \\ \lim_{t\to\infty} (\mathrm{Im}\,\nu + (\mathrm{Im}\lambda\rho^2(\infty))/C_2)(C-C_0)t , & (C < C_0) \qquad (49.b) \end{cases}$$

Thus we have a stable limit cycle of amplitude $\rho(\infty)$ when $C < C_0$ and an unstable limit cycle otherwise. It comes as a onesided bifurcation to oscillation which is stable provided $\mathrm{Re}\lambda$ is negative, and $C < C_0$ in (45). In our case the stability condition is satisfied for the relevant range of parameter values $0 < G < 1$, (5.b)

The asymptotic form ($t \to +\infty$) of the stable limit cycle up to terms of order $\varepsilon^2 = |(C_0-C)/C_2|$ is obtained as

$$\begin{bmatrix} x(t) \\ n(t) \end{bmatrix} = \begin{bmatrix} x_0 \\ n_0 \end{bmatrix} + 2|(C_0-C)/C_2|^{1/2} \rho(\infty) \begin{bmatrix} \cos\tilde{\omega}t \\ -\cos\tilde{\omega}t - \{(1-G)/G\}^{1/2} \sin\tilde{\omega}t \end{bmatrix}$$
$$+ |(C_0-C)/C_2|\rho^2(\infty)\{(1-G)/G \begin{bmatrix} 2G \\ -1/2 \end{bmatrix} +$$
$$+ 1/(3G(1-G)) \begin{bmatrix} -2(1-G)^3 \cos 2\tilde{\omega}t - (1+2G) \sin 2\tilde{\omega}t \\ (1-G)(3G^2-15G/2+9/2)\cos 2\tilde{\omega}t + (8-3G-2/G)\sin 2\tilde{\omega}t \end{bmatrix} \qquad (50)$$

where the first term arises as we go back to the origin (0,0) in x,n space, and the angular frequency is

$$\tilde{\omega} \equiv \omega + \{\mathrm{Im}\,\nu + (\mathrm{Im}\lambda\rho^2(\infty))/C_2\}(C-C_0) \qquad (51)$$

which is to order $(C-C_0)$ slightly different from ω .

iv. Numerical estimates

The stable limit cycle solution is of the form

$$x(t) = x_0 + x_1\cos\tilde{\omega}t + x_{21}\cos 2\tilde{\omega}t + x_{22} \sin 2\tilde{\omega}t$$

$$n(t) = n_0 - n_{11}\cos\tilde{\omega}t - n_{12} \sin\tilde{\omega}t + n_{21}\cos 2\tilde{\omega}t + n_{22}\sin 2\tilde{\omega}t$$

where $x_0, x_1, x_{21}, x_{22}, n_0, n_{11}, n_{12}, n_{21}$ and n_{22} are constants to be indentified from (50). For (51) to be a good approximation to a limit cycle solution, the second harmonic terms must be small compared with the first harmonic terms. This is satisfied when

$$x_{22} \sim x_{21} << x_1 \qquad (52.a)$$

and

$$|n_{22}| \sim |n_{21}| << n_{11} \sim n_{12} \qquad (52.b)$$

From (50) it is seen that (52) amounts to

$$\left[C_0-C/C_2^2\right] << G^4(1-G)^2 / \left[\left|(1-G)^4(269G^3/4-118G^2-118G-359/4G-39/2)\right|\right] \qquad (53)$$

When (53) breaks down one must consider 3rd and higher order harmonics corresponding to terms of the order of $|(C_0-C)/C_2|^{3/2}$ and higher in (50) in order to get a good approximation to the limit cycle solution. Thus assuming (53) and keeping only the first harmonic terms, the limit cycle is approximated by

$$(x-x_0) = x_1 \cos \tilde{\omega} t \qquad (54.a)$$

$$(n-n_0) = -n_1 \sin(\tilde{\omega} t+\phi) \qquad (54.b)$$

where

$$x_1 = 2\rho(\infty) \; |(C_0-C)/C_2|^{1/2} \qquad (55)$$

$$n_1 = 2\rho(\infty) \; |(C_0-C)/C_2|^{-1/2} \qquad (56)$$

and

$$\text{tang}\, \phi = \{G/(1-G)\}^{1/2} \qquad (57)$$

For general values of ϕ ,(54) represents a tilted ellipse centred around the steady state (x_0,n_0).It is interesting to calculate the relative changes in exciton (x) and electron-hole concentrations (n) as the limit cycle is traced.

This is given by

$$(x_1/x_0)/(n_1/n_0) = (x_1 n_0)/(n_1 x_0) = 2G^{-3/2} > 1$$

since by (5.b) $G < 1$. The numerical value of G in a realistic semiconductor is $\sim 10^{-3}$. In an experiment one would have to look for oscillating light emission due to the oscillating exciton concentration, and this should be accompanied by oscillating electron concentrations as revealed,e.g.,by the Hall effect. One cycle would be described in each case in a time order $\omega^{-1} \sim 1\mu$ sec . The expected fractional changes in concentrations are of the order given in Table 1 . It is hoped that experiments will be made to find this effect, which, as far as we are aware, has not been considered before.

Table 1. Numerical values for a limit cycle oscillation(units in main text)

Photogeneration rate, G	$\lvert(C_o-C)/C_2\rvert$	*Angular frequency*, ω	*Fractional changes in concentrations.* x_1/x_o	n_1/n_o
10^{-3}	$0.1023.10^{-14}$	0.03158	0.08174	$0.132.10^{-2}$

N.B.: Other values and a series of pictures can be found in Pimpale *et al.*(1981).

4.- EXAMPLE OF SPACE DISTRIBUTION AND MULTIPLE STEADY STATES

i. Homogeneous and primary solution

For purposes of illustration we take Degn-Harrison (1969) model as a mathematical exercise disregarding its origin and its relevance to the oscillating respiration of *Klebsiella Aerogenes* . Moreover we shall concentrate on a description of the possible nonuniform steady states of such model problem in a one-dimensional geometry under the simplifying assumption taht one of the components, say Y, diffuses much faster than the other: $D_y >> D_x$, $D_y \to \infty, D_x < \infty$.

It is convenient to recast (2.2) into dimensionless form. Introduction of

$$x = k_5^{1/2} X, \quad a = k_1 k_3^{-1} k_5^{1/2} A$$

$$b = k_2 k_3^{-1} k_5^{1/2} \left(B - \frac{k_1 A}{k_2} \right), \quad y = k_4 k_3^{-1} Y$$

$$\tau = k_3 t, \quad D = D_x k_3^{-1} L^{-2}, \quad D' = D_y k_4^{-1} k_5^{1/2} L^{-2} \tag{1}$$

yields

$$\frac{\partial x}{\partial \tau} = b + a - x - \frac{xy}{1+x^2} + D \frac{\partial^2 x}{\partial z^2} \tag{2.a}$$

$$k_3 k_4^{-1} k_5^{1/2} \frac{\partial y}{\partial \tau} = a - \frac{xy}{1+x^2} + D' \frac{\partial^2 y}{\partial z^2} \tag{2.b}$$

when only spatial variations in the z-direction are considered. Here the parameters a anc b can be freely chosen; the physical values correspond to $a > 0$ and $b \gtrsim -a$. L denotes the dimension of the container and is taken as the unit of length. The equations (2) possess precisely one homogeneous steady state

$$x_s = b \tag{3}$$

and

$$y_s = a(1+b^2)/b \tag{4}$$

We will use these values as boundary conditions in our problem whether the steady state is homogeneous or not. This choice eliminates the problem of having boundary layers at the extreme ends of the 1d "reactor". We set

$$x(0) = x(1) = b \tag{5}$$

$$x(0) = y(1) = a(1+b^2)/b \tag{6}$$

The limit $D' \to \infty$ in Equation (2.b) leaves the equation $\partial^2 y(z)/\partial z^2 = 0$, with the solution

$$y(z) = a(1+b^2)/b \tag{7}$$

in view of the boundary condition (6).Insertion into Equation (2.a) yields

$$\frac{\partial x}{\partial \tau} = b - x + a - a\,\frac{(1+b^2)x}{b(1+x^2)} + D\,\frac{\partial^2 x}{\partial z^2} \tag{8}$$

ii. Stability of Homogeneous Solution

For small deviations $\psi = x - b$ from the homogeneous solution (8) is linearized to read (Hemmer and Velarde, 1977)

$$\frac{\partial \psi}{\partial \tau} = L\psi = \{-1+\frac{a(b^2-1)}{b(b^2+1)} + D\,\frac{\partial^2}{\partial z^2}\}\,\psi \tag{9}$$

Stability is decided by the sign of the eigenvalues

$$\lambda_k = -1+\frac{a(b^2-1)}{b(b^2+1)} - D\pi^2 k^2 \qquad k = 1,2,3... \tag{10}$$

of the operator L, corresponding to the eigenfunctions

$$\psi_k(z) = \sin(\pi k z) \tag{11}$$

which fulfill the boundary condition (5). (The alternative boundary condition $\partial x/\partial z = 0$ gives $\cos(\pi kz)$ instead, with the same eigenvalues λ_k, but with $k = 0$ inclued. We do not consider this case).

When $a(b^2-1)/b(b^2+1) < 1+D\pi^2$, all λ_k are negative, and the constant concentration profile is therefore stable. Having thus disposed of this case we turn to the more interesting unstable situation

$$\frac{a(b^2-1)}{b(b^2+1)} > 1 + D\pi^2 \tag{12}$$

Since $a > 0$ and the boundary condition (5) requires $b > 0$, (12) implies

$$b > 1 \quad \text{and} \quad a > b(b^2+1)/(b^2-1) \tag{13}$$

When (13) is fulfilled, the number K of unstable modes is given by

$$K = \| \{-1+a(b^2-1)/b(b^2+1)\}^{1/2} D^{-1/2}\,\pi^{-1} \| \tag{14}$$

where $\| x \|$ is the integer part of x

iii. The Non-Uniform Steady States (Hamilton's equations).

For steady states the basic equation (8) can be integrated, yielding

$$(a+b)x-x^2/2-a(2b)^{-1}(1+b^2)\ln(1+x^2)+D(dx/dz)^2/2 = E \tag{15}$$

where E is the integration constant. With z considered as a time variable this defines a Hamiltonian flow in phase plane (v,x), with

$$v = D^{1/2}dx/dz \tag{16}$$

acting here as "momentum" variable and E as the "energy".

The flow pattern is determined by the equations

$$D^{1/2}dx/dz = v$$

$$D^{1/2}dv/dz = a(1+b^2)x \;/b(1+x^2)-a+x-b \qquad (17)$$

The right-hand side of (17) has a zero at x = b, and can be rewritten as

$$D^{1/2}dv/dz = \{(x-b)(x-x_+)(x-x_-)\}/(1+x^2) \qquad (18)$$

with

$$x_\pm = a/2\pm(a^2-4-4a/b)^{1/2}/2 \qquad (19)$$

For the concentrations a,b that the instability criterion (12) is fulfilled, we have

$$x_- < b < x_+ \qquad (20)$$

which is necessary if the homogeneous solution (x,v) = (b,0) shall be an unstable center. Figure 1 shows the phase-plane flow for concentrations corresponding to a=10,b=2. The phase trajectories may be labeled by the value of the constant E. By (15) E increases with increasing distance from the x-axis. Three topologically different phase-plane flows are possible. They arise when $E_+ > E_-$, $E_+ = E_-$, $E_+ < E_-$, respectively. Here

$$E_\pm(a,b) = (a+b)x_\pm - x_\pm^2/2 - ab^{-1}(1+b^2)\ln(1+x_\pm^2) \;/\; 2 \qquad (21)$$

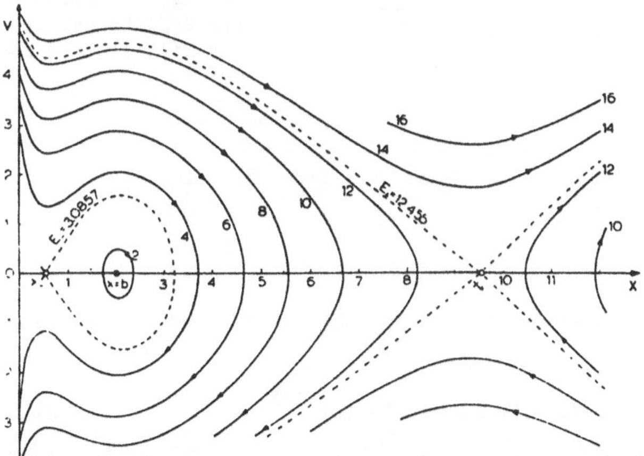

Fig.1.: Phase-plane flow for the concentration values a=10,b=2. The unstable center(b,0) and the two saddle points ($X_\pm$,0)are marked with filled and open circles,respectively. The separatrices through the saddles are dashed. Trajectories are labelled according to their "energy".

are the values corresponding to the two saddlepoints. Using (19) one finds that $E_+=E_-$ can be realized when one of the concentration parameters is carefully adjusted to a definite value depending on the value of the other variable. Figure 2 describes the situation.

The trajectories or trajectory that satisfy the boundary condition (5) are those that start at time $z=0$ at a point (b,v_0) in phase space and at the later time $z=1$ are back to the same value of x, i.e., at (b,v_0) or at $(b,-v_0)$. They may wind around the center any number of times.

The times of flight are obtained by integrating (15). The time for travelling from (b,v_0) to $(b,-v_0)$, with $v_0 > 0$, is

$$T_+(v_0) = 2(D/2)^{1/2} \int_b^{X_M} \{E-(a+b)x+x^2/2+a(2b)^{-1}(1+b^2\ln(1+x^2))\}^{-1/2}\, dx \quad (22)$$

and with $v_0 < 0$

$$T_-(v_0) = 2(D/2)^{1/2} \int_{X_m}^{b} \{E-(a+b)x+x^2/2+a(2b)^{-1}(1+b^2)\ln(1+x^2)\}^{-1/2}\, dx \quad (23)$$

Here X_m and X_M are the two zeros of the radical, $X_m < b < X_M$. The parametrization can be in terms of v_0 or E, which are connected via $E=E(x=b,v=v_0)$, or in terms of X_M, X_m.

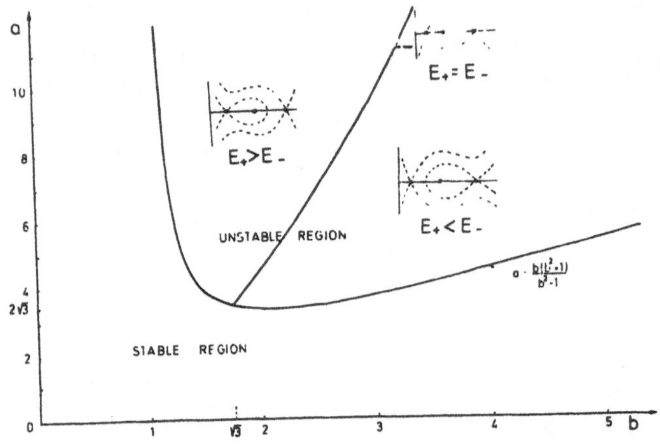

Fig.2.: Linear stability portrait of the homogeneous steady state at vanishing D; a and b denote concentrations. The 3 different types of flow correspond to the "energy" relations. At $(\sqrt{3},2\sqrt{3})$ the center merges with the two saddle points.

For trajectories close to the center one evaluates (22) and (23) easily and finds in limit $v_0 \to 0$

$$T_+(0) = T_-(0) = T_0 = D^{1/2}\{[a(b^2-1)/\, b(b^2+1)] - 1\}^{-1/2} \quad (24)$$

Moreover, one can show that $T_\pm$ diverge when the trajectory passes through one of the saddle points. Figure 3 shows the behaviour of $T_\pm$ as function of v_0. We also plot $T_+ + T_-$, corresponding to a complete loop around the origin; $2T_+ + T_-$ and $T_+ + 2T_-$, which correspond to one-and a-half loop starting from the upper or the lower side, respectively. In general, any number of loops $(T_+ + T_-)$ added to the T_+, T_- and $T_+ + T_-$ curves is of interest.

The solution of the boundary value problem is now found by equating these times of flight to unity. This yields the corresponding value for the initial momentum v_0.

For the special case $D=0.01$, $a=10$, $b=2$, Figure 3 shows that there are precisely *eight* steady-state space trajectories and the actual concentration profiles are given in Figure 4.

In the general case, we note that precisely one allowed trajectory lies between the separatrix through x_- and the one through x_+.The trajectories that wind a half-integer number of times around the center correspond to symmetric concentration profiles. The profiles corresponding to trajectories winding an integer number of times are neither symmetric nor antisymmetric around z=1/2.For each of these trajectories there are two profiles, one corresponding to a positive initial value v_0, the other to a negative v_0.

To the next approximation the time of flight equals

$$T_\pm = T_0 \{ 1+\alpha v_0+O(v_0^2) \}$$

with

$$\alpha = 4a(b^2-3)/3\pi(1+b^2)^2\{a(b^2-1)/b(b^2+1)-1\}^{3/2}$$

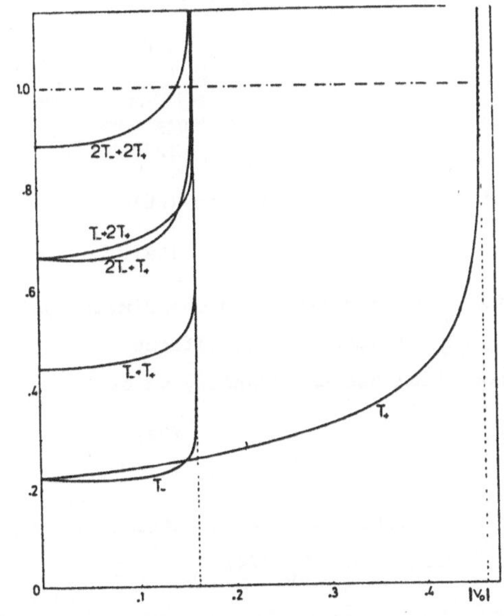

Fig.3.: The "time of flight" as a function of the initial "momentum", v_0. Case D=0.01,a=10,b=2. There are eight stationary concentration profiles.

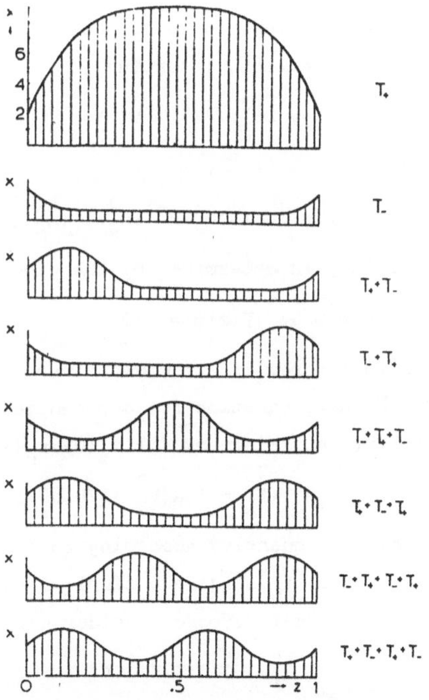

Fig.4.: The eight stationary concentration profiles for the case described in Fig.3.

Using the condition (13) we find for the slope of the $T_\pm$ versus v_0 curves at origin

$$\text{sgn} \{\partial T_\pm/\partial v_0\}_{v_0=0} = \pm \text{sgn}(b-\sqrt{3})$$

Together with the fact that $T_\pm$ tend to $+\infty$ when v_0 approaches a separatrix this shows that the function $T_+(v_0)$ has a minimum when $b<\sqrt{3}$, the function $T_-(v_0)$ a minimum when $b>\sqrt{3}$; in both cases for a non-zero value of v_0.

When the homogeneous state is unstable according to the criterion given by linear analysis, the value of T_0, Equation (24) is less than unity,and we have therefore at least two non-uniform steady states. Moreover,if the concentration of A,say, is increased new unstable modes $\psi_k(z)$ appear, one at a time. When a new unstable mode appears there are two new inhomogeneous states available, and it is tempting to assume a one-to-one correspondence between the stationary structures and the initial infinitesimal perturbation,depending upon the sign of the latter. This is not correct, however,as we now show.

iv. Stability of the multiple nonuniform steady states (Schrödinger's equation)

Consider small deviations

$$\psi(z,\tau) = x(z,\tau) - x_s(z) \tag{25}$$

from one of the stationary non-uniform concentration profiles $x_s(z)$ determined in the previous section. Inserting (25) into Equation (8) and linearizing in ψ we find

$$\partial\psi/\partial\tau = D\partial^2\psi/\partial z^2 - V(z)\psi \tag{26}$$

with

$$V(z) = 1-a(1+b^2)b^{-1}(x_s^2-1)(x_s^2+1)^{-2} \tag{27}$$

Stability is determined by the eigenvalues λ of the r.h.s. operator in (26)

$$-D\partial^2\psi/\partial z^2+V(z)\psi = \lambda\psi \tag{28}$$

and is ensured if all eigenvalues λ are positive. Equation (28) is a one-dimensional Schrödinger equation for a potential $V(z)$ on $[0,1]$ with boundary conditions $\psi(0)=\psi(1)=0$. Since by (5) $x_s(0)=x_s(1)=b$, the potential has the boundary values

$$V(0) = V(1) = 1-a(b^2-1)/b(b^2+1) \tag{29}$$

a negative quantity according to (12).

Let us first give a simple demostration that unstable dissipative structures may occur.For that purpose consider parameters a,b,D such that $1/T_0$ (24), is slightly larger than an integer N,

$$T_0 \lesssim N^{-1} \tag{30}$$

Then the construction in Figure 3 shows that there are precisely 2N steady concentration patterns. One of these states (one of the two with most oscillations) will correspond to a trajectory very close to the center (i.e. to $|v_0|$ very small). For this state the potential $V(z)$ will ,therefore, be closely approximated by the constant (29). Hence the lowest eigenvalue λ_0 is obtained for a ground-state wave function close to $\psi_0 = \sin(\pi z)$. Thus

$$\lambda_0 \approx V(0) + D\pi^2 = D\pi^2(1-T_0^{-2}) \approx -D\pi^2(N^2-1) \tag{31}$$

when (29 and (24) are used. Hence this state is definitely unstable provided $N > 1$, i.e., if there are more than two dissipative structures.

Having thus demostrated the possibility of instability, we now proceed with a complete discussion of the same particular case as used in Figure 3 and 4, viz, $a = 10$, $b = 2 > \sqrt{3}$ and $D = 0.01$. Some of the corresponding potentials $V(z)$ are sketched in Figure 5. A numerical evaluation of the ground-state eigenvalue λ_0 gives that the profile labeled T_- in Figure 4-corresponding to a concentration depletion in the middle of the container-is most stable ($\lambda_0 = 5.2$). The structure with concentration increase in the middle (label T_+) is also stable ($\lambda_0 = 0.7$) while the other six structures are all unstable. Computer studies of the full time-dependent equation (8) confirm this picture.

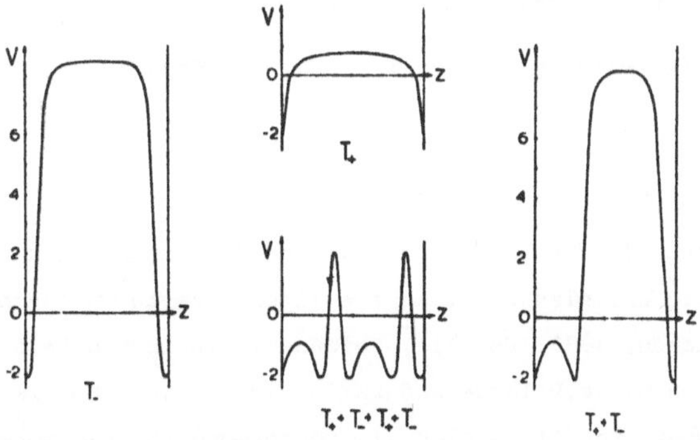

Fig.5.: A nonequilibrium reaction-diffusion problem was transformed into a conservative dynamical system that later on led to a one-dimensional Schrödinger equation. Here the "potentials",V(z),that enter the stability analysis of some of the dissipative structures shown in Fig.4 are drawn(case D=0.01,a=10,b=2).The labels T_+*,etc. correspond to the notation introduced in the main text and on Figs.3 & 4.Note that for* T_+ *and* T_- *these potentials have negative values in a small range of z; these two potentials have positive ground-state energies(stability).*

For the general case it happens that the same result holds, viz: When two dissipative structures with a single extremum occur in one-dynamical-variable problems, one corresponding to a concentration minimum the other to a concentration maximum, these are both stable. All other profiles are unstable (Fairén and Velarde,1979; Jetschke,1979,1980. Further details on a related model problem can be found in Velarde,1978; Ibañez and Velarde,1978).

Further details about multiple steady states and localized structures in model problems involving the Michaelis-Menten kinetics, etc. can be found in the papers by Bonilla and Velarde (1979,1980). In the latter paper the description is obtained by using the J-WKB method. For a general account see the monograph by Nicolis and Prigogine (1977).

5. EXAMPLE OF LIMIT CYCLE IN SYSTEMS WITH TIME DELAY, DIFFUSION AND ADVECTION

i. Introduction

Time-delay problems arise in many areas of physics, physical chemistry, chemical engineering, biochemistry, ecology, economy, etc. We consider here a model problem where time-delay, diffusion and advection compete. It will be seen that dissipation can play a *dual* role in the stability of a limit cycle and that slowly modulated variations of its amplitude obey a Landau-Ginsburg equation. The model is

$$\partial N/\partial t - rN(1-BN/K) + bN^2/(H^2+N^2) = \partial/\partial x \{D(N)\partial N/\partial x\} - \partial/\partial x\{ V(N)N \} \qquad (1)$$

where N is the unknown. N may represent some population density. BN accounts for a time-delay. This is either

i) discrete or punctual (with time-delay, T)

$$BN \equiv N^n(x,t-T) \quad n \geq 1 \quad (n=1 \text{ is the standard case}) \qquad (2.a)$$

or ii) continuous, although peaked around $t = T$

$$BN \equiv \int_0^\infty (\sigma/T^2)\ \exp(-\sigma/T)N(x,t-\sigma)d\sigma \qquad (2.b)$$

D and V denote diffusion and velocity coefficients which are taken density dependent.

Specific examples exist in the literature to which the model could apply (Bonilla and Velarde, 1981; Bonilla, Fernández Cancio and Velarde, 1981; Bonilla and Velarde, 1981; Bonilla, Velarde and Parisi, 1981). One case is the spruce budworm-forest ecosystem where the time-delay, T, corresponds to the seven to ten year time interval that balsam fir trees take to complete their refoliage. Worms, however, reproduce yearly with a birth rate r. K is the carrying capacity of the ecosystem. Another example has been recently suggested by E. Parisi to account for some features of Calcium transport in sea-urchin eggs.

The depletion term in (1) is Holling's S-shaped functional law, and it accounts for predation of the worms by birds. H determines the scale of budworm densities at which saturation begins to take place. b is the rate of worms consumption by the birds. H and K are quantities usually taken proportional to the available branch surface in the forest, say.

For certain values of the parameters in (1) the time-delay equation has two homogeneous steady states: $N = 0$ (extinction) and $N = N_0$ where N_0 is some constant whose actual value depends on the given values of the parameters of the problem. In the absence of time-delay and dispersion (no spatial effects) N_0 can be *stable* steady state of (1) whereas $N = 0$ becomes *unstable*. On the other hand, provided the time-delay, T, is large enough, N_0 can be unstable to time-periodic oscillatory disturbances (Hutchinson, 1948; Hale, 1978).

We call $\tilde{D} \equiv D(N_0) \neq 0$, $\tilde{V} = V(N_0)$, and define

$$u = N/H, \quad \tau = t/T, \quad x = X(\tilde{D}T)^{1/2}, \quad \beta = rT, \quad R = rH/b$$

$$Q = K/H, \quad c = V(T/\tilde{D})^{1/2}, \quad \tilde{V}c(u) = V(N), \quad \tilde{D}d(u) = D(N)$$

With the new scales the equation (1) becomes in dimensionless form

$$\frac{\partial u}{\partial \tau} - \beta u\,\{1 - \frac{Bu}{Q} - u/R(1+u^2)\} = \frac{\partial}{\partial x}\,\{d(u)\;\partial u/\partial x\} - c\,\frac{\partial}{\partial x}\{c(u)u\} \tag{3}$$

where

$$d(u_0) = c(u_0) = 1, \quad u_0 = N_0/H \tag{4}$$

and

$$Bu = u(x,\tau-1) \tag{5.a}$$

or

$$Bu = \int_0^\infty \sigma\exp(-\sigma)u(x,\tau-\sigma)\,d\sigma \tag{5.b}$$

ii. Linear Stability Analysis

ii.1 Dispersion-free case

In the absence of space-dependent phenomena linearization of eq (3) around the homogeneous steady state u_0 (which is stable in the absence of delay) leads to the following equation

$$du/dt = -\{a_{10}(\alpha)u + \alpha Bu\} \tag{6.a}$$

with

$$a_{10} = -\beta u_0(u_0^2-1)/R(u_0^2+1)^2 \equiv -\alpha Q(u_0^2-1)/R(u_0^2+1)^2 \tag{6.b}$$

and

$$\alpha = \beta u_0/Q \tag{6.c}$$

The characteristic equation for the eigenvalues λ is

$$\lambda + a_{10}(\alpha) + \alpha\tilde{B}(\lambda) = 0 \tag{7.a}$$

where

$$\tilde{B}(\lambda) = \exp(-\lambda) \tag{7.b}$$

for discrete delay, and

$$\tilde{B}(\lambda) = (1+\lambda)^{-2} \tag{7.c}$$

for continuos delay.

It happens that for $\alpha < \alpha_0$ all roots of (7) have negative real parts, with

$$\nu_0\cot\nu_0 = -a_{10}, \text{ with } 0 < \nu_0 \leqq \pi \tag{8.a}$$

$$\nu_0\operatorname{cosec}\nu_0 = \alpha_0 \tag{8.b}$$

$$a_{10} + \alpha_0 > 0 \tag{8.c}$$

for discrete time delay , and

$$\alpha_0 = 2(1+a_{10})^2 \tag{9.a}$$

$$\nu_0 = (1+2a_{10})^{1/2} \qquad (9.a)$$

for continuous time delay (Routh-Hurwitz criterion).

At $\alpha = \alpha_0$ there are two purely imaginary roots, $\pm\, i\nu_0$, that cross axis with positive speed, Re $\lambda_1 > 0$, where

$$\lambda_1 = \lambda'(\alpha_0) = -\{a'_{10}(\alpha_0) + B(i\nu_0)\} / \{1+\alpha_0 B'(i\nu_0)\}$$

Here the prime denotes derivative with respect to the argument. Thus, we have a Hopf bifurcation to a limit cycle whose stability is exchanged with the trivial solution(Hale,1977,1978;Bonilla and Velarde,1981). α_0 corresponds to a critical β_c.

ii.2. Purely diffusive dispersion (no advection).

If we set $a = a_{10} + b$, $b > 0$ (not depending of α by assumption) instead of a_{10} in (6.a) and again perform the stability analysis, it turns out that b is always a stabilizing parameter for the homogeneous steady state. Therefore, when adding diffusion here restricted to dimension one, $\partial^2 u/\partial x^2$, we obtain the following results:

(i) In unbounded media or in bounded media with Neumann zero-flux b.c. there is bifurcation to homogeneous limit cycle. The most unstable Fourier mode in (6) has zero wavenumber.

(ii) In bounded media (length, L) with Dirichlet zero-data conditions there is bifurcation to a space modulated limit cycle with wavelength 2L. Thus in dimensional units time-delay and length act destabilizing and diffusion stabilizing the trivial state.

ii.3 General dispersion with Dirichlet zero-dat b.c. (diffusion and advection).

Here

$$b = \pi^2/L^2 + c^2/4$$

or

$$b = \pi^2 DT/L^2 + V^2 T/4D$$

in dimensional units. Diffusion plays a dual role:

(i) If the dimensionless number

$$Re \equiv VL/D$$

is less than 2π , diffusion tends to stabilize the trivial (homogeneous) solution.

(ii) If Re > 2π diffusion can play a destabilizing role of that solution.

The critical value $Re^c = 2\pi$ comes from the geometry and b.c. of the problem. Here Re is the Reynolds or Péclet number of fluid dynamics (Normand et.al,1977) *iii. Results of the nonlinear analysis*

Search for direction of bifurcating solutions is carried out by taking into account the specific nonlinearity of the model. This can be done with a method as described earlier in Section 3 (see also Bonilla and Velarde,1979,1982) to derive from (1) the following Landau-Gin burg equation for the slowly varying complex nonlinear amplitude of the limit cycle:

$$\partial a/\partial\tau = \delta\partial^2 a/\partial\xi^2 + (\alpha_2\lambda_1 + \mu|a|^2)a \qquad (10)$$

where $\alpha = \alpha_0 + \alpha_2\varepsilon^2 + \theta(\varepsilon^3)$; λ_1, δ, μ are complex quantities, ξ, τ are rescaled space and time variables. Homogeneous time periodic solutions of (10) represent those of (1) which are stable to homogeous disturbances if and only if $\mathrm{Re}\,\mu < 0$.

iii.1. Dispersion-free case. Here $\delta = 0$ in (10) and there is a Hopf bifurcation which is supercritical (stable) if $\mathrm{Re}\,\mu < 0$ and subcritical (unstable) if $\mathrm{Re}\,\mu > 0$.

iii.2. Bounded media. The b.c. impose an exponential decay to all Fourier modes except one whose complex amplitude verifies (10) with $\delta = 0$. As μ depends on the nonlinear dispersion,then an appropriate density-dependent diffusion can change the direction of bifurcation and the stability of the limit cycle.

iii.3. Purely diffusive dispersion in unbounded media. We have (10) with $\delta \neq 0$. Even limit cycles which are stable to homogeneous disturbances can be destabilized by diffusion, if a number, κ, (essentially the nonlinear frequency correction ν_1), is negative.

According to the sign of

$$\kappa = \mathrm{Re}\,\delta\,\{1 + (\mathrm{Im}\,\delta\,\mathrm{Im}\,\mu)/(\mathrm{Re}\,\delta\,\mathrm{Re}\,\mu)\}$$

either local spatial oscillators can be sinchronized through diffusion if $\kappa > 0$ (Kuramoto and Tsuzuki,1976; Kuramoto,1977), or if $\kappa < 0$ a *chaotic(turbulent)* state (chemical *turbulence*) appears(Fujisaka and Yamada,1977;Kuramoto and Yamada,1976; Yamada and Kuramoto,1976).

iv. Applications

iv.1. Budworm-forest ecosystem.

In this particular case, some relevant data are: $r = 1.52$/year, and $T \sim 7$-10 years, $R = 0.994$, $Q = 302$ and $u_0 = 300.991$. Then $\mathrm{Re}\,\mu < 0$ and $\nu_1 < 0$.Thus in the dispersion-free case there is bifurcation to a stable limit cycle, whereas a turbulent state is expected in the purely diffusive case if the space is unbounded.

In the dispersion-free case, at larger supercritical values ($\beta > \beta c$) the limit cycle appears like a relaxation oscillation. Figure 1 corresponds to the case $\beta = 14$. For about 2/3 of the period ($4T \sim 37$ years) there are few worms (endemic state) whereas in the later third of the period there is the outbreak which is expected to repeat periodically some forty years later. The population ratio between the densities at the two states is about 25 x 10^5. Such predicted values (period and ratio) agree satisfactorily with data given in the literature (Clark et al,1978). One can numerically test the stability of the limit cycle to various illustrative distubances: u u_0, $u = 5u_0$, $u = u_0 + e^{\tau+1}$ $(-1 \leq \tau \leq 0)$. It takes about $\tau \sim 120$ for such initial conditions to decay into the stable limit cycle within less than one per cent error bar.

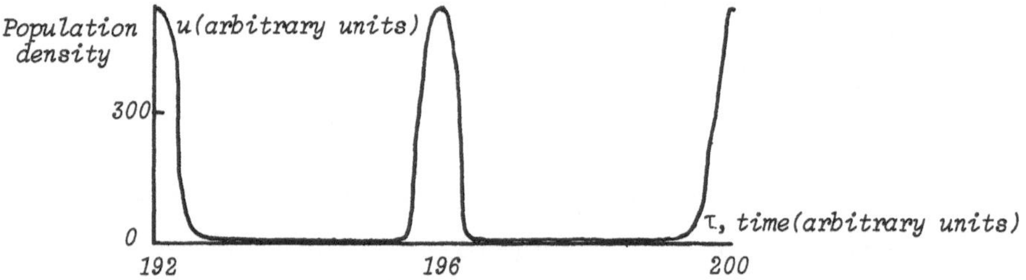

Fig.1: Stable limit cycle(oscillations in population density) for β=14,and $u(\tau)$=300.991,$-1<\tau<0$. Note that the minimum of u is 0.0002(non vanishing).Ratio of maximum to minimum is $u_{max}/u_{min} \sim 25.10^5$. The period is about 37 years.

iv. 2 Time-delay in an *ionic pump* of sea-urchin eggs.

The equation

$$dN/d\tau = 1 - \alpha N \{\beta + \gamma N^n(\tau-1)\} \quad , n > 1 \qquad (11)$$

is a model recently proposed by E. Parisi (private communication; Bonilla et al,1981) to account for the Calcium ionic pump in sea-urchin eggs. As for (1), here (11) has a unique physically observable steady state $N_0 > 0$ (α, β, γ positive numbers) and there is a Hopf bifurcation for suitable values of the parameters if $n > 1$. It appears that for n = 1 the steady state can only be destabilized for $\alpha\gamma = \infty$, $N_0 = 0$. $\nu_0 = \pi$.

As an illustration, for $\beta = 0$, $n = 2$, $N_0 = \Gamma^{-1/3}$ with $\Gamma = \alpha\gamma$ there is a supercritical Hopf bifurcation for $\Gamma > \Gamma_0 \equiv (2\pi\sqrt{3}/9)^3$. The nonlinear correction to the linear frequency $\nu_0 = 2\pi/3$ is $\nu_1 > 0$ (the period is slightly less than 3 minutes). If diffusion is added to the model, this homogeneous limit cycle is stable to inhomogeneous spatial disturbances.

6. ROLE OF EXTERNAL NOISE

Years ago,Landauer (1962) and Prigogine and collaborators (Horsthemke and Malek Mansour 1976; Horsthemke and Lefever,1977; Arnold et al,1978) realized that phase transitions can be induced in a nonequilibrium system by the sole effect of external noise. Their finding was not a specific consequence of the model problem they considered. Quite similar results were also found in other models (de la Rubia and Velarde,1978; see also the book edited by Arnold and Lefever,1981). We illustrate now the possibility of noise induced transitions in the model already used in Section 4, without diffusion. It refers to the macroscopic evolution given by the equation

$$dX/dt = \alpha - X - \beta X/(1+X^2) \qquad (1)$$

where α and β are some given *externally varying* parameters.

For $\alpha \geq \alpha_c = 27^{1/2}$, and varying β, the steady problem of (1) has a triple root (for β = 8, $X = 3^{1/2}$, say) wherwas for $\alpha < \alpha_c$ there is a single steady state whatever the values of β. To (1) we associate a Fokker-Planck equation

$$\partial P(X,t)/\partial t = \partial\{P(X,t)(\alpha-X-\beta X/(1+X^2))\}/\partial X + (\sigma^2/2)\partial^2\{P(X,t)X^2/(1+X^2)^2\}/\partial X^2 \qquad (2)$$

for which the origin and infinity are *natural* boundaries (Gihman and Skorohod,1972, Arnold,1974).

Fig.1 corresponds to *mean* values of α and β such that (1) has a unique steady state, $X_0 = 1$. For $\sigma^2 = 1$ the maximum of probability is located over X_0, as it should be expected from the macroscopic description or from a Langevin-type of description. But for $\sigma^2 = 5$ the distribution shows *three* peaks of which two are maxima and one minimum. The latter is precisely located over X_0 and its depth goes further deep upon increase of the variance of the external parameter, σ^2. Thus, merely increasing this variance, or *external noise*, the system can be forced to operate in a state not predicted from the deterministic descripition.

Figure 2 corresponds to α and β such that (1) has three steady soluions: $X_1 = 1$ (at least *locally* stable), $X_2 = 2$ (unstable), and $X_3 = 3$ (stable). For $\sigma^2 = 1$ the maximum located over X_3 is higher than the maximum over X_1 with the property that upon decreasing $\sigma^2 < 1$ this difference of height decreases.This is already a clear consequence of the variable external noise. For $\sigma^2 = 5$ the maximum located over X_1 goes to a minimum and at the same time a new maximum shows up near the origin. This latter maximum gets higher upon increasing the value of the variance. On the ohter hand the probability looses extremality at X_2.

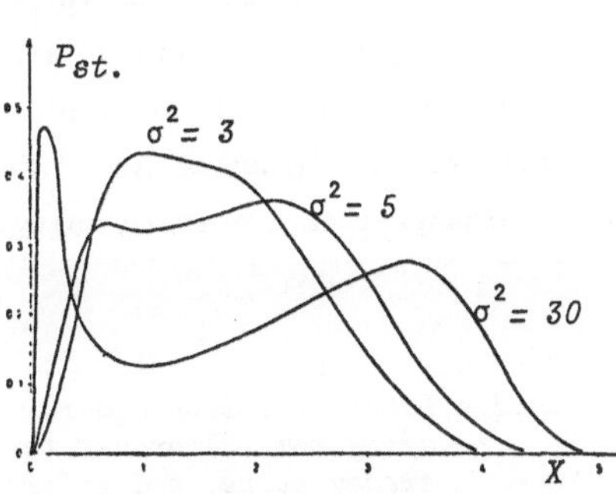

Fig.1.:Stationary probability density for α=5, β =8, and three values of the variance when, according to deterministic theory there is only ONE stable steady state, X=1.

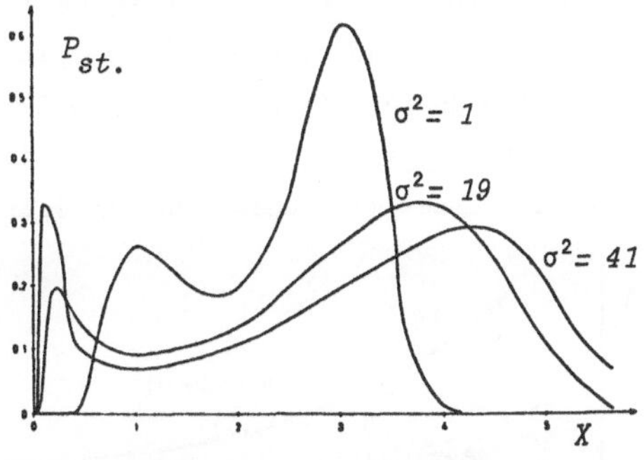

Fig.2.:Stationary probability density in the case of three deterministic steady states: X_1*=1(stable),* X_2*=2(unstable) and* X_3*=3(stable). Case α=6, β =10, and three values of the variance.*

APPENDIX: *Autocatalysis is not enough to sustain limit cycle oscillations.*

Consider the following autocatalytic reaction scheme

$$A \longrightarrow Y \quad (1.a)$$

$$Y + qX \longrightarrow (q+1)X \quad (1.b)$$

$$X \longrightarrow P \quad (1.c)$$

in which all three steps are taken to be irreversible, i.e., all the reverse reaction's constants are set at zero. We take $q > 0$ but otherwise, and for illustration of mathematical behaviour of (1), q is considered here as a real continuosly varying parameter. If $q = 1$ (viz.2) the autocatalytic step (1.b) is bimolecular (viz.trimolecular). It is to be noted that model (1) contains as a particular case the essential trimolecular ($q=2$) step of the *Brusselator* model extensively studied by I. Prigogine and collaborators (Nicolis and Prigogine, 1977). We use model (1) to illustrate with a computer-aided representation, a result proved by Sel'kov(1968), Hanusse(1972) and Tyson(1973), for the case $q = 1$.

In the absence of diffusion, and after a suitable adimensionalization, model (1) is described by the following set of ordinary differential equations

$$dX/d\tau = X^q Y - X; \quad dY/d\tau = A - X^q Y \quad (2)$$

System (2) has the following solutions: (i) the *unbounded* solution

$$X(\tau) = 0; \quad Y(\tau) = A\tau + Y_0 \quad (3)$$

for a given initial concentration $Y(\tau=0) = Y_0$. (ii) the critical point (i.e., a stationary solution)

$$X_s = A; \quad Y_s = A^{1-q} \quad (4)$$

For different values of A and q the local stability of the critical point (4) is given by the real parts of the roots of the characteristic equation

$$\lambda^2 - (q-1-A^q)\lambda + A^q = 0 \quad (5)$$

The results are depicted in Figure 1.

Regions Ia and IB are of local asymptotic stability of (4) with characteristics of stable focus (IA) and stable node (IB), respectively. In region II the steady state is unstable with characteristics of unstable focus (IIA), and unstable node (IIB), respectively. If $q \leq 1$, the steady state (4) is stable for all values of $A > 0$. The two instability areas (II) described in figure 1 just dissapear as q goes to unity.

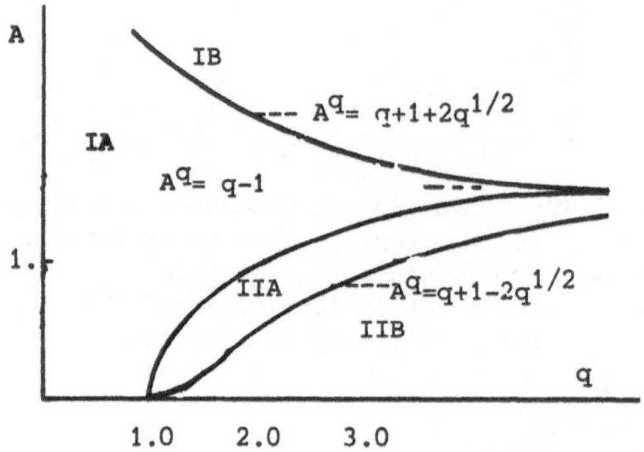

Fig.1.: Local stability diagram of system (2) around the steady state. q=1 refers to a bimolecular step and q=2 to a trimolecular one (Brusselator).

REFERENCES AND OTHER RELATED BIBLIOGRAPHY.

1.Andronov,A.A.,E.A. Leontovich,I.I.Gordon and A.G.Maier,1973(English transl.)THEORY OF BIFURCATIONS OF DYNAMIC SYSTEMS ON A PLANE,Wiley, New York.

2.Aris,R.,1976,THE MATHEMATICAL THEORY OF DIFFUSION AND REACTION IN PERMEABLE CATALYSTS(2 vols.),Clarendon Press,Oxford.

3.Arnold,L.,1974,STOCHASTIC DIFFERENTIAL EQUATIONS,Wiley,New York.

4.Arnold,L.,L.,W.Horsthemke and R.Lefever,1978,Z.Phys. B29,367.

5.Arnold,L. and R.Lefever,1981(editors),STOCHASTIC NONLINEAR SYSTEMS,Springer-Verlag New York.

6.Arnold,V.,1978(English transl.) ORDINARY DIFFERENTIAL EQUATIONS,M.I.T.Press, Cambridge,Mass.

7.Arnold,V.,1980(French transl.)CHAPITRES SUPPLÉMENTAIRES DE LA THÉORIE DES EQUATIONS DIFFERENTIELLES ORDINAIRES,Mir, Moscow.

8.Balslev,I. and H. Degn, 1975,J.Theor.Biol.49,173.

9.Bonilla,L.L. and M.G.Velarde,1979,J.Math.Phys.,20,2692.

10.Bonilla,L.L. and M.G.Velarde,1980,J.Math.Phys.,21,2586.

11.Bonilla,L.L. and M.G.Velarde,1981,J.interdisciplin.Cycle Res.,to appear.

12.Bonilla,L.L.,A.Fdez.Cancio and M.G.Velarde,1981,J.interdiciplin.Cycle Res.in press.

13.Bonilla,L.L. and M.G.Velarde,1981, J.Math.Phys. submitted por publication.

14.Bonilla,L.L.,M.G.Velarde and E.Parisi,1981(report in preparation).

15.Bonilla,L.L.,M.G.Velarde,A.Pimpale and P.Landsberg,1981(report in preparation).

16.Boyce,W.E. and R.C.DiPrima,1977,ELEMENTARY DIFFERENTIAL EQUATIONS AND BOUNDARY VALUE PROBLEMS,Wiley,New York.

17.Clark,W.C.,D.D.Jones and C.S.Holling,in SPATIAL PATTERNS IN PLANKTON COMMUNITIES (J.H.Steele, editor)Plenum Press, New York.

18.Coddington,E.A. and N.Levinson,1955, ORDINARY DIFFERENTIAL EQUATIONS,McGraw-Hill, New York.

19.Cushing,J.M.,1977,INTEGRODIFFERENTIAL EQUATIONS AND DELAY MODELS IN POPULATION DYNAMICS, Springer-Verlag,New York.

20.Degn,H. and D.E.F. Harrison,1969,J.Theor.Biol.22,238.

21.de la Rubia,J.F. and M.G.Velarde,1978,Phys.Lett.A69,304.

22.Degtyarenko,N.N.,V.F.Elesin and V.A.Furmenov,1974,Sov.Phys.Semicond.7,1147.

23.Ebeling,W.,1976,STRUKTURBILDING BEI IRREVERSIBLEN PROZESSEN,Teubner Verlag,Leipzig.

24.Eigen,M.,1971,Die Naturwiss.,58,465.

25.Eigen,M. and P.Chuster,1979, THE HYPERCYCLE, Springer-Verlag,New York.

26.Fairén,V. and M.G.Velarde,1979,J.Math.Biol.,8,47.

27.Fairén,V. and M.G.Velarde,1979,Prog.Theor.Phys.61,801.

28.Fairén,V. and M.G.Velarde,1979,Rep.Math.Phys.16,421.

29.Faraday Symposia (Chemical Society),1974,nº9,PHYSICAL CHEMISTRY OF OSCILLATORY PHENOMENA,London.

30.Fife,P.C.,1979,MATHEMATICAL ASPECTS OF REACTING AND DIFFUSING SYSTEMS, Springer-Verlag,New York.

31.Frank,G.M. et al.,1967 (editors),OSCILLATORY PROCESSES IN BIOLOGICAL AND CHEMICAL SYSTEMS,Nauka,Moscow(in Russian).

32.Fujisaka,H. and T.Yamada,1977,Prog.Theor.Phys.55,734.

33.Gihman,I.I. and A.V.Skorohod,1972 (English transl.) STOCHASTIC DIFFERENTIAL EQUATIONS,Springer-Verlag,New York.

34.Gross,E.F.,D.A.Permogorov,V.V.Travnikov and A.V.Silin,1972,Sov.Phys.Solid-St. 14,1193.

35.Haken,H.,1977,SYNERGETICS(2nd. edition),Springer-Verlag, New York.

36.Hale,J.,1977,THEORY OF FUNCTIONAL DIFFERENTIAL EQUATIONS,Springer-Verlag,New York.

37.Hale,J.,1978, in NONLINEAR OSICLLATIONS IN BIOLOGY,Lect.Appl.Math.,17,157.

38.Hanusse,P.,1972,Comptes Rendus Acad .Sci.Paris,C274,1247.

39.Harrison,D.E.F. and S.J.Pirt,1967,J.Gen.Microbiol.46,193.

40.Harrison,D.E.F. and H.H.Topiwàla,1974,Adv.Biochem.Eng.,vol.3,Springer-Verlag,Berlin.

41.Hemmer,P.C. and M.G.Velarde,1977,Z.Phys.B 31,111.

42.Higgins,J.,1967,Ind.Eng.Chem.59,19.

43.Hirsch,M.W. and S.Smale ,1974,DIFFERENTIAL EQUATIONS,DYNAMICAL SYSTEMS AND LINEAR ALGEBRA, Academic Press,New York.

44.Horsthemke,W. and M. Malek-Mansour,1976,Z.Phys.B24,307.

45.Horsthemke,W.and L.Brenig,1977,Z.Phys.B27,341.
46.Horsthemke,W. and R.Lefever,1977,Phys.Lett.A64,19.
47.Hulin,D.,A.Mysyrowicz and C.Benoit,1980,Phys.Rev.Lett.45,1980.
48.Hutchinson,G.E.,1948,Ann.N.Y.Acad.Sci.,50,221.
49.Ibañez,J.L.,V.Fairén and M.G.Velarde,1976,Phys.Lett.A58,364.
50.Ibañez,J.L.and M.G.Velarde,1977,J.Physique Lett.38,L465.
51.Ibañez,J.L. and M.G.Velarde,1978,Phys.Rev.A18,750.
52.Ibañez,J.L. and M.G.Velarde,1978,J.Non-equilib.Therm.,3,63.
53.Iooss,G.and D.D. Joseph,1980,ELEMENTARY BIFURCATION THEORY,Springer-Verlag,New York.
54.Jetschke,G.,1979,Phys.Lett.A72,265.
55.Jetschke,G.,1980,Report N.80.17 F.Schiller Univ.Jena (in German).
56.Klingenstein,W. and W.Schmid,1979,Phys.Rev.B20,3285.
57.Kuramoto,Y. and T.Tsuzuki,1976,Prog.Theor.Phys.55,356.
58.Kuramoto,Y and T.Yamada,1976,Prog.Theor.Phys.56,679.
59.Kuramoto,Y. , 1977, in SYNERGETICS (H.Haken,editor)Springer-Verlag,New York.
60.Kurtz,T.G.,1971,J.Appl.Prob.8,344.
61.Landauer,R.,1962,J.Appl.Phys.33,2209.
62.Landsberg.,P.T. and A.Pimpale,1976,J.Phys.C9,148.
63.Landsberg,P.T.,1978,THERMODYNAMICS AND STATISTICAL MECHANICS, Clarendon Press, Oxford.
64.Landsberg,P.T.,1980,European J.Phys.1,31.
65.La Salle,J. and S.Lefchetz,1961,STABILITY BY LYAPUNOV'S DIRECT METHOD WITH APPLICATIONS,Academic Press,New York.
66.MacDonald,N.,1978,TIME LAGS IN BIOLOGICAL MODELS,Springer-Verlag,New York.
67.Mahr,H. and C.L.Tang,1972,J.Appl.Phys.43,1818.
68.Margalef,R.,1980,LA BIOSFERA,Omega,Barcelona.(in Spanish).
69.May,R.M.,1976,Nature(London)459.
70.Maynard-Smith,J.,1974,MODELS IN ECOLOGY,University Press, Cambridge.
71.Minorsky,N.,1962,NON-LINEAR OSCILLATIONS,Van Nostrand,New York.
72.Murray,J.D.,1977,LECTURES ON NONLINEAR-DIFFERENTIAL-EQUATION MODELS IN BIOLOGY, Clarendon Press,Oxford.
73.Nicolis,G. and I.Prigogine,1977,SELF-ORGANIZATION IN NONEQUILIBRIUM SYSTEMS, Wiley,New York.
74.Normand,Ch.,Y.Pomeau and M.G.Velarde,1977,Rev.Mod.Phys.,49,581.
75.Okubo,A.,1980,DIFFUSION AND ECOLOGICAL PROBLEMS,Springer-Verlag,New York.
76.Oppenheim,I.,K.E.Shuler and G.H.Weiss,1977(editors) STOCHASTIC PROCESSES IN CHEMICAL PHYSICS,M.I.T.Press,Cambridge,Mass.
77.Pikovsky,A.S.,1981,Phys.Lett.,A85,13.
78.Pimpale,A. and P.Landsberg,1977,J.Phys.C10,1447.
79.Pimpale,A.,P.Landsberg,L.L.Bonilla and M.G.Velarde,1981,J.Phys.Chem.Solids,42,873.
80.Pomeau,Y.J.C.Roux,A.Rossi,S.Bachelart and C.Vidal,1981,J.Phys.(Paris)Lett.42,L271.
81.Prigogine,I.,1967,(published in 1969), in FROM THEORETICAL PHYSICS TO BIOLOGY (1st. Int. Symposium;M.Marois,editor)North Holland,Amsterdam.
82.Prigogine,I.,1974,(Spanish transl.) INTRODUCCION A LA TERMODINAMICA DE LOS PROCESOS IRREVERSIBLES, Seleciones Científicas,Madrid.
83.Prigogine,I. and I.Stengers,1979,LA NOUVELLE ALLIANCE,Gallimard,Paris.(in French).
84.Rabinovich,M.I.,1978,Sov.Phys.Usp.21,443.
85.Rayleigh,Lord, 1921,Proc.Roy.Soc.A99,372.
86.Sel'Kov,E.E.,1968,European J.Biochem.,4,79.
87.Turner,J.S.,J.C.Roux,W.D.McCornick and H.L.Swinney,1981,Phys.Lett.A85,9.
88.Tyson,J.J.,1973,THE BELOUSOV-ZHABOTINSKII REACTION,Springer-Verlag,New York.
89.Van Kampen,N.G.,1976,Adv.Chem.Phys.,34,245.
90.Vasil'ev,V.A.,Yu.M.Romanovsky and V.G.Yakhno,1979,Sov.Phys.Usp.22,615.
91.Velarde,M.G.,1981,in NONLINEAR PHENOMENA AT PHASE TRANSITIONS AND INSTABILITIES (T.Riste,editor),Plenum Press,New York.
92.Waltman,P.,1974,DETERMINISTIC THRESHOLD MODELS IN THE THEORY OF EPIDEMICS, Springer-Verlag,New York.
93.Winfree,A.T.,1980,THE GEOMETRY OF BIOLOGICAL TIME,Springer-Verlag,New York.
94.Yamada,T. and Y.Kuramoto,1976,Prog.Theor.Phys.,56,681.
95.Zhabotinskii,A.M.,1964,Doklady Akad.Nauk USSR,157,392.

FLUCTUATIONS IN ELECTROMAGNETIC SYSTEMS

J.M. RUBI
Departamento de Termología
Universidad Autónoma de Barcelona
Bellaterra (Barcelona) Spain*

1. INTRODUCTION.

Fluctuation phenomena have been the subject matter of many recent investigations [1]. Since matter is not exactly a continuum, field variables can fluctuate around average values. One of the most fruitful procedures to deal with fluctuations in continuum systems is to assume that the dissipative currents occurrring in the system split in systematic and random parts [2]. Systematic parts constitute the linear or phenomenological laws of thermodynamics of irreversible processes [3] whereas random parts are responsible for fluctuations and satisfy fluctuation-dissipation theorems. These theorems and techniques based on the coupling of diverse modes, allow to calculate correlation functions of field variables and constitute the so-called fluctuation-dissipation [2] and mode-mode coupling theories [4].

In this paper we are dealing with fluctuations in electromagnetic systems. When matter interacts with electromagnetic fields, many new interesting phenomena occur[5]. Fluctuation of the electric and magnetic

* Part of this work was done at Instituut-Lorentz (Leiden University, The Netherlands)

field strengths are the starting point of theories of molecular attractive forces between solids [6]. Many efforts have been devoted to study fluctuations in conducting media [7] and around stationary states of superconductors [8]. The latter systems are of special interest since there are still opened questions referent to phenomenological coefficients [9], [10]. Finally we can quote the unsolved problem about the origin of 1/f noise in charged systems [11], [12].

The text is divided in two main parts. In the first one we shall briefly indicate a procedure to study linear fluctuations around stationary states of one-dimensional conductors. The method is based on assuming that at the steady state, fluctuation-dissipation theorems have the same form than their equilibrium counterpart, but the coefficients of delta functions depend on quantities calculated at the steady state [17], [18], [19]. The systems under study are described by means of a common one-dimensional equation. As an example we shall discuss fluctuations around stationary states in the ballast resistor [13] and near the Gunn instability [14]. The evolution of a kind of plancton (see Margalef seminar) belongs also to the type of equation treated in this part. In the second part we shall deal with non-linear fluctuations in a charged two-component fluid in equilibrium. In the Landau-Lifshitz spirit, it is possible to arrive to non-linear stochastic differential equations and from them to a Fokker-Planck equation involving the distribution function of the physical fields

2. FLUCTUATIONS AROUND STATIONARY STATES OF ONE-DIMENSIONAL CONDUCTORS

A number of one-dimensional systems can be described by means of the equation

$$\partial\Psi(x,t)/\partial t = \alpha_1\partial^2\Psi(x,t)/\partial x^2+\alpha_2\partial\Psi/\partial x+\alpha_3 d\phi(\Psi)/d\Psi \qquad (1)$$

where $\Psi(x,t)$ is a physical field as temperature or density. The right hand side of (1) consists of three contributions: the first term is a diffusive contribution, α_1 being a diffusion coefficient, the second term is a "convective" term, and the third involves the potential $\phi(\Psi)$ and contains the remaining terms responsible for the temporal variations of $\Psi(x,t)$, as are for example external source terms or terms proportional to $\Psi(x,t)$. In (1), α_1 and α_3 are unspecified coefficients. Stationary states are the solutions of

$$\alpha_{1,st} d^2\Psi_{st}(x)/dx^2+\alpha_{2,st} d\Psi_{st}(x)/dx+\alpha_{3,st} d\phi(\Psi_{st})/d\Psi_{st} = 0, \qquad (2)$$

which, of course, depend on the form of the potential $\phi(\psi)$

In the linear case, it is possible to study the stability of the stationary solutions $\Psi_{st}(x)$ by solving the eigenvalue problem associated to (2) (normal mode analysis), namely

$$H\,(x;\beta_i)\,\Psi(x,k) = \alpha\Psi(x,k) \tag{3}$$

where $H(x;\beta_i)$ is the "hamiltonian" of the system, which in general may depend on a set of parameters β_i, α is the eigenvalue and k is a function of α. The stationary states $\Psi_{st}(x)$ will be stable if the real part of the lowest eigenvalue is positive. If this real part is negative the state will be unstable while the state of marginal(indifferent)stability will be reached for a zero eigenvalue. A non-linear stability analysis can be achieved by using more sophisticated methods as for example the energy method [15]) or perturbative expansions [16]).

We are interested in studying linear fluctuations around the stationary states $\Psi_{st}(x)$. In the linear case, one gets from (1) the evolution equation for the perturbations $\Delta\Psi(x,t) = \Psi(x,t)-\Psi_{st}(x)$ in the form

$$\partial\Delta\Psi(x,t)/\partial t = H\,(x;\beta_i)\Delta\Psi(x,t) \tag{4}$$

where, if the coefficients α_1, α_2 and α_3 are constant, the "hamiltonian" is given by

$$H(x;\beta_i) = (\alpha_1\partial^2/\partial x^2+\alpha_2\partial/\partial x+\alpha_3 d^2\phi(\Psi)/d\Psi^2)_{st} \tag{5}$$

Fluctuations of physical fields can be introduced by adding a stochastic force $F_R(x,t)$ to the right hand side of (4) in such a way that this equation becomes a Langevin-like equation

$$\partial\Delta\Psi(x,t)/\partial t = H(x;\beta_i)\Delta\Psi(x,t)+F_R(x,t) \tag{6}$$

In order to arrive at a closed description of the system we must specify the properties of the random force $F_R(x,t)$. To this end it is necessary to realize that although equation (6) is formally analogous to the Langevin equation of brownian motion, our reference state is not an equilibrium state. It is possible, however, to follow a procedure close to the one used when the reference state is an equilibrium state. Thus we shall assume that $F_R(x,t)$ has zero mean

$$\langle F_R(x,t)\rangle_{st} = 0 \tag{7}$$

and the correlations are given by

$$\langle F_R(x,t)\ F_R(x',t')\rangle_{st} = F(\Psi_{st}(x))\ \delta(x-x')\ \delta(t-t') \tag{8}$$

where now the averages $\langle ....\rangle_{st}$ are taken at the steady state. This assumption has been successfully employed in the treatment of fluctuations near the Bénard [17]), ballast resistor [18]) and Gunn [19]) instabilities.

By means of the fluctuation-dissipation theorem (8) it is possible to calculate correlation functions of the type

$$\langle \Delta\Psi(x,t)\ \Delta\Psi(x',t')\rangle_{st} \tag{9}$$

or

$$\langle \Delta\Psi(x,t)\ \Delta\Psi(x',t')\rangle_{st} \tag{10}$$

where $\Delta\Psi(x,t) = L(x,t)\ \Delta\Psi(x,t)$, $L(x,t)$ being an operator. To arrive to (9) or (10) one must formally solve equation (6) in Fourier space or in the space of the eigenvalues of $H(x;\beta_i)$. Hence one gets

$$\langle \Delta\Psi(k,\omega)\ \Delta\Psi(k',\omega')\rangle_{st} \tag{11}$$

which by inversion leads to (9). In (11) k or k' stands for either an element of Fourier space or an element of the eigenvalues space.

2.1 FLUCTUATIONS IN THE BALLAST RESISTOR.

The ballast resistor [13]) consists of a thin conducting wire surrounded by a gas which is kept to constant temperature. Both the electrical current $I(t)$ and the temperature of the bath T_G can be externally controlled The quantity Ψ introduced above is in this case the temperature of the wire and the diffusion coefficient α_1 is nothing but the thermal diffusivity $a=\lambda/c$, λ and c being the heat conductivity of the wire and the heat capacity per unit lenght, respectively. In the case in which λ is constant and the Thomson coefficient is zero, the balance equation of the internal energy of the wire and the linear laws lead to an equation as (1)

$$\partial T(x,t)/\partial t = a\ \partial^2 T(x,t)/\partial x^2 + (\lambda c)^{-1} d\phi(T)/dT \tag{12}$$

where the potential $\phi(T)$ is given by

$$\phi(T) = \int_0^T dT\lambda[I^2R - q(T-T_G)] \tag{13}$$

Here R(T) is the isothermal resistance per unit lenght and $q(T,T_G)$ a coefficient which modulates the energy transfer between the wire and the gas.

The stationary states are the solutions of the equation

$$ad^2T_{st}(x)/dx^2+(\lambda c)^{-1}d\phi(T_{st})/dT_{st} = 0 \tag{14}$$

Obviously these solutions will depend on the form of R(T), $q(T,T_G)$ and also on boundary conditions. The Langevin-like equation (6) is in this case

$$\partial\Delta T(x,t)/\partial t = H(x;\beta_i)\ \Delta T(x,t)+F_R(x,t) \tag{15}$$

where the random force contains the random parts of the heat current, the electric field and the energy transfer between the gas and the wire. From the properties of the random parts of the currents one concludes that F_R has zero mean, (see eq. (7)), and the correlation at the stationary state is given by

$$\langle F_R(x,t)F_R(x',t')\rangle_{st} = \frac{2K_B}{c^2}[T^2(x)q + T(x)RI^2 + +\lambda(\partial/\partial x)\ T^2(x)\ (\partial/\partial x')]_{st}\ \delta(x-x')\delta(t-t') \tag{16}$$

where K_B is the Boltzmann constant and the derivatives act on all factors at their right. When the reference steady state is homogeneous and the wire is infinite, equation (15) can be formally solved in Fourier space. One has

$$\Delta T(k,\omega) = G(k,\omega)\ F_R(k,\omega) \tag{17}$$

where the operator $G(k,\omega)$ is given by

$$G(k,\omega) = [-i\omega + ak^2- (\lambda c)^{-1}d^2\phi/dT^2]_{st} \tag{18}$$

Because T_{st} is constant, equation (16) can be immediatly Fourier-transformed and used to calculate the temperature correlation function at

Fourier space. One gets

$$\langle \Delta T(k,\omega)\Delta T(k',\omega')\rangle_{st} = H(k,\omega)\delta(k+k')\delta(\omega+\omega') \tag{19}$$

where the operator $H(k,\omega)$ involves the parameters of the system. The temperature correlation function at equilibrium can be calculated from (19) by considering that this state is reached when $I=0$ and $T=T_G$. At the critical point the temperature correlation function in (k,t) and (x,t) space obtained by inversion of (19) diverges and it is possible to define a relaxation time and a correlation lenght for the decay of fluctuations.

To calculate the voltage correlation function one can use the equation

$$V(x,t) = -\int_0^x E(x',t)\, dx' \tag{20}$$

where $V(x,t)$ is the voltage difference between 0 and x and $E(x,t)$ the electric field. In view of (20), voltage fluctuations are related to electric field fluctuations in the manner

$$\Delta V(x,t) = -\int_0^x \Delta E(x',t)\, dx' \tag{21}$$

According to the linear law obtained from the entropy production, the field fluctuations around homogeneous steady states depends on tempera ture fluctuations and write

$$\Delta E(x,t) = [I\, dR/dT - \eta\, \partial/\partial x]_{st}\, \Delta T(x,t) + E_R(x,t) \tag{22}$$

Here η is the thermo-electric power. Using (22) one has in (k,ω) space and therefore the voltage correlation function is

$$\langle \Delta V(x,\omega)\Delta V(x,\omega')\rangle_{st} = (2\pi)^{-1}\int_0^x dz\int_0^x dy\int dk\; C'(k,\omega,\omega')\,.$$

$$\exp[ik(z-y)] \tag{24}$$

which at the critical point and at low frecuencias behaves as

$$\langle \Delta V(x,\omega)\Delta V(x,\omega')\rangle_{st} \sim \omega^{-3/2} \tag{25}$$

If the reference steady state is not homogeneous it is necessary to solve formally (15) in the space of the eigenvalue of H. In view of (5) and (13), the "hamiltonian" and its eigenfunctions and eigenvalues will depend on the functional form of R and q (models). A rather simple and interesting case is the "hot-spot" model [20)22] in which the wire is superconductor below a critical temperature T_c and q and λ are constant. In this model, the resistance is equal to $R_o\theta(T-T_c)$ where θ is the Heaviside function and R_o is a constant. The potential (13) is now given by

$$\phi(T) = (q/c) \int_0 dt\,[T_h\theta(T-T_c)-T] \tag{26}$$

where $T_h = I^2R_o/q$ and T_G has been, for simplicity'sake, taken zero. The stationary solutions of the model, corresponding to eq.(2), contain cold $(T>T_c)$ and hot $(T<T_c)$ sections and depends on boundary conditions at both extremes of the wire (Neumann and Dirichlet boundary conditions) [21]). Due to the fact that the wire is superconductor below T_c, the voltage difference between the end points of the wire is proportional to the lenght of the hot section.

2.2 FLUCTUATIONS NEAR THE GUNN INSTABILITY

In some semiconductors (Ga As for example) when electric current passing through the sample exceeds a critical value, microwave oscillations occur. This electrodiffusive effect called Gunn effect [14]) has been the subject of intensive studies due to its importance as microwaves generator. The fundamental equations governing the system, which is assumed again one-dimensional, are the balance equation for the charge of electrons

$$z_e\partial\rho_e(x,t)/\partial t = -\partial I(x,t)/\partial x \tag{27}$$

where ρ_e is the density of electrons and z_e its constant charge per unit mass, and the Maxwell equation (Poisson equation)

$$\varepsilon\partial E(x,t)/\partial x = z_e\rho_e(x,t)+z_i\rho_i \tag{28}$$

where ε is the dielectric constant, ρ_i the density of ionized donors and z_i its charge. Moreover we must use the linear law

$$E(x,t)-z_e^{-1}\partial\mu_e(x,t)/\partial x = RI(x,t) \tag{29}$$

derived from the entropy production by considering that the diffusion flux is measured with respect to ionized fonors. In (29), $\mu_e(x,t)$ is the chemical potential of electrons. Equations (27)-(29) can be combined to get the differential equation

$$\partial E(x,t)/\partial t = D\partial^2 E(x,t)/\partial x^2 - \varepsilon^{-1}\sigma E(x,t) + \varepsilon^{-1} I_{ex}(t) \tag{30}$$

where D is the diffusion coefficient, $\sigma = R^{-1}$ the electrical conductivity and $I_{ex}(t)$ the external electric current. The diffusion coefficient D may depend on the electric field, moreover σ is often related to the mobility μ through $\sigma = \rho_e z_e \mu(E)$. In function of the velocity of electrons $v(E) = \mu(E)E$, (30) reads

$$\partial E(x,t)/\partial t = D\partial^2 E(x,t)/\partial x^2 - v\partial E/\partial x + \varepsilon^{-1}\rho_i z_i v + \varepsilon^{-1} I_{ex}(t) \tag{31}$$

The latter equation is again an equation of (1) type. The potential $\phi(\psi)$ is in this case

$$\phi(E) = \int_0^E dE[\ \rho_i z_i v + I_{ex}] \tag{32}$$

Electric field steady solutions are the solutions of

$$D_{st} d^2 E_{st}(x)/dx^2 - v_{st} dE_{st}/dx + \varepsilon^{-1} d\phi_{st}\ /dE_{st} = 0 \tag{33}$$

As we saw in the ballast resistor, to get $E_{st}(x)$, particular models and convenient boundary conditions should be taken into account. In this case models involve the form of D(E) and V(E) [23].

To study linear fluctuations around stationary states $E_{st}(x)$ we must work with the Langevin-like equation

$$\partial \Delta E(x,t)/\partial t = H(x,\beta_i)\Delta E(x,t) + F_R(x,t) \tag{34}$$

where $\Delta E(x,t) = E(x,t) - E_{st}(x)$. Here the "hamiltonian" linearized at the stationary state is

$$H = [D\partial^2/\partial x^2 + vD^{-1}\partial D/\partial x - (\varepsilon D)^{-1}(dD/dE)(d\phi/dE)\ - v\partial/\partial x$$

$$-\partial v/\partial x + \varepsilon^{-1} d^2\phi/dE^2]_{st} \tag{35}$$

Since the only dissipative flux appearing in the entropy production is the electric current, $F_R(x,t)$ is only related to the random part $I_R(x,t)$

in the form, $F_R = -\varepsilon^{-1} I_R$. Therefore one has

$$\langle F_R(x,t)\rangle_{st} = 0 \tag{36}$$

$$\langle F_R(x,t)F_R(x',t')\rangle_{st} = -2\varepsilon^{-2} D_{st}\rho_{e,st} z_e^2\ \delta(x-x')\ (t-t') \tag{37}$$

An equation similar to (35) but involving voltage fluctuations can be obtained from (21). In this equation, the stochastic force is obtained from $F_R(x,t)$ by simple integration and the fluctuation-dissipation theorem is related to (37).

Electric field and voltage fluctuations [19]) can be easily calculated when $\rho_{e,st}$ is homogeneous. One finds that the equal frecuency voltage correlation function behaves as

$$\langle \Delta V(\omega)\Delta V(\omega')\rangle_{st} \sim (a+b\omega^2)^{-1} \tag{38}$$

where a and b involve the parameters of the system and depend on voltage. Voltage fluctuations have been measured in [24]) and experimental results agree with (38).

To calculate electric field or voltage fluctuations when the reference steady state is inhomogeneous it is necessary, first of all, to find a stationary solution (model) and afterwards to solve the eigenvalue problem as discussed above.

3. ELECTROMAGNETIC FLUCTUATIONS IN FLUIDS.

When the system under consideration is a fluid, it is necessary to work with the whole set of conservation laws, namely, the conservation laws of mass, charge, momentum and energy. To arrive at a closed description of the system one also needs Maxwell equations, linear laws, derived from the entropy production, and equations of state [3]). Even though simplifications can be made, charged fluids are more complicated than one-dimensional conductors. Moreover due to the presence of convective nonlinearities, it is difficult (if not impossible) to find analytic solutions of the steady state. Remember that in the pure hydrodynamic case only a few stationary flows can be exactly solved[2]). The presence of nonlinearities also leads to use fluctuation theories different from the one described in the first part.

In this part we are dealing with charged (globally uncharged) two-component fluids in absence of polarization and magnetization phenomena.

The balance equations derived from the conservation laws mentioned above are

$$\partial\rho/\partial t = -\nabla\cdot\rho\underline{v} \tag{39}$$

$$\partial\rho_e/\partial t = -\nabla\cdot\rho_e\underline{v}-\nabla\cdot\underline{i} \tag{40}$$

$$\partial\underline{g}/\partial t = -\nabla\cdot(\underline{j}\underline{v}+p-\underline{\underline{T}}) - \nabla\cdot\underline{\underline{\Pi}} \tag{41}$$

$$\partial e_v/\partial t = -\nabla\cdot\{(e_v-\tfrac{1}{2}(E^2+B^2)+p)\underline{v}+c'\underline{E}\times\underline{B}-\nabla\cdot(\underline{J}+\underline{\underline{\Pi}}\ \underline{v})\} \tag{42}$$

Here ρ is the mass density, ρ_e the charge density, $\underline{g}$ the total momentum density of matter and field, $\underline{v}$ the velocity, $\underline{j}$ the momentum density of matter, p the dydrostatic pressure, $\underline{\underline{T}}$ the Maxwell tensor, e_v the total energy of matter and field per unit volume, $\underline{E}$ and $\underline{B}$ the electric and magnetic field strengths respectively and c' the velocity of light. The irreversible phenomena occurring in the fluid: heat conduction, viscous phenomena and electric conduction are linked to dissipative currents: $\underline{J}$ the heat current, $\underline{\underline{\Pi}}$ the viscous pressure tensor and $\underline{i}$ the conduction current. Tn absence of polarization and magnetization $\underline{g}$ and $\underline{\underline{T}}$ are univocally defined as

$$\underline{g} = j + c'^{-1}\underline{E} \times \underline{B} \tag{43}$$

$$\underline{\underline{T}} = \underline{E}\,\underline{E} + \underline{B}\,\underline{B} - c'^{-1}(E^2+B^2)\,\underline{\underline{U}} \tag{44}$$

$\underline{\underline{U}}$ being the unit tensor.

The electromagnetic field will evolve according to the Maxwell equations

$$\partial\underline{E}/\partial t = c'\nabla \times \underline{B} - \rho_e\underline{v} - \underline{i} \tag{45}$$

$$\partial\underline{B}/\partial t = -c'\nabla \times \underline{E} \tag{46}$$

$$\nabla\cdot\underline{E} = \rho_e \tag{47}$$

$$\nabla\cdot\underline{B} = 0 \tag{48}$$

Balance equations (39)-(42) and Maxwell equations (45)-(48) are not independent. In fact (47) and (48) can be obtained from (40), (45) and (46). Thus (47) and (48) can be viewed as initial conditions to be satisfied for $\underline{E}$, $\underline{B}$ and ρ_e.

3.1. NON-LINEAR FLUCTUATIONS.

To study fluctuations in the case in which nonlinearities appears in (39)-(42), we shall make use of a theory recently developed in [25], [26]) and we shall extend it to the case in which matter interacts with electromagnetic fields [27]). As we did with electrothermal and electrodiffusive systems, we shall assume that the dissipative currents split in systematic and random parts in the form

$$\underline{J} = \underline{J}^S+\underline{J}^R \quad ; \quad \underline{\underline{\Pi}} = \underline{\underline{\Pi}}^S+\underline{\underline{\Pi}}^R \quad ; \quad \underline{i} = \underline{i}^S+\underline{i}^R \tag{49}$$

The random parts have zero mean

$$\overline{\underline{J}^R(\underline{r},t)}=0 \quad ; \quad \overline{\underline{\underline{\Pi}}^R(\underline{r},t)}=0 \quad ; \quad \overline{\underline{i}^R(\underline{r},t)}=0 \tag{50}$$

and the non-zero correlations are

$$\overline{J_j^R(\underline{r},t)\ J_k^R(\underline{r}'t')}= 2k_B l_{qq}\delta_{jk}\delta(\underline{r}-\underline{r}')\delta(t-t') \tag{51}$$

$$\overline{J_j^R(\underline{r},t)\ i_k^R(\underline{r}',t')}= 2k_B L_{qi}\delta_{jk}\delta(\underline{r}-\underline{r}')\delta(t-t') \tag{52}$$

$$\overline{i_j^R(\underline{r},t)\ i_k^R(\underline{r}'t')} = 2k_B L_{ii}\delta_{jk}\delta(\underline{r}-\underline{r}')\delta(t-t') \tag{53}$$

$$\overline{\Pi_{jk}^R(\underline{r},t)\Pi_{lm}^R(\underline{r}',t')} = 2k_B L_{jklm}\delta(\underline{r}-\underline{r}')\delta(t-t') \tag{54}$$

where the tensor L_{jklm} is defined by

$$L_{jklm} = L(\delta_{jk}\delta_{km}+\delta_{jm}\delta_{kl}-\frac{2}{3}\delta_{jk}\delta_{lm}) + L_v\delta_{ij}\delta_{kl} \tag{55}$$

In (50)-(54) an upper bar stands for the average over an ensemble of random forces of systems with the same initial conditions for the variables. Furthermore the phenomenological constants L, L_v, L_{ii},L_{qq} and L_{qi} depend only on the equilibrium quantities of the system.

Because we are dealing with fields we must work with functional derivatives. To avoid difficulties inherent to the definition of such derivatives and following [25]) we shall make use of discretization rules. The space occupied by the fluid will be divided into small cubic cells each one of size Δ^3. The cells will be located at $\underline{r}=\underline{n}\Delta$, $\underline{n}$ being a vector of which the components are integer numbers. We shall introduce the following vector in the space of the discretized physical quantities

$$\alpha_{\underline{n}} \equiv (\rho_{\underline{n}},\ \rho_{e,\underline{n}},\ e_{\underline{v},\underline{n}},\ \underline{g}_{\underline{n}},\ \underline{E}_{\underline{n}},\ \underline{B}_{\underline{n}}) \tag{56}$$

The balance equations (39)-(42), the Maxwell equations (45)-(48) and the stochastic properties of the dissipative currents can be rewritten in the discretized space. When inserting the discretized dissipative currents into the discretized balance and Maxwell equations one arrives to stochastic differential equations of the form

$$\partial\alpha_{\beta,\underline{n}}/\partial t = F^{rev}_{\beta,\underline{n}} + F^{irrev}_{\beta,\underline{n}} + \sum_{\underline{n}'}\sum_{\gamma\delta} M_{\beta\gamma\delta,\underline{nn}'} F_{\gamma\delta,\underline{n}'} \tag{57}$$

The terms $F^{rev}_{\beta,\underline{n}}$ and $F^{irrev}_{\beta,\underline{n}}$ make reference to the reversible and irreversible parts of the discretized conservation laws and Maxwell equations, respectively. The terms $f_{\gamma\delta,\underline{n}'}$ are related to the random parts of the dissipative currents while $M_{\beta\gamma\delta,\underline{nn}'}$ is a matrix whose terms involve discretized gradients $\nabla_{\underline{n}}$, delta functions $\delta_{\underline{nn}'}$ and the velocity $\underline{v}_{\underline{n}}$. As an example, the discretized stochastic differential equation associated to the balance equation of charge is obtained from (57) by putting

$$\begin{aligned}
&\alpha_{2,\underline{n}} = \rho_{e,\underline{n}}\\
&F^{rev}_{2,\underline{n}} = -\nabla_{\underline{n}}\cdot\rho_{e,\underline{n}}\, v_{\underline{n}} \quad ; \quad F^{irrev}_{2,\underline{n}} = -\nabla_{\underline{n}}\cdot\underline{i}^{s}_{\underline{n}}\\
&f_{4j,\underline{n}} = i^{R}_{j,\underline{n}}\\
&M_{24\,j,\underline{n}\,\underline{n}'} = -\nabla_{j,\underline{n}}\,\delta_{\underline{n}\,\underline{n}'}
\end{aligned} \tag{58}$$

Since our stochastic quations are non-linear (they contain the term $\nabla\cdot(\underline{\Pi}^{R}\cdot\underline{v})$), we should interpret them in the Itô or Stratonovich sense[28]. In our case, however, both interpretations can be shown to be equivalent [26] [27]. Then the Fokker-Planck equation derived from the stochastic differential equarions interpreted in the Stratonovich sense is

$$\partial P(\{\alpha_{\underline{n}}\},t)/\partial t = \sum_{\underline{n}} \sum_{\beta} \partial[- F^{rev}_{\beta,\underline{n}} - F^{irrev}_{\beta,\underline{n}} + \quad + \sum_{\underline{n}'} \sum_{\beta'} \partial D_{\beta\beta',\underline{n}\underline{n}'}/\partial\alpha_{\beta',\underline{n}'}]/\partial\alpha_{\beta,\underline{n}} P(\{\alpha_{\underline{n}}\}t) \tag{59}$$

Here $P(\{\alpha_{\underline{n}}\},t)$ is the distribution probability and $D_{\beta\beta',\underline{n}\underline{n}'}$ is the matrix of diffusion coefficients defined by

$$D_{\beta\beta',\underline{n}\underline{n}'} = \sum_{\underline{n}''\underline{n}''} \sum_{\gamma\delta\gamma'\delta'} M_{\beta\gamma\delta,\underline{n}\underline{n}''} M_{\beta'\gamma'\delta',\underline{n}'\underline{n}''} \Lambda_{\gamma\delta\gamma'\delta',\underline{n}''\underline{n}''} \tag{60}$$

where the matrix $\Lambda_{\gamma\delta\gamma'\delta',\underline{n}''\underline{n}''}$ is defined through

$$\overline{f_{\beta\gamma,\underline{n}}(t)\, f_{\delta\varepsilon,\underline{n}'}(t')} = 2\Lambda_{\beta\gamma\delta\varepsilon,\underline{n}\underline{n}'}\, \delta(t-t') \tag{61}$$

The Fokker-Planck equation (59) can be transformed by using the identities

$$\sum_{\beta} \partial F^{rev}_{\beta,\underline{n}} / \partial\alpha_{\beta,\underline{n}} = 0 \tag{62}$$

$$\sum_{\underline{n}'} \sum_{\beta'} \partial D_{\beta\beta',\underline{n}\underline{n}'} / \partial\alpha_{\beta',\underline{n}'} = 0 \tag{63}$$

which have been shown in [26] and [27]. One finally gets

$$\partial P(\{\alpha_{\underline{n}}\},t)/\partial t = \sum_{\underline{n}} \sum_{\beta} [-F^{rev}_{\beta,\underline{n}}\, \partial/\partial\alpha_{\beta,\underline{n}} + \partial(-F^{irrev}_{\beta,\underline{n}} + \quad + \sum_{\underline{n}'} \sum_{\beta'} D_{\beta\beta',\underline{n}\underline{n}'}\, \partial/\partial\alpha_{\beta',\underline{n}'})/\partial\alpha_{\beta,\underline{n}}]P(\{\alpha_{\underline{n}}\},t) \tag{64}$$

3.2 THE FORM OF THE DISSIPATIVE CURRENTS

Graham and Haken [29] derived the conditions under which the principle of detailed balance is satisfied for the Fokker-Planck equation (c4). These conditions, that involve the equilibrium distribution pro bability $p^{eq}(\{\alpha_{\underline{n}}\})$, are

$$\sum_{\underline{n}} \sum_{\beta} F^{rev}_{\beta,\underline{n}} \; \partial p^{eq}(\{\alpha_{\underline{n}}\})/\partial\alpha_{\beta,\underline{n}} = 0 \tag{65}$$

$$F^{irrev}_{\beta,\underline{n}} = \sum_{\underline{n}'} \sum_{\beta'} D_{\beta\beta',\underline{nn}'} \; \partial \ln p^{eq}(\{\alpha_{\underline{n}}\})/\partial\alpha_{\beta',\underline{n}'} \tag{66}$$

which in the continuum limit, defined as

$$\lim_{\Delta\to 0} \alpha_{\underline{n}} \equiv \alpha(\underline{r}) \tag{67}$$

transforms in

$$\sum_{\beta} \int d\underline{r} \; F^{rev}_{\beta}(\underline{r}) \; \delta p^{eq}(\{\alpha(\underline{r})\})/\delta\alpha_{\beta}(\underline{r}) = 0 \tag{68}$$

$$F^{irrev}_{\beta}(\underline{r}) = \sum_{\beta'} \int d\underline{r}' \; D_{\beta\beta'}(\underline{r},\underline{r}')\delta \ln P^{eq}(\{\alpha(\underline{r})\}/\delta\alpha_{\beta}(\underline{r}') \tag{69}$$

Equation (68) is satisfied for the Einstein equilibrium distribution

$$P^{eq}(\{\alpha(\underline{r})\}) \sim e^{S(\{\alpha(\underline{r})\})/K_B} \tag{70}$$

where $S = \int d\underline{r} \; s_v(\underline{r})$ is the total entropy. The derivatives of the logarithm of equation (69) can be computed making use of (70) and of the expression of the entropy per unit volume s_v

$$s_v = s_v(u_v,\rho,\rho_e) \tag{71}$$

where the internal energy per unit volume u_v is

$$u_v = e_v - (\underline{g} - c'^{-1}\underline{E} \times \underline{B})^2/2\rho - (E^2+B^2)/2 \tag{72}$$

As an example we shall explore the consequences derived from (69) in the case in which $\beta=2$. In such a case, the derivative of the logarithm is

$$\delta \ln P^{eq}/\delta\alpha_2(\underline{r}) = -\{\mu_1(\underline{r})-\mu_2(\underline{r})\}/K_B T(r)(z_1-z_2) \tag{73}$$

where $\mu_K(\underline{r})$ and z_K, K=1,2, are the chemical potentials and the char ges per unit mass of both components respectively. One gets the condition

$$\underline{i}^{s} = L_{qi}\nabla T^{-1} - L_{ii}\{\nabla[(\mu_1-\mu_2)/T(z_1-z_2)] - (\underline{E}+c'^{-1}\underline{v} \times \underline{B})/T\} \quad (74)$$

which is nothing but the phenomenological equation for the conduction current derived in thermodynamics of irreversible processes [3]. The constants L_{qi} and L_{ii} which were introduced in the correlation functions of the dissipative currents (52) and (53) are the phenomenological coefficients of thermodynamics of irreversible processes [3]. The Fokker- Planck equation (64) allows the study of the linear response of the charged two-component fluid to an external electric field. Green-Kubo relations have been given in [27]

ACKNOWLEDGEMENTS

This work has been parcially supported by a grant of the Comisión Asesora de Investigación Científica y Técnica of the Spanish Goverment. It is a pleasure to acknowledge stimulating discussions with Profs. P. Mazur and D. Bedeaux and Dr. W.van Saarloos.

REFERENCES

1) See for example, D. Forster, Hydrodynamic Fluctuations, Broken Symmetry and Correlation Functions (W.A. Benjamin, Inc. London, 1975, and Fluctuations, Instabilities, and Phase Transitions (Plenum Press, New York, 1975).

2) L. Landau and E. Lifshitz, Mécanique des Fluides, (Mir, Moscou, 1971)

3) S.R. de Groot and P. Mazur, Non-Equilibrium Thermodynamics (North-Holland, Amsterdam, 1962).

4) Y. Pomeau and P. Résibois, Physics Reports 19C (1975) 63.

5) L. Landau and E. Lifshitz, Electrodynamique des milieux continus (Mir, Moscou, 1969).

6) E. Lifshitz, Soviet Physics 2 (1955) 73.

7) B.V. Felderhof, Physica 89A (1977) 205 and references quoted therein.

8) A.M.S. Tremblay, in Non-Equilibrium Superconductivity, Phonons and Kapitza Boundaries, NATO Advanced Study Institute, ed. K.E. Gray, (Plenum, New York, 1981) and references quoted therein.

[9]) A.A.J. Matsinger, R. de Bruyn and H. van Beelen, Physica 93B (1978) 63.

[10]) A. Schmid, J. Low Temp. Phys. 41 (1980) 37.

[11]) R.F. Woss and J. Clarke, Phys. Rev. B 13(2) (1976) 556.

[12]) F.N. Hooge, T.G.M. Kleinpenning and L.K.J. Vandame, Rep. Prog. Phys. 44 (1981) 479.

[13]) D. Bedeaux, P. Mazur and R.A. Pasmanter, Physica 86A (1977) 355.

[14]) J.B. Gunn, Solid State Commun. 1 (1963) 88.

[15]) D.D. Joseph, Stability of Fluid Motion I (Springer Verlag, Berlin 1976).

[16]) L.A. Segel in Non-Equilibrium Thermodynamics Variational Techniques and Stability, ed. Donnelly, Herman and Prigogine (University of Chicago Press, Chicago 1965)

[17]) V.M. Zaitsev and M.I. Shliomis, Sov. Phys. JETP 32 (1971) 866.

[18]) R.A. Pasmanter, D. Bedeaux and P. Mazur, Physica 90A (1978) 151.

[19]) J. Keizer, J. Chem. Phys. 74(2)(1981) 1350.

[20]) W.J. Skocpol, M.R. Beasley and M. Tinkham, J. Appl. Phys. 45 (1974) 4054.

[21]) P. Mazur and D. Bedeaux, J. Stat. Phys. 24 (1981) 215.

[22]) D. Bedeaux and P. Mazur, Physica 105A (1981) 1.

[23]) P.N. Butcher, Rep. Progr. Phys. 30 (1967) 97.

[24]) S. Kabashima, H. Yamazaki and T. Kawakubo, J. Phys. Soc. Japan 40 (1976) 921.

[25]) W. van Saarloos, D. Bedeaux and P. Mazur, Physica A 107A (1981) 109.

[26]) W. van Saarloos, D. Bedeaux and P. Mazur, Physica A (in press).

[27]) J.M. Rubí and W. van Saarloos, Physica A (in press).

[28]) L. Arnold, Stochastic Differential equations, Wiley, New York 1974.

[29]) R. Graham and H. Haken, Z.f. Physik 243 (1971) 289.

INSTABILITIES IN ECOLOGY

Ramón Margalef
(Ecology, University of Barcelona)

Attempts to uncover connections between biology and thermodynamics have a dismal history. Nevertheless the subject never losses its positive powers of seduction. Here I want to comment on a number of processes and situations that provide subjects for thinking about the ways in the organic world, in areas more or less related to thermodynamics. The subject focusses on ecological systems or ecosystems. In them the fundamental source of instability is the quantification provided by the distinction of individuals. The quantification concerns size and life span. Individuals are always out of equilibrium, around their boundaries and at the interfaces with other individuals. Both, these discontinuities and the process of reproduction keep the system reactive.

Asymmetries in history and the "laws" of succession and evolution.

Ecologists of certain persuasions are fond of singling out processes of change in the ecosystems and attaching arrows with definite directions to them. A pertinent example is succession, typified by the way an abandoned old field changes slowly ito a forest, increasing biomass, usually increasing complexity, and always decreasing the energy exchanged in unit time per amount of preserved biomass. Another example is the pattern of gradual evolution, as shown by the fossil record in several groups of organisms, as foraminifera, ostracoda, ammonites, horses and elephants, among others. The observed regularities have been acknowledged by many as formal "laws", concerning increase of size, specialization and irreversibility.

A complete picture of nature should be larger, and besides the continuous processes of succession and regular evolution, should include also more or less opposite processes leading to the destruction or regression of the plant cover, as well as to provide, in the field of evolution, some explanation of what happens when and where a larger component of impredictability and inventiveness has to be assumed.

The direction along our arrow marked succession or progressive evolution, might be more attractive, because it is slow, seems predictable up to a certein point, characteristic of the organic world, and may be easier to rationalize, probably for deeper psychological reasons. The opposite direction of change looks unpredictable, catastrophic, and too short or too rapid, to provide a satisfactory object of contemplation. Nature is seen as crisscrossed by processes going alternative ways, and only by the reason that one direction of change is fast and the other slow, the slow one is easier to follow and their trajects cover a larger share of the scenario.

I propose to have a closer look at a number of selected examples.

a) The growth of the population of a bacterium in a culture flask. It starts with a rapid use of available nutrients, that can be considered as wasteful or not highly efficient. Then the biological and biochemical processes slow down, and a measure of differentiation in space and time appears and increases, specially if the contents of the culture flask is not stirred, and the available clones are many.

b) Life in the droppings of cattle in a mountain pasture. Where the dropping will fall is unpredictable; its use by the organisms already present in the excrement or available in the local soil, is fast and can be considered inefficient, but we can be sure that the appropriate organisms will come in time, grow in sequence, and develop a very regular pattern of utilization, until the local instability initiated by the dropping, is assimilated in the process of persistence and succession of the whole meadow.

c) Take again the development of a forest. The destruction of the precedent arboreal vegetation was a consequence of energy from outside - wind, bulldozers - or of putting into work, in a different and catastrophic mode , energy accumulated in the ecosystem, and triggered from outside. Fire is the pertinent example; its equivalent is the burning of fossil fuels at the scale of the biosphere. The eventual persistence of a crop field requires sustained exploitation or to be kept free of the invasion by weeds and bushes. If the field is abandoned, the necessary diaspores are available, and a forest is reconstructed, according to the limitations imposed by the local climate.

d) Phytoplankton. Succession of a forest requires centuries and the life of each individual tree is counted in decades. The generation time of the organisms of phytoplankton is from less than one to a few days, and in a few weeks it is possible to observe successions equivalent to the process of reconstruction of a forest. Winter cooling or any gush of wind mixes the water and uniformizes vertical distributions of nutrients and of seed organisms. A succession starts with the dominance of small and fast multiplying organisms. Turbulence is slowly decaying and water stratifies , larger and usually motile organisms take over, and their grow metabolism diminishes as nutrients are being used.

e) Sedimentation. The regularities are by no means restricted to systems composed essentially by living organisms. Cycles of sedimentation are commonly recognized in any thick pack of sediments. Each cycle starts with coarse material, slowly and gradually passing to more fine grained material. The next transition from fine to coarse material is sharp. The rationalization is related to the one that applies to the sequences of phytoplankton. An input of energy from outside mixes water or transports large grains of solid materials; decay of turbulence is reflected in a planktonic succession that allows an ecological explanation in terms of competition under changing conditions, or else, in the case of sediment, in the gradual decrease of the individual size of the settling particules. Similar explanations are applicable to other geological differentiation processes.

In each one of the precedent examples there is a turning point between a period of rapid processes driven from outside and unpredictable from inside, and a period of slow and internally controlled change, much dependent on the existing organization. A description, short and emphatic, would be to say that the system makes history during the first period, and records part of the history during the second period. Things are difficult to sort out in nature, because of the overlapping and combination, at different scales, of a number of processes that run at different speeds. As a telling example it suffices to remember the succession in the droppings of cattle, as it becomes assimilated in the succession of the whole pasture.

Ecosystems are subjected to inputs of energy, which intensity and timing can be considered at random. Each input filters through the organization and generates a response that is relatively slow and rather long, according to the organization. Ecosystem fluctuations, in consequence, are not symmetrical, but their representation appears saw toothed. This fact diminishes the usefulness of the analysis of change in ecosystems through the summation of sinusoidal waves. When the stochastic inputs are frequent, as compared with the size and length of life of important organisms, they can be added more easily into a regular output. A capable organization is able to turn into a source of more organization what could be an event potentially generator of a catastrophe in a differently organized ecosystem. The transfer of energy between different scales is quite diverse in a fluid medium and in an organized system. And symmetrical oscillations could occur only in systems made of pure energy, out of the real world.

Systems large enough can analyze past fluctuations and recognize rhythms. If the regularity is maintained and future inputs can be anticipated more or less, a way has been found to make smaller the asymmetries in the responses relative to the inputs. As soon as life has learned the tune, the unexpected is only partially so. Rhythms anticipate events and provide an obvious advantage in competition and selection. On a more personal and psychological terrain, this may help to explain the tranquilizing effects of music not deprived of rhythm.

The more general model of succession and change allows a very simple description of the two periods. The first stage of catastrophic change, followed by the rapid colonization of an ecologically empty space by all sorts of opportunists, with wasteful use of the resources, can be described as the result of a rapid and throughout mixing of reactants in a situation that allows for a maximum and rapid increase of entropy, much like a gas that expands and mixes in a reservoir, or a bomb that explodes. The second period, or standard succession, is marked by local differentiation and increase of the number of discernible elements of structure. Then, each one of the component subsystems that have differentiated, behaves as an open system, and according to theory, the amount of energy exchanged actually, or the corresponding increase of entropy, decreases steadily per unit of total preserved biomass. This biomass is a measure of the information forwarded by the system.

In the initial state, no distinctions can be drawn. As time advances, turnover decreases, but not in the same degree in the different compartments, and the result is an enrichment of the structure. Information or organization is directly related to the possibility of making distinctions or division in the biomass, a condition for its differentiation. The historical process of internal differentiation is essential to the concept of succession, as is recognized by the conventional wisdom of naturalists. But the respectable wisdom of the theorists has followed another path and has produced mathematical models that pretend to link inputs and outputs by a rigid sort of mechanism. Consideration of natural populations, for example, of exploited fisheries, has made clear how much the functional structure of the ecosystem shifts in dependence of the degree of stress or of exploitation to which the system is subjected, and of the historic development of the same system. The models in use require another degree of complication, to accomodate the consequences of historical change in a persisting organization or, either, its destruction. Perhaps it is not inconsiderate to say that present day models are "first law" models, and that we are much in need of "second law" models. Whithout such step much modelling excercises appear futile, prompting the quip that in ecology, any expression over four inches long is wrong.

Some of the relations can be summarized in the following form:

$$\frac{P_1}{B_1} > \frac{\sum P_{i2}}{\sum B_{i2}} \quad , \quad \frac{\Delta S_1}{B_1} > \frac{\sum \Delta S_{i2}}{\sum B_{i2}} \quad , \quad \frac{T_1}{I_1}\Delta S > \frac{T_2}{I_2}\Delta S$$

P, biological production
B, biomass
S, entropy
T, temperature
I, information
1, 2, time or stage
..i,j,k.., compartments

Organization as the result of the segregation of potential reactants.

The succession of phytoplankton outlined in precedent section, d), is apt to be a subject for further discussion. Succession starts when mixing uniformizes the vertical distribution of nutrients and disperses the organisms through the whole mass of water. Then nutrients start to be used in the illuminated layers. Any atom of an element that passes from being in solution in water to be a part of an organism or of any other solid particule, increases its probability to be moved downwards. Vertical transport along the direction of gravity is pasive (sedimentation) or active. It is also a result of the migrations of the planktonic animals. Although zooplankton contributes to the recirculation of elements inside the photic zone, usually the animals defecate in layers below the level at which they feed. Moreover, copepods and euphausids produce compacted fecal pellets that sink rapidly before decomposing. The result of these an other sorts of interactions internal to the ecosystem is a general damping of the dynamics of plankton, through the segregation of the elements of production: Where there is light, nutrients have been exhausted, where nutrients accumulate, it happens because there is no light. Productivity decreases steadily, down to a limit defined by the vertical transport based on molecular diffusion and on the remaining turbulent diffusion. Any increase of turbulence sets back the whole process, and this happens rather abruptly, in the frame of the regularity of higher order, that allows for random events, but postulate an asymmetric pattern for each single event: a rapid disorganization followed by a slower change towards local segregation.

Riley, Stommel and Bumpus (1949) provided an useful theoretical frame for the study of phytoplankton. Their expression can be applied both to phytoplankton and to the nutrients, and the results, combined (Margalef, 1978), led to a new expression that may be worthy of further comment:

$$dB/dt + V'(dB/dz) = V(dN/dz) + A((d^2N/dz^2) + (d^2B/dz^2)) + a(d^2B/dz^2)$$

net increase	loss of cells	use of nutrients (advection)	effect of the lack of conformity in the distribution of nutrients and plants (diffusion)	effect of active motion or anchoring cells in large eddies
(production)				

In it, B = biomass (may allow for internal cycling), N = substrate or nutrient concentration, V, V', speed along z, A, A' = A + a , turbulent diffusion; V', A', refer to organisms; V, A, refer to water and nutrients; z, vertical coordinate.

This expression, reduced to its bare bones, is:

production = (supply from outside) + (turbulence) x (covariance of the distributions)

	supply from outside	(turbulence) x (covariance of the distributions)
	non renewable resources dependence from outside	renewable resources internally controlled

The distinction between renewable and non renewable resources is a matter of scale. If the space of reference is large enough, connected compartments are considered as a single compartment, the cycling of elements is internalized, and no surplus of production is exported. This strong dependence on scale agrees with the spectral quality of both, turbulent diffusion, and covariance in the distributions, the two most important parameters that precise the behavior of the system:

biological production = (turbulent diffusion) x (covariance in the distributions of reactants: light, cells, nutrients)

Both, turbulent diffusion and covariance in the distribution of reactants, refer to a definite size of cell of measurement. Each one of the material components is subjected to an equation of continuity that includes (biological) change. The covariance has to be studied over the resulting distributions in space and time.

The inferior limit of A (turbulent diffusion) is given by the molecular diffusion or the viscosity; the maximum covariance in the distribution of the factors of production would correspond to the model of a mixture of gas molecules inside a container and according to the elementary presentations of thermodynamics.

It could be suspected that in the real world of the plankton, the covariance in the distribution of reactants is proportional to turbulent diffusion, that depends on available kinetic energy. Production, then, would be proportional to A squared. Actually it is not so, because no amount of mechanical energy in water can displace photons and made uniform the availability of light. The vertical distribution of available radiation, as an essential factor of production, sets a limit to the covariance among the factors of biological production. This fact emphasizes the special position of radiation in natural systems. The consequence is that production is maximal for intermediate values of A, and does not increase monotonously with A. Suppose that man, through an appropiate contraption, provides light to the deep layers of the ocean. This would require more external energy, and the model is simply extended.

When an ecosystem changes in the direction of succession, the organization being generated leads to the parsimonious interaction between the candidates for reaction, and to the effective segregation of them. We cannot escape the feeling that the world tends to be made of misplaced things, in the sense that the structures are preserved only because of spatial isolation of reactants that could change or annihilate them. This is very far of the usual model of particules or of gases that tend to fill uniformly or in a random way the inside of ideal containers. The real world seems to be much more a result of the "instability of the homogeneous", a phrase included somewhere in the writings of Herbert Spencer.

As said, potential reactants are separed in nature by the quarantine of space, and what exists is the leftover of previous reactions. This seems to be generally accepted. I was impressed reading recently about the possibility that matter and antimatter were created in equal amounts, shaking them in the cosmic cocktail mixer, where most of the stuff reacted, disappearing altogether, leaving some matter where we are, and presumably antimatter in some faraway corner, space providing the isolation.

There is a steady conversion of energy into organization. Structures grow as a result of function. Making the system more complicated slows down the rates of change. Cells "burn" slowly, because enzymes actually provide a brake to many reactions that otherwise would run wild. In a slighly metaphorical extension, we could add that sex slows down reproduction, thought slows down action, and bureaucracy slows down social metabolism. This set of properties shown by the complex systems and leading to a further increase of complexity find an uneasy accomodation in the frame provided by an elementary or pedestrian view of themodynamics. I am not quite sure, neither, if much of the discussion going on on instabilities and open systems is relevant to the point. The avidity with which nature turns available energy into organization is well exhi bited in the couplings between subsystems of different degree of organization, like plankton and benthos, if it were not obvious enough in the endogenous changes of any system, in which case the transfer is done along de dimension of time: the energy degraded in the organization of to day is reflected in the organization of tomorrow.

Returning to the description of the dynamics of plankton, the expressions proposed before can be compared with other ways of representation perhaps more in the tradition of ecological theory. They purport to describe the change in numbers of each one of the components or species i (N_i) in relation with the numbers of each one of the other species with which the first one interacts. Coefficients of interaction (a_{ij}) should not be considered as intrinsec properties of the system, but rather as statistical averages describing a past history. I think reasonable to propose following comparison:

biological production	=	supply from outside	+	A x (covariances)
		non renewable resources		renewable resources
dN_i/dt	=	$\sum_{k=1} b_{ik} R_k N_i$	+	$\sum_{j=1} a_{ij} N_i N_j$
		stochastic terms (R_k)		deterministic interaction

Ecologists like to work with models reduced to the second and deterministic block of terms, but it is obvious that much change is induced from outside and in a preliminary way can be introduced in the form of stochastic inputs (R's). Enlarging the size of the system or the frame of reference, stochastic inputs turn more determinable, at least theoretically. The resulting change in the presentation is the same in both ways of looking at the ecosystem. It can be remembered that the sum of products $N_i N_j$ is a measure of covariance, scaled by the coefficients of interaction a_{ij}.

Before leaving the subject, I propose to have another look at the expression

$$\text{biological production} \quad = \quad A. \text{ (cov.)}$$

and introduce the consideration of changes in time

$$dP/dt \quad = \quad (dA/dt).(\text{cov.}) \quad + \quad (d(\text{cov.})/dt).A$$

that is helpful to state some general differences between planktonic and terrestrial ecosystems. Benthic ecosystems share some properties with terrestrial ones. Terrestrial ecosystems have a more rigid organization and any decrease in the value of turbulent diffusion is less important; in consequence, decrease in productivity along succession is more related to changes in localization, growth, movement and transport by organisms. In planktonic systems, the consequences of the shift of energy to smaller scales of turbulence is much more important; the covariance of distribution decreases in a particular way when active (and non coincident) movement takes over pasive transport.

Differences between vertical and horizontal axes.

Let us have another look at the plankton ecosystems. Solar radiation in the illuminated layers of the water provides reducing power to the chloroplasts of the primary producers, and reduced compounds of biogenetic elements (C, N, S) appear inside the cells. Material structures of the organisms sink in the water and bring reducing power in deep water, where organic matter and reduced materials (ammonia, methane, hydrogen sulfide) are set free. The combination of light and gravity define a principal axis of organization, that is close to the vertical, and is marked, among other chemical and biochemical properties, by a gradient of exydation-reduction potential and by a flux of electrons in the water, opposite to the net flux of organic matter.

Also in terrestrial ecosystems light and gravity define and single out a vertical axis in the organization. A tree can be compared to a planktonic ecosystem. The productive units, here the leaves, are held together and linked to a transport system provided by the organisms.

When an ecosystem works under a certain charge or stress from outside, two responses are quite general: elements cycle faster, and a fraction of the elements that were in cycle are led out or discharged at the boundaries, or along external paths, sometimes immobilized inside the own system. Eutrophic lakes, upwelling marine systems, as well as stressed agricultural systems, lend oxygen and nitrogen to the atmosphere, and organic carbon and phosphate to the sediment or to the soil. The same mechanism has been effective during the whole history of life, providing a source of the atmosphere of the Earth, out of the primaeval biosphere.

Light and gravity are two kinds of force that are active in the biosphere and not only lead to the vertical differentiation, but are also the base of a mechanism of transport. In this sense, the whole biosphere, that may be thought of as a pellicle covering the solid Earth, has many of the functions of a broadly generalized model of a biological membrane.

In aquatic ecosystems, succession leads to a subdivision of the whole vertical column by a certain number of discontinuities. The process of differentiation of a thermocline provides an excellent example of selfincreasing structures. In a lake a relatively complicated, but common, situation includes following compartments: atmosphere, epilimnion, hypolimnion, monimolimnion and sediment. Such structure, as it develops, slows down general turnover, increases segregation of species and richness of small detail in organization. Presumably, the development of such organization decelerates the change in the ratio between the increase of entropy and the amount of information or of organization forwarded into the future.

Along the vertical axis we have a strong gradient, expression of the energy available to create and maintain a rather rich organization, in the form of a dissipative system. The horizontal plane is essentially different, and it is appropriate to model or conceive the whole ecosystem as formed by a bunch of comparable and parallel columns. In any model, if it is to remain realistic, the section of each vertical column cannot be infinitely small. The input of external energy requires a minimum surface, relativelly small in the case of waves, much larger in the patterns of marine circulation associated with upwelling. Perhaps it could be illustrative to compare the ecosystem to a bunch of parallel test tubes in each of which a Zhabotinskii reaction is going on; the whole set can fall into a rhythm. In the case of oceanic ecosystem, the daily cycle originates, at least, a wave in the intensity of primary production, that is reflected on the distribution of many properties along each local column. Another limitation in the size of the unit columns in the model, depends on the presence of animals, able to migrate horizontally and generators of an horizontal transport and differentiation, besides other factors.

Each vertical elementary column in an ecosystem can be compared to a well tuned tense string. When it vibrates, waves are propagated around and through any horizontal plane. Daily, yearly rhythms in the ecosystem generate waves in different properties -chemical, biotic, etc.- that travel around. Internal waves in the aquatic ecosystems are not restrictedto temperature and water movement. But the horizontal heterogeneity that develops through lateral interaction appears always less determined than the vertical organization. An aquatic ecosystem is equivalent to a Zhabotinskii experiment as conducted in a dish. The relative lack of determination along the horizontal plane, in contrast with the more rigid vertical organization, explains the fugitive and apparently stochastic patterns that make so difficult -and, at the same time, so exciting- the analysis of horizontal transects across the waters of a lake or of the ocean. Often, the heterogeneity observed over an horizontal plane allows to recognize a pattern in which two topologically diverse parts are recognized. One part consists of discontinuous patches of more mixed or turbulent water, populated by oportunistic organisms that may be considered as representing the earlier stages of a succession. The reticulum or honeycomb around these patches has a more stable character in relation to water, and a more advanced character in relation to succession, with richer structure and more regular movements of organisms. As is usual in the organization of the biosphere, structures of different size, combine and overlap at different scales, from the smaller patches to the largest gyres of the marine circulation.

Summary.

Change in ecosystems is twofold. One kind of change proceeds under important inputs of energy from outside and leads to distributions rather uniform, a result that perhaps could be expected from the point of view of elementary thermodynamics. In other paths of change, energy degrades as well, but at the same time the segregation of reactants, or of what is left of them, increases, and the system becomes more heterogeneous. Biologists are delighted with this modality of change, that is relatively slow, and decelerating. Oddly they have associated this sort of change with the idea of progress. Perhaps because it appears to run against vague tendencies assumed for physical systems. In any case, it falls in the area of study of thermodynamics of open systems out of equilibrium. The main way to be out of equilibrium is to cut any organization into discontinuous individuals and populations, and this life has done from the very beginning. The study of plankton offers very instructive examples and suggests positive approaches. It shows also that vertical and horizontal dimensions are not comparable in what concerns the organization of the ecosystem. Vertical organization can respond to stresses through oscillatory behavior. If the ecosystem is ideally decomposed in a nomber of adjacent vertical columns, horizontal change propagates as waves from column to column. The natural tendency is to local segregation over the horizontal plane. But inputs of energy can run against this tendency, can enhance horizontal transportation, and result in systems highly reactive.

REFERENCES

R. MARGALEF , Oceanologica Acta, 1, 493-509 (1978).

G.A. RILEY, H. STOMMEL and D.F. BUMPUS , Bull. Bingham Oceanogr. Coll., 12, 1-169 (1949).

STRANGE ATTRACTORS

Carles Perelló
Facultat de Ciències. Secció de Matemàtiques
Universitat Autònoma de Barcelona
Bellaterra, Barcelona (Spain)

1. INTRODUCTION.

In the last few years the use of dynamical systems as models of processes in physics, chemistry, biology, economy, etc., has grown very fast. The origin of this, let's call it success, lies in the fact that in many cases the fundamental laws governing these processes take the shape of differential equations. When, given the initial state of the system, the solution exists, is unique and depends continuously on these initial condition, we have what is named a dynamical system: a space of states (each point of this space represents a possible state) and, through each of the points a path describing the change of the state with the time.

One of the most important problems of applied science is to be able to say something about the long time behaviour of the processes being studied, i.e. to what states do the system tends when the time grows to infinity, if we know the initial state. Still more practical is the problem of knowing what is the limit behaviour when the initial state is known only approximately.

These two problems are at the root of the concepts of limit set and of attractor.

The theory of dynamical systems was put in shape by Poincaré and his followers, who considered only finite dimensional spaces. The first systems studied were, quite naturally, the simplest ones, for instance in the plane, where the only limit sets and attractors are points of equilibrium (rest points), periodic solutions (closed orbits) or other things not much more complicated.

In studying tridimensional systems there appeared limit sets of a structure not so simple. See for instance the papers of Cartwright, Littlewood, and Levinson ([1]), ([2]), ([3]).

For these systems there is not any theorem telling us what the limit sets or the attractors should look like. With the help of the modern fast digital computers it has been possible to integrate approximately equations which were intractable before. With such help orbits of very complicated behaviour in the limit have been put in evidence. See the papers of Lorenz ([4]), Rikitake ([5]), Rössler ([6]).

With the help of functional analysis, also some processes modelled by parabolic differential equations have been liable to be treated as dynamical systems in spaces of functions (of infinite dimension). That is the case of the Navier-Stokes equations and also of reaction-difusion equations. The list of contributions to these developements is very large and we thing the best is to refer to the book of Henry ([7]) for further references.

The interest in complicated attractors has grown further with the conjecture far from proven by Ruelle and Takens ([8]) that turbulence in hydrodynamics is due to the existence of some "strange attractor" in the dynamical system defined by the Navier-Stokes equations.

In this lecture we shall give the basic definitions and some examples of systems with orbits showing a complicated limit behaviour atributable to a possible "strange attractor". We shall try to say something about the most relevant of these examples, without any pretension of covering, even coarsely, what is known or what problems are open today on this subject.

2. BASIC CONCEPTS.

A dynamical system consists of a topological space H and of a continous map T defined for every non-negative real t (time) and every point x of H and taking the value $T(t,x)$ in H. This map has to fulfill the conditions $T(o, x)=x$, and $T(s+t,x)= T(x,T(t,x))$. If moreover T is differentiable as a function of (t,x) we say that our dynamical system is differentiable.

For instance a system of n autonomous ordinary differential equations of first order $x'=f(x)$, with f continuously differentiable, (that is, a differentiable vector field in $\mathbb{R}^n$), defines through its solutions to the initial value problem, a dynamical system in $\mathbb{R}^n$, if the solutions are defined for all values of $t>0$ (which is the same as saying that no solution tends to infinity with t tending to a finite value, as is the case with $x'=x^2$).

Also the solutions of the Navier-Stokes equations in a bounded region of the plane define a dynamical system in a function space (as a matter of fact, a Sobolev space, which is infinite dimensional). (Let's remind that it is not yet proven that if the bounded region is three-dimensional the solution can not escape to infinity in finite time, see Ladyshenskaya ([9])). The mapping $T(.,x):\mathbb{R}^+\to H$, which for a given $x\in H$ associates to each t the value $T(t,x)$ is named the "trajectory" or "solution" by x, and the set of points which lie on such a trajectory with the orientation given by growing t is called the "orbit" by x, and is denoted by $\gamma(x)$.

For further convenience we shall use the notation $T(t)x:=T(t,x)$.

The limit set $\Lambda(x)$ of $\gamma(x)$ consists of the points y such that there is a sequence of points $T(t_n,x)$ in $\gamma(x)$ tending to y with t_n tending to infinity.

It can also be defined as the intersection of the closure of the orbits of $T(\pi,x)$ for all $\pi>0$. It turns out that if H is finite dimensional and γ is bounded, then Λ is non-empty, is invariant (which means that with x it contanins $\gamma(x)$), and is closed.

If our dynamical system is a model of some practical process, and we expect to be able to deduce from its study the long term behaviour of the state of the process, we have to ask some kind of stability of $\Lambda(x)$ with respect to small changes in x and in the parameters defining the system itself. We cannot make exact measuments of the initial conditions!

The fact that $\Lambda(x)$ can change a lot by varying x as little as we want is illustrated by any system in the plane having a saddle point 0. If x lies in one of the oncoming separatrices $\Lambda(x)$ is 0. If we move x slightly away from this separatrix, $\Lambda(x)$ is certainly very different from 0. For instance, in the figure, the limit instead of 0 would be the closed orbit C_1 or the closed orbit C_2 , if we more x above or under the separatrix in which it lies.

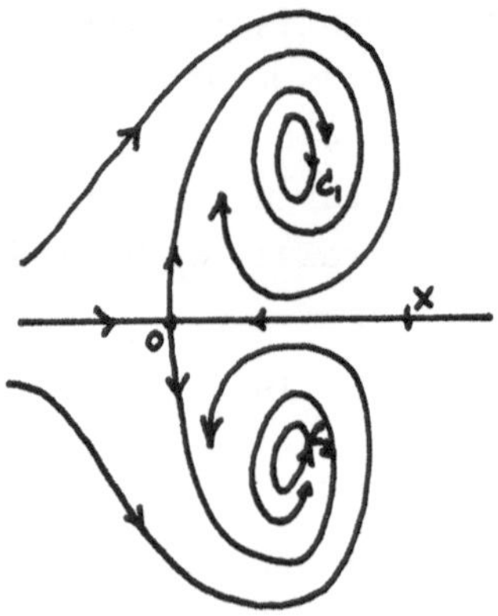

$\Lambda(x)$ would not change much (in some sense) should it be stable, that is, if the orbits beginning close enough to Λ do not leave a prescribed neighborhood of Λ. The orbits C_1 and C_2 in the figure have such property, and if a point y has them as $\Lambda(y)$, then a small change on y does not change Λ at all. The fact is that with more complicated systems it is quite hard, not to say impossible, to ascertain the stability of limit sets. Or, to be more sincere, there are cases of practical importance, like the Lorentz system for instance, where the limit sets do not fulfill the stability requirement (however there is some qualitative behaviour shared by the orbits).

Owing to this facts one becomes interested in sets which are invariant, closed and bounded and which contain all the limit sets of the points in some of their neighborhoods. This is the concept of attractor: a set A which is invariant, closed and bounded and which has a neighborhood U such that if $y \in U$, then $T(t)\,y$ tends to A as t tends to infinity.

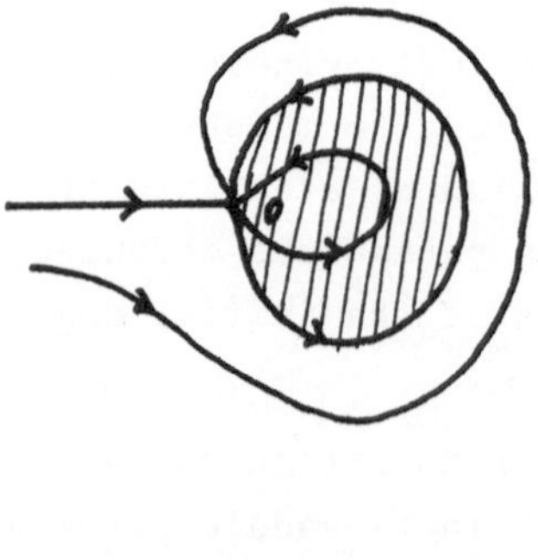

This concept is still not suficient to insure some stability:there are unstable attractors, as the example shown in the figure, due to Mendelson : the point 0 is an attractor,(as a matter of fact it is the Λ of all the points of the plane),but it is not stable because as close as we wish to it there are orbits starting which go quite far before returning.

Hence we have to strenghten the concept of attractor asking for its stability: given any neighborhood V of A we ask for a neighborhood U of A such that if x is in U then $T(t)x$ is in V.

Notice that the shaded set in the figure above is an attractor in the sense just given. Notice also that it could be that an attractor is not minimal, in the sense that it may contain a subset which is an attractor.

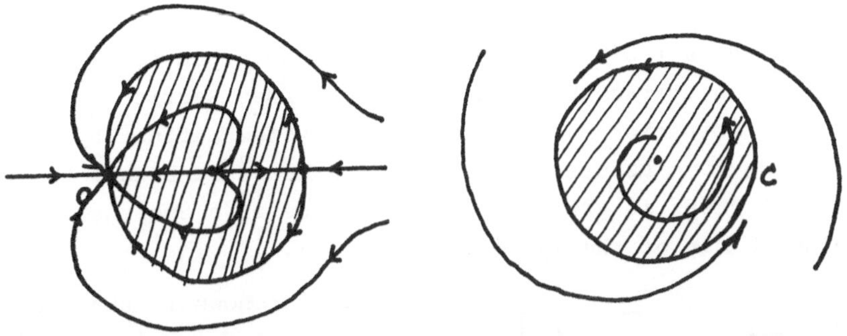

Two simple possibilities are shown in the figures above where the shaded sets are attractors as are also the rest point 0 and the closed orbit C.

Among the possible attractors we are more interested in those which have some qualitative properties preserved under small changes of the system. We say that they are structurally stable with respect to those properties. Of course it is not always easy to decide which are the properties preserved (see later the description of de Lorentz attractors, for instance).

Some authors furnish their attractors with more conditions and this allows, if not full characterizations, at least the possibility of saying more things about them. For instance one could ask that the set of periodic orbits be dense in A, or that there is a hyperbolic structure (which means that the differential of T(t) splits in a expanding and a contracting part) or that A has a dense orbit (topological transitivity). For more details see Williams ([10]).

One way of studying some qualitative properties of dynamical systems is through the use of what is called the "Poincaré map". Supose that we can find a transversal section to the "flow" of the dynamical system: that is, a manifold Σ (possibly with boundary), of dimension one less than the space H (if H is three dimensional, Σ is a piece of a surface), and which the orbits cross from one side to the other. If for all $x \in \Sigma$ we can find $t>0$ such that $T(t)x$ is in Σ, that is, if all orbits after crossing Σ, cross again after a time, we have a mapping from Σ into itself which is called the Poincaré map ϕ. If $\phi^n(x)=x$ for some positive integer n, we say that x is a periodic point, and if this is the least n, it is an n-periodic point. No need to say that all periodic points correspond to periodic orbits.

If we have such a Poincaré map defined, the intersection of $\Lambda(x)$ with Σ ($x\varepsilon\Sigma$), is the set of limit points of the sequence $\{\phi^n(x)\}$. Likewise, if there is an attractor of which Σ is a cross section, its intersection is an "attractor" of the "discrete" system defined by ϕ and its iterates ϕ^n on Σ. A beautiful example of some weird behaviour in three-dimensional cases is given by the existence of a Smale's "horseshoe" in a Poincaré map of the system (see ([11])).

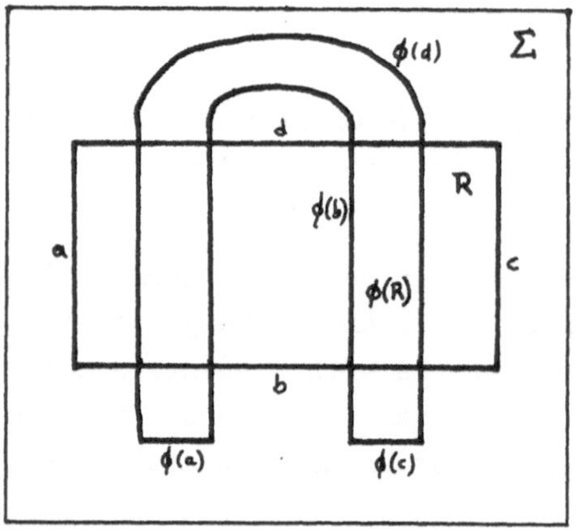

This is shown in the figure: the rectangle R in Σ is mapped on the "horseshoe" $\phi(R)$. Under these conditions there are periodic points of all the periods, which shows that there are limit sets quite complicated. If we have such a horseshoe, then there must exist are attractor seccioned by Σ in the set A (as a matter of fact $\lim\{\phi^n\Sigma\}$ is an attractor under the discrete system. The problem may lie in this attractor not being minimal, i.e., that there can exist in A a proper subset which is itself an attractor. For instance one of the periodic points in A could be a stable attractor by itself. On the other hand there are complicated orbits which do not tend to any periodic point, but the points with such complicated limits could occupy a nowhere dense set, while the rest tend to periodic points. This happens in the standard horseshoe map.

In trying to ascertain the existence of "strange" attractors, what one generally does is to establish a Poincaré map with some of the features of the horseshoe above, and then, through some other condition, as hyperbolicity, for instance, try to prove that A does not have proper sets which are attractors.

We are then forced to distinguish among two cases according to the fact of having periodic points of all periods in A, in which case we speak of "chaos" and having a minimal attractor which is not a periodic point , in which case we speak of "strange" or "complicated" attractors. In practice chaos is much easier to establish than the existence of strange attractors, as we shall see in the examples of next section.

3. EXAMPLES OF SYSTEMS WITH STRANGE ATTRACTORS.

In the paper by Lorenz ([4]) a system of three first order ordinary differential equations is studied. The system is obtained after considering Saltzman's convexion equations for Bénard's problem. Some solutions which are assumed periodic in two of the space variables are expanded in double Fourier series with time depending coefficients.

Plugging these series in the equations and equating coefficients, an inifinite set of ordinary differential equations is obtained. From these a finite number is chosen by a procedure of Lorenz himself, and which in some cases is three. The system to consider is then:

$$x' = -\sigma x+\sigma y$$
$$y' = rx-y-xz$$
$$z' = -bz+xy$$

(of a rather simple appearance).

When this system is solved by numerical procedures for certain values of the parameters σ, r and b one observes that the solutions in $\mathbb{R}^3$ behave in the following way: first they turn several times, n_1, around a fixed point A in the octant $(x>0,\ y>0,\ z>0)$, then they switch to the octant $\{x<0,\ y<0,\ z>0\}$ where they turn around a fixed point B n_2 times, to go back to turn n_3 times around A, etc.. In appearance the sequence $n_1, n_2, n_3, \ldots$ is aleatory and depends of the initial point in a very unstable way.

The interesting thing about such a behaviour, besides being of a purely mathematical interest is that it could be the explanation of what turbulence is (see Ruelle ([12])). A good and very concise account of what the behaviour is for $\sigma=10$, $b=8/3$ and r going from 0 to about 100 can be found in Marsden ([13]). (See also ([14])).

In order to explain what is happening Guckenheimer ([15]) considers a system which looks like the Lorenz one, but is somehow indealized, in the sense that he assumes that certain Poincaré map contracts in a given direction. He is able to show that in this case we have a strange attractor in the strong sense given before, i.e., a minimal attractor which is not a closed orbit. As a matter of fact he proves that this attractors are not structurally stable, in the sense that changing - the system very little, the topological type of the attractor may change (Williams has later proved in ([16]) that there are uncountably many topologically distruct Lorenz attractors). Anyhow, even being different, the attractors have in common many features, in the sense that through numemerical analysis we would not be able to distinguish among them. The model of Guckenheimer admits a Poincaré map which is sketched in the following figure.

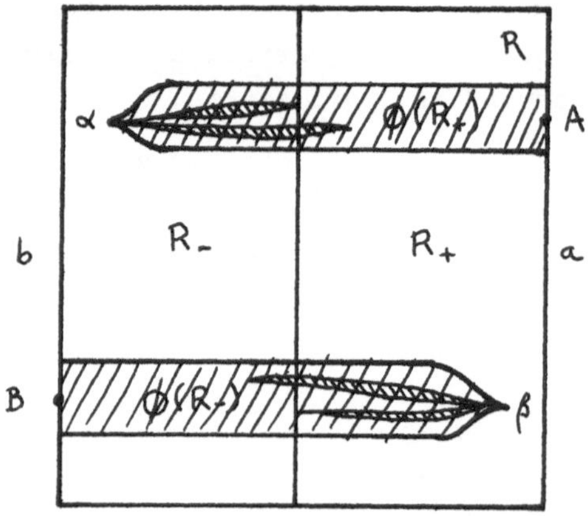

The map is defined in R_- and R_+ (not in the line separating them), and has the images shown by the shaded regions. Vertical lines go into vertical lines, and in particular a and b go into themselves and α,β are the (limit) images of L. Vertically the map experiments a uniform contraction.

The double-hatched V sets are the image under ϕ^2, and we observe some features of the horseshoe. Under the iterations of ϕ we obtain in the limit the intersection of R with the attractor, which has a book-like structure, with no interior points; a sort of Cantor set.

In ([17]) an explanation of the behaviour of the Lorenz attractor is intended through the scrolling and intertwining of some invariant surfaces (separatrices). In the next figure we picture how the intertwining scrolls, plus the barrier posed by the unstable manifolds of the fixed points A and B force the turning-around-a-point and rebouncing features of the Lorenz attractor.

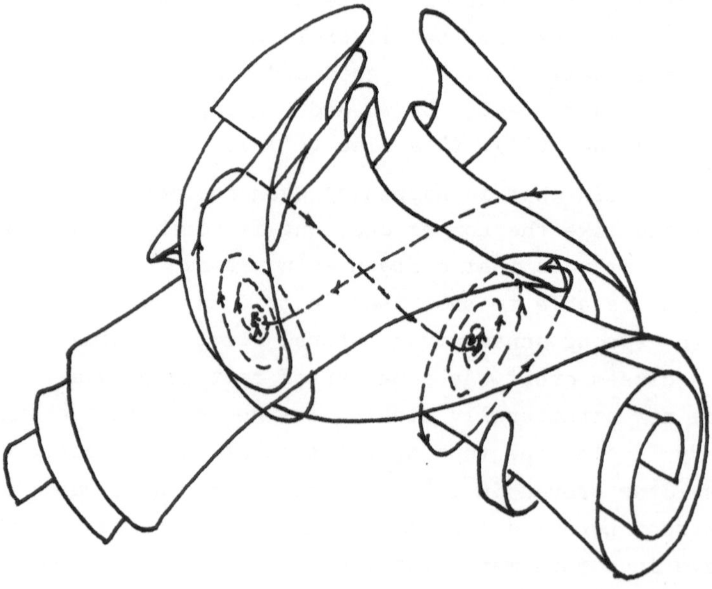

Another system showing a "strange attractor" is the one proposed by Rikitake ([5]) to explain the apparently aleatory changes of polarity of the terrestrial magnetic field.

Cook and Roberts ([18]) study the system using mainly numerical integration (they use some analysis to ascertain the unstability of the fixed points). The result is that apparently there is an attractor, with features quite similar to the Lorenz one.

The Rikitake system is given by the equations

$$x' = -\mu x+y(z+\alpha)$$
$$y' = -\mu y+x(z-\alpha)$$
$$z' = 1-xy$$

It admits two fixed points A and B and the bahaviour of the solutions is very similar to the Lorenz's ones: they turn alternatively around A and B, switching from one to the other in an apparently aleatory way. The main difference seems to lie in the fact that here the orbits don't need to be bounded.

Valero in ([19]) makes a study of the system using numerical computations only to choose among a few cases when necessary, but otherwise purely analytical, and he is able to prove the existence of chaos and quasi-aleatority (which means that the sequence $n_1, n_2, n_3, \ldots$ of turns around A and B seems to respect any arbitrarily given sequence by choosing the initial conditions conveniently). He is able to say, further, that the attractor is minimal, and strange (in the sense that it is not a periodic orbit), if there are no attracting periodic orbits (or, what is the same, if the attractor has no interior points). He does not obtain the estimates to insure this.

In order to prove this the system is mapped in the interior of the unit ball and extended to its surface. Then a Poincaré map which shows some of the horseshoe features is studied, using, among others, the invariant surfaces of orbits tending to the fixed points in infinity (the surface of the ball).

The following figure, borrowed from the work just mentioned, shows a piece of an orbit of the Rikitake system.

Besides the two systems just reviewed, there are others, let's call them "artificial" systems in three and four dimensional space which exhibit the same kind of behaviour (See Rössler ([20])).

It appears as if some chemical systems would show some chaotic behavior. This has been already observed experimentally in open Belousov-

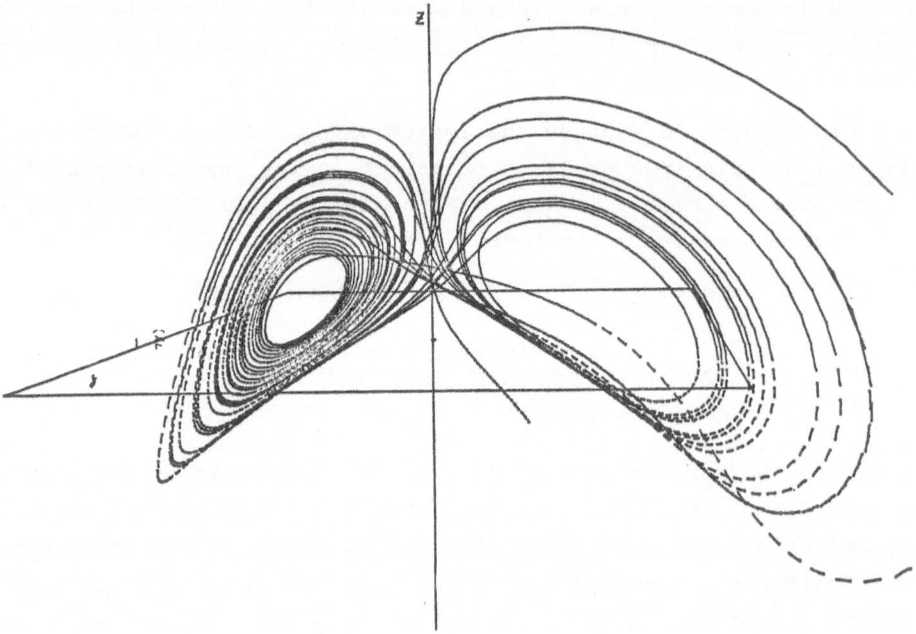

Zhabotinsky reactions. These reactions are generally modelled by systems with variables much slower than others. That could make profitable to study, instead of the original system, what we could call the reduction to a "relaxation" system, that is, one where we have split the fast behavior from the slow one. For instance, if we have the system

$$x' = f(x,y,z)$$
$$y' = g(x,y,z)$$
$$\varepsilon z' = h(x,y,z)$$

with small ε, we can assume that the z' component is infinity in all points outside the surface $h(x,y,z)=0$, and that, in this surface (the "slow" manifold), the system behaves according to a tangent field whose x' and y' components are given by the first two equations.

We can consider with Rössler ([21]) the system despicted in the figure, if the segment AB, goes into A'B' through transportation along the orbits of the relaxation system, we will have chaos.

One may wonder if there is a complicated behavior for the system for positive and small ε when we are in the previous situation for the relaxation system. This is the case under some hypothesis, which are analysed by Bonet in ([22]). The method he uses is, as usual, the consideration of a horseshoe-like Poincaré mapping. These equations are not very far from the classical Belousov-Zhabotinsky equations, only there is no

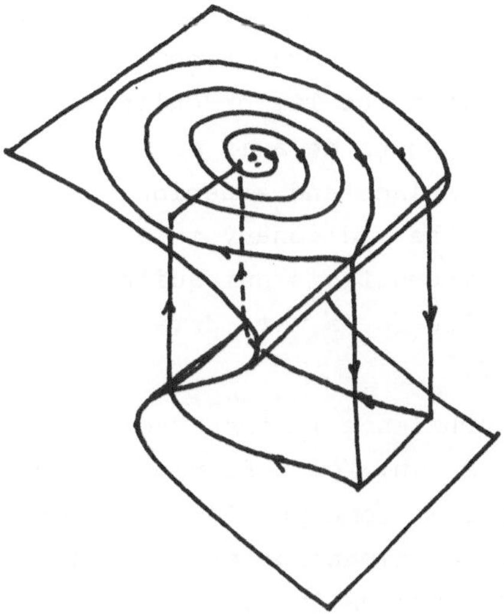

S folding there, but only a $\int$ slow surface. It could be that chaos appear when this folding appears (through change of parameters of the equation).

4. POSSIBILITY OF FINITE DIMENSIONAL STRANGE ATTRACTORS IN INFINITE DIMENSIONAL SYSTEMS AND ITS POSSIBLE RELATION TO TURBULENCE.

As we have mentioned in the introduction, one of the basic reasons to study strange attractors, is the conjecture that turbulence in hydrodinamics, say, may be due to the presence of these attractors in the infinite dimensional space of states of the system. (see Ruelle-Takens ([8]). This conjecture is supported by some experimental evidence. Consider for instance the Couette flow of a viscous fluid between two coaxial cylinders rotating one with respect to the other. For small Reynold numbers the flow observed is stationary and all the fluid particles move in circles which are coaxial with the cylinders. When we increase the Reynolds number (for instance making larger the relative speed of the cylinders or decreasing the viscosity), there is a critical value at which a bifurcation takes place and Taylor cells appear. This step has been mathematically explained in ([23]) under some idealizing hypothesis on the upper and lower lids of the cylinders which compactify conveniently the problem. The flow with Taylor cells is still stationary, that is, corresponds to an equilibrium point in H, the space of states. The mathematical explanation of the bifurcation runs through the computation of the spectrum of the infinitessimal generator of the linear part of the system at the Couette equilibrium state: if a simple real eigenvalue passes from being negative to being positive, the original stationary state becomes unstable and new stationary states, two of them owing to symmetry corresponding to Taylor flows, become attractors. If we keep increasing the Reynolds number the Taylor cells loose their shape to become undulating and turn around de axis of the cylinders: apparently a new bifurcation has taken place and a periodic solution has inherited the attracting property of the stationary solution. After this there can still appear few new bifurcations but quite suddenly we observe the appearance of turbulence. See for a more detailed description the paper by Swinney and Gollub ([24]).

This sudden onset of turbulence is observed in most of the experiences made to study the fenomenon, for instance in the flow past a cylinder. The working hypothesis of some people goes then to say that as the Reynolds number increases the dimension of invariant manifolds to which all orbits of the dynamical system tend, also increases: in the Couette cases it goes from 0 to 1 in passing to the Taylor cells, to 2 in becoming periodic and may be to more in becoming turbulent.

This is very suggestive: we know that in submanifolds of dimension one only stationary points can be structurally stable attractors (for any worthwhile structure), and that in dimension two only stationary points and periodic orbits. But in dimension three we know already there can be strange attractors, which will appear in any Fourier analysis as having a broad band of frequencies, as in the experiments of Swinney and Gollub.

The numerical study of attractors in dimension three, show that in some sense all the observed strange attractors share some common behavior. In his conference at the Journées Fermat in Toulouse in april - 1979, Rössler (see ([25])) presented several examples of three and four dimensional chaotic systems.

Among other things, once he had the solution being computed being delived to a graphic screen at sufficiently large speed, he introduced a voltage proportional to one of the components to a loudspeaker. The result was astonishing: in all three dimensional cases with chaotic behavior the noise made ressembled the one made by an exhaust of gases from a pipe. The four dimensional case presented sounded differently: as statics in radio communications. That seems to show that the ear, a very fine Fourier analyzer, is capable of detecting some common feature in the three dimensional attractors and distinguish it from the four dimensional case. It would be a nice thing to characterize this feature in a mathematical way! Moreover it suggests that in the dynamics of exhaust gases there is some three dimensional attractor **hidden** somewhere.

We are still very far from characterizing and classifying the possi ble structurally stable attractors in dimension three or superior. The first question would be what the equivalence classes of attractors would be: topological equivalence in the sense of homeomorphisms sending orbits into orbits seems far too strong to provide a satisfactory structural stability. On the other hand it seems that in the space of vector fields in dimension three or more, systems with strange attractors are not scar ce. As a matter of fact they have interior points (in some natural topo logy). Then maybe the name "strange" is not quite adequate for such a usual thing!

Inspired by the possibility of having the attractors contained in in finite dimensional submanifolds in H, a study is being conducted about the existence of finite dimensional submanifolds in H to which all the orbits tend. This is done systems defined by some parabolic equations by Mañé in ([26]). Also Mora in ([27]) intends a similar thing for the syst tems defined by reaction-diffusion equations (see Calsina ([28]) and Mora ([29]) for a detailed proof of reaction-diffusion equation defining dyna mical systems).

In spite of the evidence presented it is not quite established that turbulence is due to strange attractors and still less to strange attrac tors in finite dimensional submanifolds of H. The possibility of the Na vier-Stokes equations not being valid for large Reynold numbers or the possibility of lack of existence of smooth (strong) solutions of the equations after some time have been not totally discarded.

We don't discard also as a possibility for an explanation of turbulence, that all limit "behaviors" of the orbits are unstable. That means that the limit sets of the orbits are unstable in such a way that in - any neighborhood of the limit set there are orbits which are thrown -- away from it and behave very differently from Λ. This means that the system doesn't show a behavior determined qualitatively by an attractor but depending on the fluctuations exerted by the surrounding medium, - which will decide, when close to its limit set, in what direction will the orbit move, and hence changing its behavior in an apparently erratic way. Let's remark that this is the case in the Lorenz system, where the sequence $n_1, n_2, n_3, \ldots$ of turns of the orbit around the fixed - points A and B depend very critically from the initial state. On the - other hand this unstability of the limit sets can be present without any strange attractor. To end this discussion we mention also the possi bility of the system not having any attractor of finite dimension, - which would complicate quite a lot de description of what happens.

REFERENCES

(1) M.L. CARTWRIGHT, Almost periodic solutions of equations with periodic coefficients. In Nonlinear Problems, University of Wisconsin Press, 207-218 (1964).

(2) N. LEVINSON, Transformation theory of nonlinear differential equations of the second order, Annals of Mathematics, 45, 723-737 (1944)

(3) J.E. LITTLEWOOD, On van der Pol's equation with large k. Ibid (1), 161-165.

(4) E.N. LORENZ, Deterministic nonperiodic flow. J. Atmos.Sci. 20, 130-141 (1963).

(5) T. RIKITAKE, Proc. Cambridge Philos. Soc., 54, 89 (1958).

(6) O.E. RÖSSLER, "Different types of chaos in two simple differential equations". Z. Naturforsch. 31a, 1664-1670 (1976).

(7) D. HENRY, Geometric theory of semilinear parabolic equations. Lect. Notes in Math. 840 Springer-Verlag (1981).

(8) D. RUELLE and F. TAKENS, On the nature of turbulence. Commun. Math. Phys. 20, 167-192 (1971).

(9) O.A. LADYSHENSKAYA, The mathematical theory of viscous incompressible flow, Gordon and Breach (1963).

(10) R.F. WILLIAMS, The Lorenz attractor. Lect. Notes in Math. 615. Springer-Verlag, 94-112 (1977)

(11) Z. NITECKY, Differentiable dynamics. M.I.T. Press (1971).

(12) D. RUELLE, The Lorenz attractor and the problem of turbulence. Lecture Notes in Math. 565. Springer-Verlag, 146-158 (1976).

(13) J.E. MARSDEN, Attempts to relate the Navier-Stokes equations to turbulence. Lect. Notes in Math. 615, Springer-Verlag, 1-22 (1977).

(14) J.L. KAPLAN and J.A. YORKE, Preturbulence: a regime observed in a fluid flow model of Lorenz. Commun. Math. Phys. 67, 93-108 (1979).

(15) J. GUCKENHEIMER, A strange strange attractor. In Marsden, J.E., McCracken, M. Hopf bifurcation and its applications, Springer-Verlag, 368-381 (1976).

(16) R.F. WILLIAMS, The structure of Lorenz attractors. Preprint (1977).

(17) C. PERELLO, Intertwining invariant manifolds and the Lorenz attractor, Lect. Notes in Math. 819, Springer-Verlag, 375-378 (1980)

(18) A.E. COOK and P.N. ROBERTS, The Rikitake two-disc dynamo system. Proc. Cambridge Philos. Soc., 68, 547-569 (1970).

([19]) J. VALERO, El sistema de Rikitake. Actas III Congreso de Ecuaciones diferenciales y aplicaciones, Santiago de Compostela, 313-329, (1980).

([20]) O.E. ROSSLER, An equation for hyperchaos. Physics letters, 71A, 155-157 (1979).

([21]) O.E. ROSSLER, Chaotic behavior in simple reaction systems. Z. Naturforsch. 31a, 259-264 (1976).

([22]) C. BONET, Tesis de Llicenciatura. Facultat de Ciencies. Universitat Autònoma de Barcelona.

([23]) K. KIRCHGÄSSNER, and H. KIELHÖFER, Stability and bifurcation in fluid mechanics, Rocky Mtn. Math. J. 3, 275-318 (1973).

([24]) H.L. SWINNEY and J.P. GOLLUB, The transition to turbulence.Physics Today, August 1978, 41-49.

([25]) O.E. ROSSLER, The flueing-together principle in chaos. In nonlinear problems of analysis in geometry and mechanics, Research Notes in Math. 46. Pitman, 50-56 (1981).

([26]) R. MAÑE, Reduction of semilinear parabolic equations to finite dimensional C_1 flows, Lect. Notes in Math. 597, 361-378 (1977).

([27]) X. MORA, Finite-dimensional attracting manifolds in reaction-diffusion equations. University of Warwick Notes (1981).

([28]) A. CALSINA, Bifurcacions genèriques d'attractors en sistemes de reacció i difusió. Publ.Secc. de Mat.24, Univ.Aut.de Barcelona,73-162(1981)

([29]) X. MORA, Reaction diffusion equations define dynamical systems. University of Warwick Notes (1981).

Lecture Notes in Physics

Vol. 136: The Role of Coherent Structures in Modelling Turbulence and Mixing. Proceedings 1980. Edited by J. Jimenez. XIII, 393 pages. 1981.

Vol. 137: From Collective States to Quarks in Nuclei. Edited by H. Arenhövel and A. M. Saruis. VII, 414 pages. 1981.

Vol. 138: The Many-Body Problem. Proceedings 1980. Edited by R. Guardiola and J. Ros. V, 374 pages. 1981.

Vol. 139: H. D. Doebner, Differential Geometric Methods in Mathematical Physics. Proceedings 1981. VII, 329 pages. 1981.

Vol. 140: P. Kramer, M. Saraceno, Geometry of the Time-Dependent Variational Principle in Quantum Mechanics. IV, 98 pages. 1981.

Vol. 141: Seventh International Conference on Numerical Methods in Fluid Dynamics. Proceedings. Edited by W. C. Reynolds and R. W. MacCormack. VIII, 485 pages. 1981.

Vol. 142: Recent Progress in Many-Body Theories. Proceedings. Edited by J. G. Zabolitzky, M. de Llano, M. Fortes and J. W. Clark. VIII, 479 pages. 1981.

Vol. 143: Present Status and Aims of Quantum Electrodynamics. Proceedings, 1980. Edited by G. Gräff, E. Klempt and G. Werth. VI, 302 pages. 1981.

Vol. 144: Topics in Nuclear Physics I. A Comprehensive Review of Recent Developments. Edited by T.T.S. Kuo and S.S.M. Wong. XX, 567 pages. 1981.

Vol. 145: Topics in Nuclear Physics II. A Comprehensive Review of Recent Developments. Proceedings 1980/81. Edited by T. T. S. Kuo and S. S. M. Wong. VIII, 571-1.082 pages. 1981.

Vol. 146: B. J. West, On the Simpler Aspects of Nonlinear Fluctuating. Deep Gravity Waves. VI, 341 pages. 1981.

Vol. 147: J. Messer, Temperature Dependent Thomas-Fermi Theory. IX, 131 pages. 1981.

Vol. 148: Advances in Fluid Mechanics. Proceedings, 1980. Edited by E. Krause. VII, 361 pages. 1981.

Vol. 149: Disordered Systems and Localization. Proceedings, 1981. Edited by C. Castellani, C. Castro, and L. Peliti. XII, 308 pages. 1981.

Vol. 150: N. Straumann, Allgemeine Relativitätstheorie und relativistische Astrophysik. VII, 418 Seiten. 1981.

Vol. 151: Integrable Quantum Field Theory. Proceedings, 1981. Edited by J. Hietarinta and C. Montonen. V, 251 pages. 1982.

Vol. 152: Physics of Narrow Gap Semiconductors. Proceedings, 1981. Edited by E. Gornik, H. Heinrich and L. Palmetshofer. XIII, 485 pages. 1982.

Vol. 153: Mathematical Problems in Theoretical Physics. Proceedings, 1981. Edited by R. Schrader, R. Seiler, and D.A. Uhlenbrock. XII, 429 pages. 1982.

Vol. 154: Macroscopic Properties of Disordered Media. Proceedings, 1981. Edited by R. Burridge, S. Childress, and G. Papanicolaou. VII, 307 pages. 1982.

Vol. 155: Quantum Optics. Proceedings, 1981. Edited by C.A. Engelbrecht. VIII, 329 pages. 1982.

Vol. 156: Resonances in Heavy Ion Reactions. Proceedings, 1981. Edited by K.A. Eberhard. XII, 448 pages. 1982.

Vol. 157: P. Niyogi, Integral Equation Method in Transonic Flow. XI, 189 pages. 1982.

Vol. 158: Dynamics of Nuclear Fission and Related Collective Phenomena. Proceedings, 1981. Edited by P. David, T. Mayer-Kuckuk, and A. van der Woude. X, 462 pages. 1982.

Vol. 159: E. Seiler, Gauge Theories as a Problem of Constructive Quantum Field Theory and Statistical Mechanics. V, 192 pages. 1982.

Vol. 160: Unified Theories of Elementary Particles. Critical Assessment and Prospects. Proceedings, 1981. Edited by P. Breitenlohner and H.P. Dürr. VI, 217 pages. 1982.

Vol. 161: Interacting Bosons in Nuclei. Proceedings, 1981. Edited by J.S. Dehesa, J.M.G. Gomez, and J. Ros. V, 209 pages. 1982.

Vol. 162: Relativistic Action at a Distance: Classical and Quantum Aspects. Proceedings, 1981. Edited by J. Llosa. X, 263 pages. 1982.

Vol. 163: J. S. Darrozes, C. Francois, Mécanique des Fluides Incompressibles. XIX, 459 pages. 1982.

Vol. 164: Stability of Thermodynamic Systems. Proceedings, 1981. Edited by J. Casaş-Vázquez and G. Lebon. VII, 321 pages. 1982.

Selected Issues from

Lecture Notes in Mathematics

Vol. 728: Non-Commutative Harmonic Analysis. Proceedings, 1978. Edited by J. Carmona and M. Vergne. V, 244 pages. 1979.

Vol. 729: Ergodic Theory. Proceedings 1978. Edited by M. Denker and K. Jacobs. XII, 209 pages. 1979.

Vol. 730: Functional Differential Equations and Approximation of Fixed Points. Proceedings, 1978. Edited by H.-O. Peitgen and H.-O. Walther. XV, 503 pages. 1979.

Vol. 731: Y. Nakagami and M. Takesaki, Duality for Crossed Products of von Neumann Algebras. IX, 139 pages. 1979.

Vol. 733: F. Bloom, Modern Differential Geometric Techniques in the Theory of Continuous Distributions of Dislocations. XII, 206 pages. 1979.

Vol. 735: B. Aupetit, Propriétés Spectrales des Algèbres de Banach. XII, 192 pages. 1979.

Vol. 738: P. E. Conner, Differentiable Periodic Maps. 2nd edition, IV, 181 pages. 1979.

Vol. 742: K. Clancey, Seminormal Operators. VII, 125 pages. 1979.

Vol. 755: Global Analysis. Proceedings, 1978. Edited by M. Grmela and J. E. Marsden. VII, 377 pages. 1979.

Vol. 756: H. O. Cordes, Elliptic Pseudo-Differential Operators – An Abstract Theory. IX, 331 pages. 1979.

Vol. 760: H.-O. Georgii, Canonical Gibbs Measures. VIII, 190 pages. 1979.

Vol. 762: D. H. Sattinger, Group Theoretic Methods in Bifurcation Theory. V, 241 pages. 1979.

Vol. 765: Padé Approximation and its Applications. Proceedings, 1979. Edited by L. Wuytack. VI, 392 pages. 1979.

Vol. 766: T. tom Dieck, Transformation Groups and Representation Theory. VIII, 309 pages. 1979.

Vol. 771: Approximation Methods for Navier-Stokes Problems. Proceedings, 1979. Edited by R. Rautmann. XVI, 581 pages. 1980.

Vol. 773: Numerical Analysis. Proceedings, 1979. Edited by G. A. Watson. X, 184 pages. 1980.

Vol. 775: Geometric Methods in Mathematical Physics. Proceedings, 1979. Edited by G. Kaiser and J. E. Marsden. VII, 257 pages. 1980.

Vol. 779: Euclidean Harmonic Analysis. Proceedings, 1979. Edited by J. J. Benedetto. III, 177 pages. 1980.

Vol. 780: L. Schwartz, Semi-Martingales sur des Variétés, et Martingales Conformes sur des Variétés Analytiques Complexes. XV, 132 pages. 1980.

Vol. 782: Bifurcation and Nonlinear Eigenvalue Problems. Proceedings, 1978. Edited by C. Bardos, J. M. Lasry and M. Schatzman. VIII, 296 pages. 1980.

Vol. 783: A. Dinghas, Wertverteilung meromorpher Funktionen in ein- und mehrfach zusammenhängenden Gebieten. Edited by R. Nevanlinna and C. Andreian Cazacu. XIII, 145 pages. 1980.

Vol. 786: I. J. Maddox, Infinite Matrices of Operators. V, 122 pages. 1980.

Vol. 787: Potential Theory, Copenhagen 1979. Proceedings, 1979. Edited by C. Berg, G. Forst and B. Fuglede. VIII, 319 pages. 1980.

Vol. 791: K. W. Bauer and S. Ruscheweyh, Differential Operators for Partial Differential Equations and Function Theoretic Applications. V, 258 pages. 1980.

Vol. 792: Geometry and Differential Geometry. Proceedings, 1979. Edited by R. Artzy and I. Vaisman. VI, 443 pages. 1980.

Vol. 793: J. Renault, A Groupoid Approach to C*-Algebras. III, 160 pages. 1980.

Vol. 798: Analytic Functions, Kozubnik 1979. Proceedings. Edited by J. Ławrynowicz. X, 476 pages. 1980.

Vol. 799: Functional Differential Equations and Bifurcation. Proceedings 1979. Edited by A. F. Izé. XXII, 409 pages. 1980.

Vol. 801: K. Floret, Weakly Compact Sets. VII, 123 pages. 1980.

Vol. 802: J. Bair, R. Fourneau, Etude Géometrique des Espaces Vectoriels II. VII, 283 pages. 1980.

Vol. 804: M. Matsuda, First Order Algebraic Differential Equations. VII, 111 pages. 1980.

Vol. 805: O. Kowalski, Generalized Symmetric Spaces. XII, 187 pages. 1980.

Vol. 807: Fonctions de Plusieurs Variables Complexes IV. Proceedings, 1979. Edited by F. Norguet. IX, 198 pages. 1980.

Vol. 809: I. Gumowski and Ch. Mira, Recurrences and Discrete Dynamic Systems. VI, 272 pages. 1980.

Vol. 810: Geometrical Approaches to Differential Equations. Proceedings 1979. Edited by R. Martini. VII, 339 pages. 1980.

Vol. 816: L. Stoica, Local Operators and Markov Processes. VIII, 104 pages. 1980.

Vol. 819: Global Theory of Dynamical Systems. Proceedings, 1979. Edited by Z. Nitecki and C. Robinson. IX, 499 pages. 1980.

Vol. 821: Statistique non Paramétrique Asymptotique. Proceedings, 1979. Edited by J.-P. Raoult. VII, 175 pages. 1980.

Vol. 826: Ph. G. Ciarlet et P. Rabier, Les Equations de von Kármán. VI, 181 pages. 1980.

Vol. 827: Ordinary and Partial Differential Equations. Proceedings, 1978. Edited by W. N. Everitt. XVI, 271 pages. 1980.

Vol. 831: Representation Theory I. Proceedings, 1979. Edited by V. Dlab and P. Gabriel. XIV, 373 pages. 1980.

Vol. 832: Representation Theory II. Proceedings, 1979. Edited by V. Dlab and P. Gabriel. XIV, 673 pages. 1980.

Vol. 836: Differential Geometrical Methods in Mathematical Physics. Proceedings, 1979. Edited by P. L. García, A. Pérez-Rendón, and J. M. Souriau. XII, 538 pages. 1980.

Vol. 837: J. Meixner, F. W. Schäfke and G. Wolf, Mathieu Functions and Spheroidal Functions and their Mathematical Foundations Further Studies. VII, 126 pages. 1980.

Vol. 838: Global Differential Geometry and Global Analysis. Proceedings 1979. Edited by D. Ferus et al. XI, 299 pages. 1981.

Vol. 840: D. Henry, Geometric Theory of Semilinear Parabolic Equations. IV, 348 pages. 1981.

Vol. 841: A. Haraux, Nonlinear Evolution Equations- Global Behaviour of Solutions. XII, 313 pages. 1981.

Vol. 842: Séminaire Bourbaki vol. 1979/80. Exposés 543–560. IV, 317 pages. 1981.

Vol. 843: Functional Analysis, Holomorphy, and Approximation Theory. Proceedings. Edited by S. Machado. VI, 636 pages. 1981.

Vol. 845: A. Tannenbaum, Invariance and System Theory: Algebraic and Geometric Aspects. X, 161 pages. 1981.

Vol. 849: P. Major, Multiple Wiener-Itô Integrals. VII, 127 pages. 1981.